Frontiers in Mathematics

Jie Xiao

Geometric Q_P Functions

Birkhäuser Verlag
Basel · Boston · Berlin

Author:

Jie Xiao
Department of Mathematics and Statistics
Memorial University of Newfoundland
St. John's, NL A1C 5S7
Canada
e-mail: jxiao@math.mun.ca

2000 Mathematical Subject Classification, 28, 30C, 30F, 31, 42, 44, 46B, 46E, 47A,
47G, 49Q, 53A, 53D, 58D

A CIP catalogue record for this book is available from the
Library of Congress, Washington D.C., USA

Bibliographic information published by Die Deutsche Bibliothek
Die Deutsche Bibliothek lists this publication in the Deutsche Nationalbibliografie;
detailed bibliographic data is available in the Internet at <http://dnb.ddb.de>.

ISBN 3-7643-7762-3 Birkhäuser Verlag, Basel – Boston – Berlin

© 2006 Birkhäuser Verlag, P.O. Box 133, CH-4010 Basel, Switzerland
Part of Springer Science+Business Media
Cover design: Birgit Blohmann, Zürich, Switzerland
Printed on acid-free paper produced from chlorine-free pulp. TCF ∞
Printed in Germany
ISBN 10: 3-7643-7762-3   e-ISBN: 3-7643-7763-1
ISBN 13: 978-3-7643-7762-5

9 8 7 6 5 4 3 2 1   www.birkhauser.ch

To Xianli and Sa

Contents

Preface

The aim of the book *Geometric Q_p Functions* is to document the rich structure of the holomorphic Q functions which are geometric in the sense that they transform naturally under conformal mappings, with particular emphasis on the last few years' development based on interaction between geometrical function and measure theory and other branches of mathematical analysis, including complex variables, harmonic analysis, potential theory, functional analysis, and operator theory.

The book comprises eight chapters in which some results appear for the first time. The first chapter begins with a motive and a very brief review of the mostly standard characterizations of holomorphic Q functions presented in the author's monograph — Springer's LNM 1767: *Holomorphic Q Classes* — followed by some further preliminaries on logarithmic conformal maps, conformal domains and superpositions and harmonic majorants with an application to Euler–Lagrange equations. The second chapter gives function-theoretic characterizations by means of Poisson extension and Berezin transform with two more generalized variants. The third chapter takes a careful look at isomorphism, decomposition, and discreteness of spaces via equivalent forms of the generalized Carleson measures. The fourth chapter discusses invariant preduality through Hausdorff capacity, which is a useful tool to classify negligible sets for various fine properties of functions. The fifth chapter develops some essential properties of the Cauchy dualities via both weak factorizations and extreme points of the target function spaces. The sixth chapter shows particularly that each holomorphic Q function can be treated as a symbol of the holomorphic Hankel and Volterra operators acting between two Dirichlet spaces. The seventh chapter deals with various size estimates involving functions and their exponentials and derivatives. Finally, the eighth chapter handles how much of the basic theory of holomorphic Q functions can be carried over the hyperbolic Riemann surfaces by sharpening the area and isoperimetric inequalities and settling the limit spaces.

Although this book may be more or less regarded as a worthy sequel to the previously-mentioned monograph, it is essentially self-contained. And so, without reading that monograph, readers can understand the contents of this successor, once they are familiar with some basic facts on geometric function-measure theory and complex harmonic-functional analysis. For further background, each chapter ends with brief notes on the history and current state of the subject. Readers may consult those notes and go further to study the references cited by this book for more information.

As is often the case, the completion of a book is strongly influenced by some organizations and individuals. This book has been no exception. Therefore, the author would like to deliver a word of thanks to: Natural Sciences and Engineering Research Council of Canada as well as Faculty of Science, Memorial University of Newfoundland, Canada, that have made this book project possible; next, a number of people including (in alphabetical order): D. R. Adams (University of Kentucky), A. Aleman (Lund University), R. Aulaskari (University of Joensuu), H. Chen (Nanjing Normal University), K. M. Dyakonov (University of Barcelona), P. Fenton (University of Otago), T. Hempfling (Birkhäuser Verlag AG), M. Milman (Florida Atlantic University), M. Pavlovic (University of Belgrade), J. Shapiro (Michigan State University), A. Siskakis (University of Thessaloniki), K. J. Wirths (Technical University of Braunschweig), Z. Wu (University of Alabama), G. Y. Zhang (Polytechnic University of New York), R. Zhao (State University of New York at Brockport) and K. Zhu (State University of New York at Albany), who have directly or indirectly assisted in the preparation of this book. Last but not least, the author's family, who the author owes a great debt of gratitude for their understanding and moral support during the course of writing.

St. John's   *J. Xiao*
Fall 2005 – Summer 2006   *jxiao@math.mun.ca*

Chapter 1

Preliminaries

Our major goal in this chapter is to deal with some of the necessary preliminary results motivating a further study of the holomorphic Q classes via the following five sections:

- Background;
- Logarithmic Conformal Mappings;
- Conformal Domains and Superpositions;
- Descriptions via Harmonic Majorants;
- Regularity for the Euler–Lagrange Equation.

1.1   Background

In order to make this book self-contained and accessible, we start with some notation and terminology, provide with the readers a motive from geometric function theory, and look over several principal properties presented in the monograph: *Holomorphic Q Classes*, LNM 1767, Springer-Verlag, 2001.

First of all, let us have a look at the concept of conformal radius which plays an important role in geometric function theory. For every simply connected domain Ω on the finite complex plane $\mathbf{C}$ whose boundary $\partial\Omega$ contains at least two points, denote by $\mathcal{F}(\Omega)$ the class of all univalent functions (conformal mappings) defined on Ω whose images are subsets of the open unit disk $\mathbf{D} = \{z \in \mathbf{C} : |z| < 1\}$ with boundary $\mathbf{T} = \{z \in \mathbf{C} : |z| = 1\}$ such that $f(z_0) = 0$ for a fixed point $z_0 \in \Omega$. Then there is an $f_0 \in \mathcal{F}(\Omega)$ such that

$$\sup_{f \in \mathcal{F}(\Omega)} |f'(z_0)| = |f_0'(z_0)|.$$

This extremal function f_0 is called a Riemann mapping. Since it is unique up to a rotation, we can define

$$r_\Omega(z_0) = |f_0'(z_0)|^{-1}$$

and call it the conformal radius of Ω at the point z_0. By the superposition f_0 with a standard Möbius transform of $\mathbf{D}$:

$$\sigma_w(z) = \frac{w - z}{1 - \bar{w}z}, \quad z, w \in \mathbf{D},$$

yields that

$$f(z, w) = \sigma_{f_0(z)}(f_0(w)) = \frac{f_0(z) - f_0(w)}{1 - \overline{f_0(z)}f_0(w)}, \quad z, w \in \Omega,$$

is a Riemann mapping with $f(z, z) = 0$. Consequently, the conformal radius at an arbitrary point $z \in \Omega$ can be evaluated by

$$r_\Omega(z) = \frac{1 - |f_0(z)|^2}{|f_0'(z)|}.$$

Accordingly, if $f^{-1} : \mathbf{D} \mapsto \Omega$ stands for the inverse of f, then $f^{-1}(0) = z_0$ and

$$r_\Omega\big(f^{-1}(z)\big) = (1 - |z|^2)|(f^{-1})'(z)|, \quad z \in \mathbf{D}.$$

Even more interestingly, the conformal radius also appears in the study of geometric elliptic equations. To see this, suppose $\Omega \subset \mathbf{C}$ is a hyperbolic domain, a planar domain whose boundary $\partial\Omega$ contains at least two points. Consider the Liouville equation

$$\Delta u(z) = 4\frac{\partial^2}{\partial z \partial \bar{z}}u(z) = 4e^{2u(z)},$$

where

$$\frac{\partial}{\partial z} = 2^{-1}\left(\frac{\partial}{\partial x} - i\frac{\partial}{\partial y}\right) \quad \text{and} \quad \frac{\partial}{\partial \bar{z}} = 2^{-1}\left(\frac{\partial}{\partial x} + i\frac{\partial}{\partial y}\right) \quad \text{for} \quad z = x + iy.$$

If

$$U(z) = \sup\{u(z) : \ \Delta u(z) = 4e^{2u(z)}\}, \quad z \in \Omega,$$

stands for the maximal solution of the Liouville equation, then

$$R_\Omega(z) = e^{-U(z)}$$

is called the hyperbolic radius of Ω at z. When Ω is simply connected and $f : \Omega \mapsto \mathbf{D}$ is a Riemann mapping, the maximal solution U exists and is given by the following formula:

$$U(z) = -\log\left(\frac{1 - |f(z)|^2}{|f'(z)|}\right) = -\log r_\Omega(z), \quad z \in \Omega.$$

It is worth pointing out that $R_\Omega(z)$ provides a complete, conformally flat metric, called hyperbolic or Poincaré metric, on Ω of constant negative curvature whose line element is

$$ds = \frac{|dz|}{R_\Omega(z)} = e^{U(z)}|dz|.$$

and whose scalar curvature is determined by

$$K = -\frac{\Delta \log e^{U(z)}}{e^{2U(z)}} = -e^{-2U(z)}\Delta U(z) = -4.$$

Moreover, if $g : \mathbf{D} \mapsto \Omega$ is a universal covering map, then

$$R_\Omega(g(z)) = (1 - |z|^2)|g'(z)|, \quad z \in \mathbf{D}$$

and hence the Bloch semi-norm:

$$\sup_{z \in \mathbf{D}}(1 - |z|^2)|g'(z)|$$

is independent of the covering but dependent only on the domain.

On the other hand, we should note that any hyperbolic domain possesses Green's function. More precisely, for $z, w \in \Omega$, let $-\log|z - w|$ be the fundamental singularity of the Laplace operator Δ. Then $g_\Omega(z, w)$, the Green function of Ω, is determined by two conditions:

(i) $g_\Omega(z, w) + \log|z - w|$ is real-valued harmonic on Ω and continuous on $\Omega \cup \partial\Omega$;

(ii) $g_\Omega(z, w) = 0$ for $z \in \partial\Omega$.

Here and henceforth, log denotes an abbreviation of the logarithmic function with base e. Consequently,

$$s_\Omega(z) = \exp\left(\lim_{w \to z}\left(g_\Omega(z, w) + \log|z - w|\right)\right)$$

is called the harmonic radius of Ω at z. Note that if Ω is simply connected, then the Green function $g_\Omega(z, w) = -\log|f(z)|$ where $f : \Omega \mapsto \mathbf{D}$ is a Riemann mapping with $f(w) = 0$, and hence $s_\Omega(z) = r_\Omega(z)$. In this case, by the maximum principle and the Koebe 1/4 Theorem, we can dominate $r_\Omega(z)$ via the distance of z to $\partial\Omega$ from above and below:

$$4^{-1}r_\Omega(z) \le d(z, \partial\Omega) = \inf_{w \in \partial\Omega}|w - z| \le r_\Omega(z).$$

A further estimate of $r_\Omega(z)$ can be worked out by the following a priori estimate for $g_\Omega(\cdot, \cdot)$.

Example 1.1.1. Let $p \in [0, \infty)$, $\Gamma(\cdot)$ be the Gamma function, and dm stand for the two-dimensional Lebesgue measure element on $\mathbf{C}$. Suppose $\Omega \subset \mathbf{C}$ is a simply connected hyperbolic domain. Then

$$\pi\big(r_\Omega(z)\big)^2 \le \frac{2^p}{\Gamma(p+1)}\int_\Omega \big(g_\Omega(z, w)\big)^p dm(w) \le m(\Omega),$$

where equality occurs when Ω is a disk centered at z. Equivalently,

$$\pi\big((1 - |\zeta|^2)|f'(\zeta)|\big)^2 \le \frac{2^p}{\Gamma(p+1)}\int_\mathbf{D}|f'|^2\big(-\log|\sigma_\zeta|\big)^p dm \le \int_\mathbf{D}|f'|^2 dm$$

holds for any surjectively conformal mapping $f : \mathbf{D} \mapsto \Omega$ with $f(\zeta) = z$, where equality holds when f is the identity.

Proof. Note that if $f : \mathbf{D} \mapsto \Omega$ is conformal and onto, then

$$g_\Omega(z, w) = g_\Omega\big(f(\zeta), f(\eta)\big) = -\log|\sigma_\zeta(\eta)|, \quad z = f(\zeta), \ w = f(\eta).$$

So it suffices to check the second part of Example 1.1.1. To this end, let

$$f(\eta) = \sum_{j=0}^{\infty} a_j \eta^j, \quad \eta \in \mathbf{D},$$

be the Taylor expansion of f at 0. Then

$$
\begin{aligned}
\big(E_p(f, 0)\big)^2 &= \int_{\mathbf{D}} |f'(\zeta)|^2 (-\log|\eta|)^p dm(\eta) \\
&= 2\pi \sum_{j=1}^{\infty} j^2 |a_j|^2 \int_0^1 r^{2(j-1)}(-\log r)^p r dr.
\end{aligned}
$$

This formula easily implies

$$\big(E_p(f, 0)\big)^2 \geq 2\pi|a_1|^2 \int_0^1 r(-\log r)^p dr = 2^{-p}\pi\Gamma(p+1)|f'(0)|^2.$$

Meanwhile, integrating by parts, we derive

$$
\begin{aligned}
\big(E_p(f, 0)\big)^2 &= \pi \sum_{j=1}^{\infty} j|a_j|^2 \int_0^1 (-\log r)^p dr^{2j} \\
&\leq \pi p \sum_{j=1}^{\infty} j|a_j|^2 \int_0^1 r(-\log r)^{p-1} dr \\
&= 2^{-p}\pi\Gamma(p+1) \sum_{j=1}^{\infty} j|a_j|^2 \\
&= 2^{-p}\Gamma(p+1) \int_{\mathbf{D}} |f'(\eta)|^2 dm(\eta).
\end{aligned}
$$

Accordingly,

$$
\begin{aligned}
2^{-p}\Gamma(p+1)\pi|f'(0)|^2 &\leq \int_\Omega |f'(\eta)|^2 (-\log|\eta|)^p dm(\eta) \\
&\leq 2^{-p}\Gamma(p+1) \int_{\mathbf{D}} |f'(\eta)|^2 dm(\eta).
\end{aligned}
$$

Substituting $f \circ \sigma_\zeta(\eta)$ for f in the last estimate, we see the desired assertion right away. Obviously, the equalities over there are valid for $f(\zeta) = \zeta$ and $\Omega = \mathbf{D}$. $\square$

These observations motivate the main issue of this book. From now on, besides the previously-introduced notation, we will use $\mathcal{H}$ as the class of all holomorphic mappings from $\mathbf{D}$ to $\mathbf{C}$. By

$$E_p(f, w) = \left(\int_\mathbf{D} |f'(z)|^2 \big(-\log|\sigma_w(z)| \big)^p dm(z) \right)^{\frac{1}{2}}$$

we mean the weighted energy integral associated with $p \in [0, \infty)$, $f \in \mathcal{H}$ and $w \in \mathbf{D}$. Using this term, we can symbolically write out a holomorphic Q space, as follows.

Definition 1.1.2. For $p \in [0, \infty)$ let $\mathcal{Q}_p$ be the class of all $f \in \mathcal{H}$ such that

$$\|f\|_{\mathcal{Q}_p} = \sup_{w \in \mathbf{D}} E_p(f, w) < \infty.$$

Each $\mathcal{Q}_p$ enjoys a large number of useful properties and has many applications; in particular, it is a dual Banach space under the norm $E_p(f) + |f(0)|$, and $Aut(\mathbf{D})$, the group of all conformal mappings from $\mathbf{D}$ to itself, acts on $\mathcal{Q}_p$ via the isometry: $E_p(f \circ \phi) = E_p(f)$ for any $\phi \in Aut(\mathbf{D})$. Perhaps, more important is that the spaces $\mathcal{Q}_p$, $p \in [0, \infty)$, form a nested scale of conformally invariant function spaces, but also bear natural connections to other such spaces as stated below:

First, if $p = 0$, $\mathcal{Q}_p = \mathcal{D}$, the Dirichlet space of all $f \in \mathcal{H}$ obeying

$$\|f\|_\mathcal{D} = \left(\int_\mathbf{D} |f'(z)|^2 dm(z) \right)^{\frac{1}{2}} < \infty.$$

Second, if $p = 1$, then $\mathcal{Q}_p = \mathcal{BMOA}$, the John–Nirenberg space of all analytic (holomorphic) functions of bounded mean oscillation; that is, $f \in \mathcal{BMOA}$ if and only if $f \in \mathcal{H}$ and its nontangential limit, identified with f, exists a.e. on the unit circle $\mathbf{T} = \{z \in \mathbf{C} : |z| = 1\}$ and satisfies

$$\|f\|_{\mathcal{BMOA}} = \sup_{I \subseteq \mathbf{T}} |I|^{-1} \int_I |f(\zeta) - f_I| \frac{|d\zeta|}{2\pi} < \infty,$$

where the supremum is taken over all open subarcs $I \subseteq \mathbf{T}$ with

$$|I| = \int_I \frac{|d\zeta|}{2\pi} \quad \text{and} \quad f_I = |I|^{-1} \int_I f(\zeta) \frac{|d\zeta|}{2\pi}.$$

Third, if $p > 1$, then $\mathcal{Q}_p = \mathcal{B}$, the Bloch space of those $f \in \mathcal{H}$ satisfying

$$\|f\|_\mathcal{B} = \sup_{z \in \mathbf{D}} (1 - |z|^2)|f'(z)| < \infty.$$

Obviously, the cases $p \in (0, 1)$ are of independent interest, and worth being further studied. The forthcoming essential characterizations of $\mathcal{Q}_p$, whose proofs can be found in [Xi3], will help us with our later development of the $\mathcal{Q}_p$ theory.

Theorem 1.1.3. *Let $p \in (0, 1)$ and $f \in \mathcal{H}$. Then the following statements are equivalent:*

(i) $f \in \mathcal{Q}_p$.

(ii) $\|f\|_{\mathcal{Q}_p,1} = \sup_{w \in \mathbf{D}} F_p(f, w) < \infty$, *where*

$$F_p(f, w) = \left(\int_{\mathbf{D}} |f'(z)|^2 \big(1 - |\sigma_w(z)|^2 \big)^p dm(z) \right)^{\frac{1}{2}}.$$

(iii) $\displaystyle \sup_{w \in \mathbf{D}} \left(\int_{\mathbf{D}} |f'(z)|^2 \big(|\sigma_w(z)|^{-2} - 1 \big)^p dm(z) \right)^{\frac{1}{2}} < \infty.$

(iv) $\displaystyle \sup_{w \in \mathbf{D}} \left(p \int_0^1 \left(\int_{|\sigma_w(z)| < r} |f'(z)|^2 dm(z) \right) (1 - r)^{p-1} dr \right)^{\frac{1}{2}} < \infty.$

(v) *For any natural number $n \geq 2$, i.e., $n - 1 \in \mathbf{N}$, the n-th derivative $f^{(n)}$ of f satisfies*

$$\sup_{w \in \mathbf{D}} \left(\int_{\mathbf{D}} |f^{(n)}(z)|^2 \big(1 - |\sigma_w(z)|^2 \big)^p \big(1 - |z|^2 \big)^{2n-2} dm(z) \right)^{\frac{1}{2}} < \infty.$$

(vi) *For $f(z) = \sum_{k=0}^{\infty} a_k z^k$ with $a_k = f^{(k)}(0)/k!$,*

$$\sup_{w \in \mathbf{D}} \left(\sum_{n=0}^{\infty} \frac{(1 - |w|^2)^p}{(n+1)^{p+1}} \left| \sum_{m=0}^{n} \frac{(m+1) a_{m+1} \Gamma(n - m + p)}{(n - m)!} \bar{w}^{n-m} \right|^2 \right)^{\frac{1}{2}} < \infty.$$

(vii) $d\mu_{f,p}(z) = |f'(z)|^2 (1 - |z|^2)^p dm(z)$ *is a p-Carleson measure on $\mathbf{D}$; that is,*

$$\|f\|_{\mathcal{Q}_p,2} = \sup_{S(I) \subseteq \mathbf{D}} \big(|I|^{-p} \mu_{f,p}(S(I)) \big)^{\frac{1}{2}} < \infty,$$

where the supremum ranges over all Carleson boxes $S(I) = \{r\xi \in \mathbf{D} : 1 - |I| \leq r < 1,\ \xi \in I\}$ based on open subarcs I of $\mathbf{T}$.

(viii) $\displaystyle \sup_{w \in \mathbf{D}} \left(\int_{\mathbf{D}} \left(\int_{\mathbf{D}} E_1(f, u) dm(u) \right) \big(1 - |\sigma_w(z)|^2 \big)^p \big(1 - |z|^2 \big)^{-2} dm(z) \right)^{\frac{1}{2}} < \infty.$

(ix) *There are a complex number c, an inner function I and an outer function O such that $f = cIO$ with $O \in \mathcal{Q}_p$ and*

$$\sup_{w \in \mathbf{D}} \left(\int_{\mathbf{D}} |O(z)|^2 \big(1 - |I(z)|^2 \big) \big(1 - |\sigma_w(z)|^2 \big)^p \big(1 - |z|^2 \big)^{-2} dm(z) \right)^{\frac{1}{2}} < \infty.$$

(x) *The nontangential limit $f(\xi)$ exists for a.e. $\xi \in \mathbf{T}$ and satisfies*

$$\|f\|_{\mathcal{Q}_p,3} = \sup_{I \subseteq \mathbf{T}} \left(|I|^{-p} \int_I \int_I \frac{|f(\zeta) - f(\eta)|^2}{|\zeta - \eta|^{2-p}} |d\zeta||d\eta| \right)^{\frac{1}{2}} < \infty,$$

where the supremum is taken over all open subarcs $I \subseteq \mathbf{T}$.

(xi) *There are functions $f_1, f_2 \in \mathcal{H}$ such that $f = f_1 + f_2$ and $\Re f_1, \Re f_2$ are essentially bounded on $\mathbf{T}$ and obey the finiteness condition of* (x).

Remark 1.1.4. The equivalence amongst (i), (ii), (iv), (v), (vi) and (vii) is also valid for $p \in [1, \infty)$. In addition, for $f \in \mathcal{Q}_p$, $p \in (0, 1)$, we have

$$\|f\|_{\mathcal{Q}_p} \approx \|f\|_{\mathcal{Q}_p,1} \approx \|f\|_{\mathcal{Q}_p,2} \approx \|f\|_{\mathcal{Q}_p,3}.$$

Here and thereafter, $X \approx Y$ means both $X \preceq Y$ and $Y \preceq X$; that is, there are positive constants c_1 and c_2 such that both $X \leq c_1 Y$ and $Y \leq c_2 X$ hold.

There is no general procedure known concerning the investigation of all classes $\mathcal{Q}_p$. Such a procedure seems unlikely to exist, given the rich variety of analytic, geometric, and even probabilistic phenomena which can be modeled by these spaces. However, we will try to get a better understanding of $\mathcal{Q}_p$ through $\|\cdot\|_{\mathcal{Q}_p}$, $\|\cdot\|_{\mathcal{Q}_p,k}$, $k = 1, 2, 3$ and their variations.

1.2 Logarithmic Conformal Mappings

The preceding Theorem 1.1.3 yields that $\log(1 - z)$ is a typical member of all $\mathcal{Q}_p$, $p \in (0, \infty)$, and accordingly that $\log f$ lies in $\bigcap_{p \in (0,\infty)} \mathcal{Q}_p$ for any univalent function $f : \mathbf{D} \mapsto \mathbf{C} \setminus \{0\}$. In what follows, we will obtain more about this fact via searching for the best estimate based on symmetrization.

Definition 1.2.1. Given a domain $\Omega \subseteq \mathbf{C}$. We call it circularly symmetric provided that for any $r \in (0, \infty)$ one of the following three conditions is valid:

(i) $\Omega \cap \{z \in \mathbf{C} : |z| = r\} = \emptyset$;

(ii) $\Omega \cap \{z \in \mathbf{C} : |z| = r\} = \{z \in \mathbf{C} : |z| = r\}$;

(iii) $\Omega \cap \{z \in \mathbf{C} : |z| = r\}$ is a single arc on $\{z \in \mathbf{C} : |z| = r\}$ which contains $\{r\}$ and is symmetric about the real line $\mathbf{R}$.

Moreover, the circular symmetrization of Ω is the domain Ω^* determined in the following way:

(iv) If $\{\theta \in [-\pi, \pi] : re^{i\theta} \in \Omega\} = \emptyset$, then $\Omega^* \cap \{z \in \mathbf{C} : |z| = r\} = \emptyset$;

(v) If $\{\theta \in [-\pi, \pi] : re^{i\theta} \in \Omega\} = [-\pi, \pi]$, then $\Omega^* \cap \{z \in \mathbf{C} : |z| = r\} = \{z \in \mathbf{C} : |z| = r\}$;

(vi) If $\{\theta \in [-\pi, \pi] : re^{i\theta} \in \Omega\}$ is neither $\emptyset$ nor $[-\pi, \pi]$, and if its one-dimensional Lebesgue measure is α, then $\Omega^* \cap \{z \in \mathbf{C} : |z| = r\}$ equals $\{re^{i\theta} : 2|\theta| < \alpha\}$;

(vii) If $0 \in \Omega$, then $0 \in \Omega^*$ and vice versa.

Clearly, Ω^* is a circularly symmetric domain, and Ω is circularly symmetric if and only if $\Omega = \Omega^*$. At the same time, Ω^* is simply connected whenever Ω is simply connected; see also [Hay, p. 115].

Lemma 1.2.2. *Let $\Omega \subset \mathbf{C}$ be a simply connected hyperbolic domain and fix $z_0 \in \Omega$. Let $g_\Omega(\cdot, z_0)$ and $g_{\Omega^*}(\cdot, |z_0|)$ be the Green functions of Ω and Ω^* with poles at z_0 and $|z_0|$ respectively. Set $g_\Omega(z, z_0) = 0$ for $z \in \mathbf{C} \setminus \Omega$ and $g_{\Omega^*}(z, |z_0|) = 0$ for $z \in \mathbf{C} \setminus \Omega^*$. Then:*

(i) *For any $r \in (0, \infty)$,*

$$\sup_{\xi \in \mathbf{T}} g_\Omega(r\xi, z_0) \leq \sup_{\xi \in \mathbf{T}} g_{\Omega^*}(r\xi, |z_0|).$$

(ii) *For any $r \in (0, \infty)$ and any $p \in [1, \infty)$,*

$$\int_{\mathbf{T}} \big(g_\Omega(r\xi, z_0)\big)^p |d\xi| \leq \int_{\mathbf{T}} \big(g_{\Omega^*}(r\xi, |z_0|)\big)^p |d\xi|.$$

(iii) $\Omega = \Omega^*$ *provided $z_0 > 0$ and*

$$\int_{\mathbf{T}} \big(g_\Omega(r\xi, z_0)\big)^p |d\xi| = \int_{\mathbf{T}} \big(g_{\Omega^*}(r\xi, |z_0|)\big)^p |d\xi|,$$

for some $r \in (0, \infty)$ and some $p \in (1, \infty)$.

Proof. (i) and (ii) are consequences of [Bae1, Theorem 5], and (iii) is a corollary of [EsSh, Theorem 1]. $\qquad\square$

Theorem 1.2.3. *Let $f : \mathbf{D} \mapsto \mathbf{C} \setminus \{0\}$ be univalent with $f(0) = 1$. Then:*

(i) *For $p \in (0, \infty)$,*

$$\|\log f\|_{\mathcal{Q}_p} \leq \left(8\pi \int_0^1 \left(\log \frac{1+r}{1-r}\right)^p r^{-1} dr\right)^{\frac{1}{2}}.$$

(ii) *For $p \in [1, \infty)$,*

$$\|\log f\|_{\mathcal{Q}_p} \leq \left(16 \int_0^1 \left(\int_{\mathbf{T}} \frac{|d\xi|}{|1 - r^2\xi^2|^2}\right) \left(\log \frac{1}{r}\right)^p r \, dr\right)^{\frac{1}{2}},$$

with equality if and only if $\|\log f\|_\mathcal{B} = 4$.

Proof. For $w \in \mathbf{D}$, let $f_w = \left(f(w)\right)^{-1} f \circ \sigma_w$, $\mathbf{D}_w = f_w(\mathbf{D})$, and $g_{\mathbf{D}_w}(\cdot, 1)$ be the Green function of $\mathbf{D}_w$ and be defined to equal 0 outside $\mathbf{D}_w$. Then a simple calculation with the conformal transform σ_w plus the conformal invariance of the Green function yields

$$\|\log f\|_{\mathcal{Q}_p}^2 = \sup_{w \in \mathbf{D}} \int_{\mathbf{D}} \left|\frac{f_w'(z)}{f_w(z)}\right|^2 \left(\log \frac{1}{|z|}\right)^p dm(z)$$

$$= \sup_{w \in \mathbf{D}} \int_{\mathbf{D}_w} \left(g_{\mathbf{D}_w}(z, 1)\right)^p |z|^{-2} dm(z)$$

$$= \sup_{w \in \mathbf{D}} \int_0^\infty \left(\int_{\mathbf{T}} \left(g_{\mathbf{D}_w}(r\xi, 1)\right)^p |d\xi|\right) r^{-1} dr.$$

Consider the Koebe-like map: $k(z) = \left(\frac{1+z}{1-z}\right)^2$. This map is conformal with $\Omega = k(\mathbf{D}) = \mathbf{C} \setminus (-\infty, 0]$, and consequently, the Green function $g_\Omega(\cdot, 1)$ of Ω with pole at 1 obeys

$$g_\Omega(z, 1) = \log \left|\frac{1 + \sqrt{z}}{1 - \sqrt{z}}\right|, \quad z \in \Omega,$$

which is also extended to be 0 on $(-\infty, 0]$.

If $\mathbf{D}_w^*$ stands for the circularly symmetric domain of $\mathbf{D}_w$ and $g_{\mathbf{D}_w^*}(\cdot, 1)$ is the Green function of $\mathbf{D}_w$ with pole at 1 — of course, the value of $g_{\mathbf{D}_w^*}(\cdot, 1)$ is set to be 0 outside $\mathbf{D}_w^*$, then $\mathbf{D}_w^* \subseteq \Omega$ and hence

$$g_{\mathbf{D}_w^*}(z, 1) \leq g_\Omega(z, 1), \quad z \in \mathbf{C}.$$

(i) In case of $p \in (0, \infty)$, Lemma 1.2.2 (i) is applied to imply

$$\|\log f\|_{\mathcal{Q}_p}^2 = \sup_{w \in \mathbf{D}} \int_0^\infty \left(\int_{\mathbf{T}} \left(g_{\mathbf{D}_w}(r\xi, 1)\right)^p |d\xi|\right) r^{-1} dr$$

$$\leq 2\pi \int_0^\infty \sup_{\xi \in \mathbf{T}} \left(g_{\mathbf{D}_w}(r\xi, 1)\right)^p r^{-1} dr$$

$$\leq 2\pi \int_0^\infty \sup_{\xi \in \mathbf{T}} \left(g_{\mathbf{D}_w^*}(r\xi, 1)\right)^p r^{-1} dr$$

$$\leq 2\pi \int_0^\infty \sup_{\xi \in \mathbf{T}} \left(g_\Omega(r\xi, 1)\right)^p r^{-1} dr$$

$$= 2\pi \int_0^\infty \left(\log \left|\frac{1 + \sqrt{r}}{1 - \sqrt{r}}\right|\right)^p r^{-1} dr$$

$$\leq 8\pi \int_0^1 \left(\log \frac{1 + r}{1 - r}\right)^p r^{-1} dr,$$

as desired.

(ii) When $p \in [1, \infty)$, we use Lemma 1.2.2 (ii) to deduce

$$
\begin{aligned}
\| \log f \|_{\mathcal{Q}_p}^2 &= \sup_{w \in \mathbf{D}} \int_0^\infty \left(\int_{\mathbf{T}} (g_{\mathbf{D}_w}(r\xi, 1))^p |d\xi| \right) r^{-1} dr \\
&\leq \sup_{w \in \mathbf{D}} \int_0^\infty \left(\int_{\mathbf{T}} (g_{\mathbf{D}_w^*}(r\xi, 1))^p |d\xi| \right) r^{-1} dr \\
&\leq \int_0^\infty \left(\int_{\mathbf{T}} (g_\Omega(r\xi, 1))^p |d\xi| \right) r^{-1} dr \\
&= \int_\Omega (g_\Omega(z, 1))^p |z|^{-2} dm(z) \\
&= \int_{\mathbf{D}} \left| \frac{k'(z)}{k(z)} \right|^2 (-\log|z|)^p dm(z) \\
&= 16 \int_0^1 \left(\int_{\mathbf{T}} \frac{|d\xi|}{|1 - r^2\xi^2|^2} \right) (-\log r)^p r \, dr,
\end{aligned}
$$

as required. Moreover, the above inequality becomes an equality for $f = k$.

Next we handle the situation of equality. If $\| \log f \|_{\mathcal{B}} = 4$, then $\| \log f \|_{\mathcal{Q}_p} \geq \| \log k \|_{\mathcal{Q}_p}$. In fact, since $\{ f_{w_n} \}$ is a normal family, we may assume that it converges locally uniformly to a univalent function F on $\mathbf{D}$ with $F(0) = 1$. Accordingly,

$$
\lim_{n \to \infty} (1 - |w_n|^2) \left| \frac{f'(w_n)}{f(w_n)} \right| = \lim_{n \to \infty} |f'_{w_n}(0)| = |F'(0)| = 4
$$

which implies $F(z) = k(\eta z)$ for some $\eta \in \mathbf{T}$ thanks to the uniqueness of the Koebe function. Furthermore,

$$
\begin{aligned}
\| \log f \|_{\mathcal{Q}_p}^2 &= \| \log f_{w_n} \|_{\mathcal{Q}_p}^2 \\
&\geq \int_{\mathbf{D}} \left| \frac{f'_{w_n}(z)}{f_{w_n}(z)} \right|^2 (-\log|z|)^p dm(z) \\
&\to \int_{\mathbf{D}} \left| \frac{F'(z)}{F(z)} \right|^2 (-\log|z|)^p dm(z) \quad \text{as} \quad n \to \infty.
\end{aligned}
$$

This gives $\| \log f \|_{\mathcal{Q}_p} \geq \| \log k \|_{\mathcal{Q}_p}$.

Note that $\| \log f \|_{\mathcal{Q}_p} \leq \| \log k \|_{\mathcal{Q}_p}$ as proved already. So we derive

$$
\| \log f \|_{\mathcal{Q}_p} = \| \log k \|_{\mathcal{Q}_p}.
$$

On the other hand, suppose $\| \log f \|_{\mathcal{Q}_p} = \| \log k \|_{\mathcal{Q}_p}$. Then there is a sequence $\{ w_n \}$ in $\mathbf{D}$ such that

$$
\lim_{n \to \infty} \int_{\mathbf{D}} \left| \frac{f'_{w_n}(z)}{f_{w_n}(z)} \right|^2 (-\log|z|)^p dm(z) = \| \log k \|_{\mathcal{Q}_p}^2.
$$

Observe that $\{f_{w_n}\}$ forms a normal family. Thus we may assume that it converges to a univalent function $F \in \mathcal{H}$ uniformly on compacta of $\mathbf{D}$. Accordingly, $F(0) = 1$ and

$$\| \log k \|_{\mathcal{Q}_p}^2 = \int_{\mathbf{D}} \left| \frac{F'(z)}{F(z)} \right|^2 (- \log |z|)^p dm(z).$$

Suppose that $g_{F(\mathbf{D})}(\cdot, 1)$ and $g_{F(\mathbf{D})^*}(\cdot, 1)$ are the Green functions of $F(\mathbf{D})$ and its circularly symmetric domain $F(\mathbf{D})^*$ with pole at 1 respectively, and suppose that their values are 0 on $\mathbf{C} \setminus F(\mathbf{D})$ and $\mathbf{C} \setminus F(\mathbf{D})^*$ respectively. Then using Lemma 1.2.2 (ii) again as well as $F(\mathbf{D})^* \subseteq \Omega$, we further get

$$
\begin{aligned}
\| \log k \|_{\mathcal{Q}_p}^2 &= \int_{\mathbf{D}} \left| \frac{F'(z)}{F(z)} \right| (- \log |z|)^p dm(z) \\
&= \int_0^\infty \int_{\mathbf{T}} \left(g_{F(\mathbf{D})}(r\xi, 1) \right)^p |d\xi| r^{-1} dr \\
&\leq \int_0^\infty \int_{\mathbf{T}} \left(g_{F(\mathbf{D})^*}(r\xi, 1) \right)^p |d\xi| r^{-1} dr \\
&\leq \int_0^\infty \int_{\mathbf{T}} \left(g_{\Omega}(r\xi, 1) \right)^p |d\xi| r^{-1} dr \\
&= \int_{\mathbf{D}} \left| \frac{k'(z)}{k(z)} \right|^2 (- \log |z|)^p dm(z) \\
&= \| \log k \|_{\mathcal{Q}_p}^2,
\end{aligned}
$$

which yields

$$
\begin{aligned}
\int_0^\infty \int_{\mathbf{T}} \left(g_{F(\mathbf{D})}(r\xi, 1) \right)^p |d\xi| r^{-1} dr &= \int_0^\infty \int_{\mathbf{T}} \left(g_{F(\mathbf{D})^*}(r\xi, 1) \right)^p |d\xi| r^{-1} dr \\
&= \int_0^\infty \int_{\mathbf{T}} \left(g_{\Omega}(r\xi, 1) \right)^p |d\xi| r^{-1} dr.
\end{aligned}
$$

If $p \in (1, \infty)$, then an application of Lemma 1.2.2 (iii) produces $F(\mathbf{D}) = F(\mathbf{D})^*$, and so $F(\mathbf{D}) \subseteq \Omega$ which yields $g_{F(\mathbf{D})}(r\xi, 1) \leq g_{\Omega}(r\xi, 1)$. Consequently, $g_{F(\mathbf{D})}(r\xi, 1) = g_{\Omega}(r\xi, 1)$ and hence $F(\mathbf{D}) = \Omega$. This means that there is an $\eta \in \mathbf{T}$ such that $F(z) = k(\eta z)$. Because $\{f_{w_n}\}$ is convergent to F uniformly on compacta of $\mathbf{D}$, it follows that

$$\lim_{n \to \infty} (1 - |w_n|^2) \left| \frac{f'(w_n)}{f(w_n)} \right| = \lim_{n \to \infty} |f'_{w_n}(0)| = |F'(0)| = |k'(0)| = 4.$$

Obviously, this limit ensures that $\| \log f \|_{\mathcal{B}} \geq 4$. However, it is well known that $\| \log f \|_{\mathcal{B}} \leq 4$. Therefore $\| \log f \|_{\mathcal{B}} = 4$.

If $p = 1$, then $\| \log f \|_{\mathcal{Q}_1} = \| \log k \|_{\mathcal{Q}_1}$ implies that there exists a sequence $\{w_n\}$ in $\mathbf{D}$ such that

$$\lim_{n \to \infty} \int_{\mathbf{D}} \left| \frac{f'_{w_n}(z)}{f_{w_n}(z)} \right|^2 (- \log |z|) dm(z) = \int_{\mathbf{D}} \left| \frac{k'(z)}{k(z)} \right|^2 (- \log |z|) dm(z).$$

As before, we may assume that $\{f_{w_n}\}$ converges to a nonzero univalent function F with $F(0) = 1$ uniformly on compacta of $\mathbf{D}$. With this, we get

$$\int_{\mathbf{D}} \left| \frac{F'(z)}{F(z)} \right|^2 (-\log|z|)dm(z) = \int_{\mathbf{D}} \left| \frac{k'(z)}{k(z)} \right|^2 (-\log|z|)dm(z).$$

If $g_{F(\mathbf{D})}(\cdot, 0)$ and $g_{k(\mathbf{D})}(\cdot, 0)$ stand for the Green functions of $F(\mathbf{D})$ and $k(\mathbf{D})$ with pole at 0, and being 0 outside of $F(\mathbf{D})$ and $k(\mathbf{D})$, respectively, then

$$\int_{-\infty}^{\infty} \int_{-\infty}^{\infty} \big(g_{F(\mathbf{D})}(x + iy, 0) - g_{k(\mathbf{D})}(x + iy, 0) \big) dy dx = 0.$$

Writing

$$H(x, s) = \int_{-s}^{s} \big(g_{F(\mathbf{D})}(x + iy, 0) - g_{k(\mathbf{D})}(x + iy, 0) \big) dy, \quad s \in [0, \infty),$$

we find $H(x, s) \geq 0$; see the proofs of Theorem 2 and Proposition 6 in [Gir1] (see (6) in [Gir2]). Accordingly, $\lim_{s \to \infty} H(x, s) \geq 0$ for $x \in (-\infty, \infty)$. So

$$0 \leq \int_{-\infty}^{\infty} \lim_{s \to \infty} H(x, s) dx = 0$$

which gives $\lim_{s \to \infty} H(x, s) = 0$ for a.e. $x \in (-\infty, \infty)$. Note that since $H(x, s)$ is increasing and continuous with $s \geq 0$, we conclude that $H(x, s) = 0$ for $s \in [0, \infty)$ and that

$$g_{F(\mathbf{D})}(x + iy, 0) - g_{k(\mathbf{D})}(x + iy, 0) = 0.$$

Accordingly, $F(\mathbf{D}) = k(\mathbf{D})$ and hence $f(z) = k(\eta z)$ for some $\eta \in \mathbf{T}$ which naturally deduces $\| \log f \|_{\mathcal{B}} = 4$. $\square$

1.3   Conformal Domains and Superpositions

The purpose of this section is two-fold. One is to give a criterion for a simply connected domain $\mathbf{\Omega} \subseteq \mathbf{C}$ to be a conformal $\mathcal{Q}_p$ domain. The other is to characterize the membership of a conformal mapping in $\mathcal{Q}_p$ by means of the geometric structure of the image domain $\mathbf{\Omega}$ and to give an application of this geometric criterion to the superposition acting between two holomorphic Q classes.

Clearly, when f is a conformal mapping from $\mathbf{D}$ into $\mathbf{\Omega}$, a proper subdomain of $\mathbf{C}$, we see that $f \in \mathcal{D}$ if and only if the area of the image of $f(\mathbf{D})$: $\int_{f(\mathbf{D})} dm$ is finite. This simple observation leads to a description of the conformal $\mathcal{Q}_p$ domains as follows.

Theorem 1.3.1. *Let $p \in [0, \infty)$ and $\Omega \subset \mathbf{C}$ be a simply connected hyperbolic domain. Then the following statements are equivalent:*

(i) Ω *is a conformal $\mathcal{Q}_p$ domain; that is, every conformal mapping $f : \mathbf{D} \mapsto \Omega$ lies in $\mathcal{Q}_p$.*

(ii) $\sup_{w \in \Omega} d(w, \partial\Omega) < \infty.$

Proof. Suppose (ii) is valid. If $f : \mathbf{D} \mapsto f(\mathbf{D}) \subseteq \Omega$ is conformal, then

$$d(w, \partial f(\mathbf{D})) \leq d(w, \partial\Omega), \quad w \in f(\mathbf{D}).$$

Note that this conformal mapping f obeys the following well-known estimate:

$$d\big(f(z), \partial f(\mathbf{D})\big) \leq (1 - |z|^2)|f'(z)| \leq 4d\big(f(z), \partial f(\mathbf{D})\big), \quad z \in \mathbf{D}.$$

So we read off $f \in \mathcal{B}$ and so $f \in \mathcal{Q}_p$. This proves (i).

    Conversely, assuming (i) holds, we verify (ii). It suffices to demonstrate that every simply connected domain $\Omega \subset \mathbf{C}$ with $\sup_{w \in \Omega} d(w, \partial\Omega) = \infty$ must contain a simply connected domain Ω^* with the same property. To this end, choose a sequence $\{w_j\}$ in Ω such that

$$|w_{j+1}| - |w_j| > 4j \quad \text{and} \quad \mathbf{D}(w_j, 2j) = \{w \in \mathbf{C} : |w - w_j| < 2j\} \subseteq \Omega, \quad j \in \mathbf{N}.$$

It is clear that the closures of the disks $\mathbf{D}(w_j, j) = \{w \in \mathbf{C} : |w - w_j| < j\}$ are mutually disjoint. Inductively, we can take distinct points:

$$\xi_1 \in \{w \in \mathbf{C} : |w - w_1| = 1\} \quad \text{and} \quad \zeta_j, \xi_{j+1} \in \{w \in \mathbf{C} : |w - w_j| = j\} \quad \text{for} \quad j \geq 2,$$

and obtain a sequence of mutually disjoint, piecewise linear, simple arcs $\mathbf{A}_j$, $j \in \mathbf{N}$, made up of segments parallel to either the real or the imaginary axis in such a way that each A_j is contained in Ω, joins ξ_j and ζ_{j+1}, and contains no other points of the closure of $\bigcup_{j=1}^{\infty} \mathbf{D}(w_j, j)$. From this construction it turns out that for each $j \in \mathbf{N}$ there is a positive ϵ_j such that $\mathbf{U}_j = \{w \in \mathbf{C} : \inf_{\xi \in \mathbf{A}_j} |\xi - w| < \epsilon_j/2\}$ are also mutually disjoint. Now setting

$$\mathbf{\Omega}_1 = \mathbf{D}(w_1, 1) \quad \text{and} \quad \mathbf{\Omega}_j = \Big(\bigcup_{k=1}^{j} \mathbf{D}(w_k, k) \Big) \cup \Big(\bigcup_{k=1}^{j-1} \mathbf{U}_j \Big), \quad j - 1 \in \mathbf{N},$$

we derive that each $\mathbf{\Omega}_j$ is not only a bounded domain whose boundary is a simple closed curve but also enjoys the following inclusion chain:

$$\mathbf{D}(w_j, j) \subseteq \mathbf{\Omega}_j \subseteq \mathbf{\Omega}_{j+1} \subseteq \Omega, \quad j \in \mathbf{N},$$

hence producing that $\mathbf{\Omega}^* = \bigcup_{j=1}^{\infty} \mathbf{\Omega}_j \subseteq \Omega$ is a simply connected domain which contains arbitrarily large disks. Due to this reason, we see that any conformal mapping from $\mathbf{D}$ onto $\mathbf{\Omega}^*$ is not in $\mathcal{B}$ and hence not in $\mathcal{Q}_p$, and so that Ω is not a conformal $\mathcal{Q}_p$ domain, contradicting the assumption (i). This completes the argument. $\square$

In order to give a geometric criterion for a conformal mapping from $\mathbf{D}$ into a simply connected hyperbolic domain $\Omega \subset \mathbf{C}$ to belong to $\mathcal{Q}_p$, we say that a function $f \in \mathcal{H}$ is of the conformally invariant Besov class $\mathcal{B}_q$, $q \in (1, \infty)$, provided

$$\|f\|_{\mathcal{B}_q} = \left(\int_{\mathbf{D}} |f'(z)|^q (1 - |z|^2)^{q-2} dm(z) \right)^{\frac{1}{q}} < \infty.$$

Here the conformal invariance of $\mathcal{B}_q$ simply means:

$$\|f \circ \sigma_w\|_{\mathcal{B}_q} = \|f\|_{\mathcal{B}_q}, \quad w \in \mathbf{D}.$$

Theorem 1.3.2. *Let $p \in [0, \infty)$ and $\Omega \subset \mathbf{C}$ be a simply connected hyperbolic domain. If $f : \mathbf{D} \mapsto \Omega$ is a surjectively conformal mapping, then:*

(i) *$f \in \mathcal{Q}_p$ when and only when $\sup_{w \in \Omega} d(w, \partial\Omega) < \infty$.*

(ii) *$f \in \mathcal{B}_{p+2}$ when and only when $\int_{\Omega} \left(d(w, \partial\Omega) \right)^p dm(w) < \infty$.*

Proof. Note again that $f(\mathbf{D}) = \Omega$ and

$$d\big(f(z), \partial f(\mathbf{D})\big) \leq (1 - |z|^2)|f'(z)| \leq 4d\big(f(z), \partial f(\mathbf{D})\big), \quad z \in \mathbf{D}.$$

So the assertion (i) follows from the equivalence $f \in \mathcal{Q}_p \Leftrightarrow f \in \mathcal{B}$ and the assertion (ii) follows from the conformal change of variable $w = f(z)$. $\square$

Corollary 1.3.3. *Let $q \in (1, \infty)$. Suppose $\{w_j\}$ is a sequence of points in $\mathbf{C}$. Assume $\{r_j\}$ and $\{h_j\}$ are sequences of positive numbers with the following property: for each $j \in \mathbf{N}$,*

$$0 \leq \arg w_j \leq \pi/4, \quad |w_j| \leq |w_{j+1}|/2, \quad r_j < |w_j|/4 \quad and \quad |j_j| < \min\{r_j, r_{j+1}\}/3.$$

Then $\Omega = \bigcup_{j=1}^{\infty} (\mathbf{D}_j \cup \mathbf{R}_j)$ is a simply connected hyperbolic domain of $\mathbf{C}$, where $\mathbf{D}_j = \{w \in \mathbf{C} : |w - w_j| < r_j\}$ and $\mathbf{R}_j$ is the rectangle whose longer symmetry axis is the segment $[w_j, w_{j+1}]$ and whose shorter side has length $2h_j$. Furthermore, a conformal mapping from $\mathbf{D}$ onto Ω belongs to $\mathcal{B}_q$ if and only if

$$\sum_{j=1}^{\infty} r_j^q + \sum_{j=1}^{\infty} h_j^{q-1} |w_{j+1} - w_j| < \infty.$$

Proof. First of all, note that the distance to the origin increases as one moves along the segment $[w_j, w_{j+1}]$ from w_j to w_{j+1}. So $\bigcup_{j=1}^{n} (\mathbf{D}_j \cup \mathbf{R}_j)$ is simply connected for each $n \in \mathbf{N}$, and consequently, Ω is simply connected.

Next, suppose $f : \mathbf{D} \mapsto \Omega$ is conformal and surjective. Then from Theorem 1.3.2 (ii) and the above-required property it is not hard to figure out that $f \in \mathcal{B}_q$

if and only if

$$\infty > \sum_{j=1}^{\infty} \left(\int_{\mathbf{D}_j} (d(w, \partial\mathbf{\Omega}))^{q-2} dm(w) + \int_{\mathbf{R}_j} (d(w, \partial\mathbf{\Omega}))^{q-2} dm(w) \right)$$

$$\approx \sum_{j=1}^{\infty} \left(\int_{\mathbf{D}_j} (d(w, \partial\mathbf{D}_j))^{q-2} dm(w) + \int_{\mathbf{R}_j} (d(w, \partial\mathbf{R}_j))^{q-2} dm(w) \right)$$

$$\approx \sum_{j=1}^{\infty} r_j^q + \sum_{j=1}^{\infty} h_j^{q-1} |w_{j+1} - w_j|,$$

as desired. $\qquad\square$

Using the previous construction of unbounded conformal mappings in $\mathcal{B}_q$, $q > 1$ we can characterize the superposition associated with $\mathcal{Q}_p$.

Theorem 1.3.4. *Let $p \in [0, \infty)$. For $f \in \mathcal{H}$ and an entire function ϕ on $\mathbf{C}$ set $\mathsf{S}_\phi(f) = \phi(f)$. Then the following statements are equivalent:*

(i) S_ϕ *sends $\mathcal{Q}_p$ to $\mathcal{B}$.*

(ii) S_ϕ *sends $\mathcal{B}_{p+2}$ to $\mathcal{B}$.*

(iii) ϕ *is linear; that is, $\phi(w) = aw + b$, where $a, w, b \in \mathbf{C}$.*

Proof. (i)$\Rightarrow$(ii) This implication is trivial thanks to $\mathcal{B}_{p+2} \subset \mathcal{Q}_p$.

(ii)$\Rightarrow$(iii) Suppose (ii) holds. If (iii) fails, then ϕ' is not a constant mapping, and hence by Liouville's Theorem on the entire functions there is a sequence of points $\{w_j\}$ in $\mathbf{C}$ such that

$$|w_1| > 2, \quad |w_{j+1}| \geq |w_j| \quad \text{and} \quad |\phi'(w_j)| \geq 2^{2j} \quad \text{for} \quad j \in \mathbf{N}.$$

Note that at least one of the eight octants: $\arg^{-1}\left([(k-1)\pi/4, k\pi/4)\right)$, $k = 1, 2, ..., 8$, contains infinitely many elements of $\{w_j\}$. Using a rotation if necessary, we may assume that $0 \leq \arg w_j < \pi/4$, and then use Corollary 1.3.3 to derive that if

$$r_j = 2^{-j} \quad \text{and} \quad h_j = 2^{-j-2} |w_j - w_{j+1}|^{-\frac{1}{p+1}},$$

and if $\mathbf{\Omega}$ is the simply connected domain given in Corollary 1.3.3 and $f : \mathbf{D} \mapsto \mathbf{\Omega}$ is a surjectively conformal mapping, then it follows that $f \in \mathcal{B}_{p+2}$ owing to Corollary 1.3.3. Choosing $z_j \in \mathbf{D}$ such that $f(z_j) = w_j$ for $j \in \mathbf{N}$, we find $\lim_{j\to\infty} |z_j| = 1$, and consequently,

$$|\phi'(w_j)||f'(z_j)|(1 - |z_j|^2) \approx |\phi'(w_j)| d_\mathbf{\Omega}(w_j, \partial\mathbf{\Omega}) \succeq r_j^{-1} \to \infty \quad \text{as} \quad j \to \infty.$$

This yields $\mathsf{S}_\phi(f) \notin \mathcal{B}$, a contradiction.

(iii)$\Rightarrow$(i) If $\phi(w) = aw + b$, where $a, w, b \in \mathbf{C}$, then ϕ' is a constant a and hence $\left(\mathsf{S}_\phi(f)\right)'(z) = af'(z)$ which implies (i) thanks to $\mathcal{Q}_p \subseteq \mathcal{B}$. $\qquad\square$

The following consequence of Theorem 1.3.4 tells us that the action of super-position between any two conformally invariant spaces from the Q families and the Besov families is trivial. The reason behind this phenomenon is that any conformal mapping from $\mathbf{C}$ onto $\mathbf{C}$ is linear, namely, of the form: $aw + b$.

Corollary 1.3.5. *Let ϕ be an entire function on $\mathbf{C}$. Suppose $\mathcal{X}$ and $\mathcal{Y}$ are any two members in $\{\mathcal{Q}_p\}_{p\in(0,\infty)}$ and $\{\mathcal{B}_{p+2}\}_{p\in(0,\infty)}$. Then S_ϕ sends $\mathcal{X}$ to $\mathcal{Y}$ if and only if ϕ is linear whenever $\mathcal{X} \subseteq \mathcal{Y}$ or ϕ is constant whenever $\mathcal{X} \not\subseteq \mathcal{Y}$.*

Proof. This follows from Theorem 1.3.4 right away.   □

1.4   Descriptions via Harmonic Majorants

In this section, we investigate $\mathcal{Q}_p$ through a solvable function determined by the integral inducing $\|\cdot\|_{\mathcal{Q}_p,1}$. To begin, we need the following notion.

Definition 1.4.1. Given a connected open proper subset Ω of the extended complex plane $\mathbf{C}^e$ with boundary $\partial\Omega$. Let u be a function from $\partial\Omega$ to the extended real number system $\mathbf{R}^e = \mathbf{R} \cup \{\pm\infty\}$. Then

(i) $\mathcal{LP}(u, \Omega) = \Big\{f: \quad f \quad$ is subharmonic and bounded above on Ω with

$$\limsup_{z\to\xi} f(z) \leq f(\xi),\ \xi \in \partial\Omega\Big\}$$

and

$$\mathcal{UP}(u, \Omega) = \Big\{f: \quad f \quad \text{is supharmonic and bounded above on } \Omega \text{ with}$$

$$\liminf_{z\to\xi} f(z) \geq f(\xi),\ \xi \in \partial\Omega\Big\}$$

are respectively called the lower and upper Perron families associated with u and Ω.

(ii) If

$$\sup\{f(z) : f \in \mathcal{UP}(u, \Omega)\} = \inf\{f(z) : f \in \mathcal{LP}(u, \Omega)\},\ z \in \Omega$$

is a finite function, then u is called solvable on $\partial\Omega$, and hence, it is the solution of the Dirichlet problem with boundary value u.

(iii) If $u \neq -\infty$, then we say that it has the least harmonic majorant $U : \mathbf{D} \mapsto \mathbf{R}$ whenever U is harmonic on $\mathbf{D}$, $u \leq U$ on $\mathbf{D}$, and $U \leq V$ on $\mathbf{D}$ for any harmonic function $V : \mathbf{D} \mapsto \mathbf{R}$ with $u \leq V$ on $\mathbf{D}$.

Definition 1.4.1, and the following proposition whose (ii) is Harnack's Principle, are classical and can be found in [Con].

Proposition 1.4.2. *Let $\Omega \subset \mathbf{C}$ be a domain with boundary $\partial\Omega$.*

(i) *If $\{u_n\}$ is an increasing sequence of solvable functions on $\partial\Omega$ and $u(z) = \lim_{n\to\infty} u_n(z)$ for $z \in \Omega$, then u is solvable on $\partial\Omega$ and $\sup\{f(z) : f \in \mathcal{UP}(u_n, \Omega)\}$ converges to $\sup\{f(z) : f \in \mathcal{UP}(u, \Omega)\}$ uniformly on compacta of Ω, or else*

$$\sup\{f(z) : f \in \mathcal{UP}(u, \Omega)\} = \inf\{f(z) : f \in \mathcal{LP}(u, \Omega)\} = \infty, \ z \in \Omega$$

and $\sup\{f(z) : f \in \mathcal{UP}(u_n, \Omega)\}$ converges to ∞ uniformly on compacta of Ω.

(ii) *If $\{u_n\}$ is an increasing sequence of harmonic functions from Ω to $\mathbf{R}$, and if $\{u_n(z_0)\}$ is bounded for some $z_0 \in \Omega$, then $\{u_n\}$ converges uniformly on compacta of Ω to a harmonic function from Ω to $\mathbf{R}$.*

Lemma 1.4.3. *Given a function $f : \mathbf{D} \mapsto \mathbf{R}$. For $(r, \xi) \in [0, 1) \times \mathbf{T}$ let $f(r, \xi) = \sup_{s\in[0,r]} f(s\xi)$. Then $f(\cdot, \xi)$ is increasing on $(0, 1)$ for each $\xi \in \mathbf{T}$. Furthermore:*

(i) *If f is continuous on $\mathbf{D}$, then $f(r, \cdot)$ is continuous on $\mathbf{T}$ for each $r \in [0, 1)$.*

(ii) *If f is continuous and bounded on $\mathbf{D}$, then*

$$F(z) = \sup_{r\in[0,1)} \int_{\mathbf{T}} \left(\frac{r^2 - |z|^2}{|r\xi - z|^2}\right) f(r, \xi) \frac{|d\xi|}{2\pi}$$

is a harmonic function on $\mathbf{D}$, and

$$\lim_{r\to 1^-} f(r, \cdot) = f(1, \cdot) = \sup_{s\in[0,1)} f(s\cdot)$$

is a solvable function on $\mathbf{T}$ with F being the solution of its Dirichlet's problem.

Proof. The fact that

$$0 \le r_1 < r_2 < 1 \Rightarrow f(r_1, \xi) \le f(r_2, \xi), \quad \xi \in \mathbf{T}$$

is a straightforward consequence of the definition of $f(r, \xi)$.

(i) In order to prove the continuity, we may assume that for a fixed $r \in [0, 1)$, $f(r, \cdot)$ is discontinuous at some point $\xi_0 \in \mathbf{T}$. Then there exists an $\epsilon_0 > 0$ such that for any $n \in \mathbf{N}$ one can find such a $\xi_n \in \mathbf{T}$ that

$$|\xi_n - \xi_0| < \frac{1}{n} \quad \text{and} \quad |f(r, \xi_n) - f(r, \xi_0)| \ge \epsilon_0.$$

Since f is continuous on $\mathbf{D}$, we can conclude that $f(r, \cdot)$ is attained. Accordingly, there are $r_0, r_n \in [0, r]$ such that

$$f(r, \xi_0) = f(r_0\xi_0) \quad \text{and} \quad f(r, \xi_n) = f(r_n\xi_n).$$

Clearly, there is a subsequence $\{r_{n_k}\}$ which converges to $s_0 \in [0, r]$. This plus the continuity of f implies $f(r_{n_k} \xi_{n_k}) \to f(s_0 \xi_0)$ and thus

$$f(r_0 \xi_0) - f(s_0 \xi_0) = |f(s_0 \xi_0) - f(r_0 \xi_0)| \geq \epsilon_0.$$

Nevertheless, there exists a natural number k_0 such that

$$k > k_0 \Rightarrow |f(r_0 \xi_{n_k}) - f(r_0 \xi_0)| < \frac{\epsilon_0}{2}.$$

Note that

$$f(r_0 \xi_{n_k}) \leq f(r, \xi_{n_k}) = f(r_{n_k} \xi_{n_k}).$$

So, if $k > k_0$, then

$$f(r_0 \xi_0) < \frac{\epsilon_0}{2} + f(r_{n_k} \xi_{n_k}) \to \frac{\epsilon_0}{2} + f(s_0 \xi_0)$$

and hence $\epsilon_0 < 0$, a contradiction.

(ii) For $r \in [0, 1)$ and $z \in \mathbf{D}$ let

$$F(r, z) = \int_{\mathbf{T}} \left(\frac{r^2 - |z|^2}{|r\xi - z|^2} \right) f(r, \xi) \frac{|d\xi|}{2\pi}.$$

Then $F(r, z)$ is harmonic in $z \in \mathbf{D}$ and hence $f(r, \cdot)$ is a solvable function on $\mathbf{T}$. Also, by the boundedness of f on $\mathbf{D}$, we obtain

$$0 \leq r_1 < r_2 < 1 \Rightarrow F(r_1, z) \leq F(r_2, z) \leq \sup_{w \in \mathbf{D}} f(w) < \infty, \quad z \in \mathbf{D}.$$

Consequently, if $\{r_n\} \subset (0, 1)$ is increasing and convergent to 1, then it follows from Harnack's Principle (Proposition 1.4.2 (ii)) that $F(z) = \lim_{n \to \infty} F(r_n, z)$ is harmonic in $z \in \mathbf{D}$. Using the boundedness of f on $\mathbf{D}$, we find that $f(1, \cdot) = \lim_{r \to 1^-} f(r, \cdot)$ is finite on $\mathbf{T}$, and so that it is a solvable function on $\mathbf{T}$ as the boundary value of F, due to Proposition 1.4.2 (i). $\qquad \square$

Returning to our principal business, we have the following result.

Theorem 1.4.4. *Let $p \in (0, \infty)$ and $f \in \mathcal{H}$. Then $f \in \mathcal{Q}_p$ if and only if $f(1, \cdot) = \lim_{r \to 1^-} f(r, \cdot)$ is bounded on $\mathbf{T}$, where*

$$f(r, \xi) = \sup_{s \in [0, r]} \int_{\mathbf{D}} |f'(z)|^2 \left(1 - |\sigma_{s\xi}(z)|^2\right)^p dm(z), \quad r \in (0, 1).$$

In this case, $f(1, \cdot)$ is solvable on $\mathbf{T}$ and the solution to its corresponding Dirichlet's problem $F : \mathbf{D} \mapsto \mathbf{C}$ is the least harmonic majorant of the family of harmonic functions $\{F(r, \cdot)\}_{r \in (0,1)}$, where

$$F(r, z) = \int_{\mathbf{T}} \left(\frac{r^2 - |z|^2}{|r\xi - z|^2} \right) f(r, \xi) \frac{|d\xi|}{2\pi}, \quad z \in \mathbf{D}.$$

Proof. The equivalence is derived immediately from (i)⟺(ii) in Theorem 1.1.3 and its extension to $p \in [1, \infty)$.

Concerning the remaining part, we first prove that if $f \in \mathcal{Q}_p$, then

$$\left(F_p(f, w)\right)^2 = \int_{\mathbf{D}} |f'(z)|^2 \left(1 - |\sigma_w(z)|^2\right)^p dm(z)$$

is a continuous and bounded function on $\mathbf{D}$. Of course, the boundedness is trivial. To see the continuity, let f be nonconstant, fix $w_0 \in \mathbf{D}$, and assume that $\delta > 0$ ensures $\{z \in \mathbf{D} : |\sigma_w(z)| \leq \delta\} \subset \mathbf{D}$. Then the mapping

$$q(z, a) = \left(\frac{1 - |a|^2}{|1 - \bar{a}z|^2}\right)^p,$$

is uniformly continuous on $\{z \in \mathbf{C} : |z| \leq 1\} \times \{z \in \mathbf{D} : |\sigma_w(z)| \leq \delta\}$. Accordingly, for any $\epsilon > 0$ there exists $s > 0$ such that $|z_1 - z_2| < s$ and $|a_1 - a_2| < s$ imply

$$|q(z_1, a_1) - q(z_2, a_2)| < \epsilon \left(\int_{\mathbf{D}} |f'(z)|^2 (1 - |z|^2)^p dm(z)\right)^{-1}.$$

Accordingly, $|w - w_0| < s$ yields

$$\left|\left(F_p(f, w)\right)^2 - \left(F_p(f, w_0)\right)^2\right| \leq \int_{\mathbf{D}} |f'(z)|^2 (1 - |z|^2)^p |q(z, w) - q(z, w_0)| dm(z)$$
$$< \epsilon.$$

Now for $(r, \xi) \in (0, 1) \times \mathbf{T}$ let $f(r, \xi) = \sup_{s \in [0, r]} \left(F_p(f, s\xi)\right)^2$. Then an application of Lemma 1.4.3 with $\left(F_p(f, \cdot)\right)^2$ implies the desired assertion. □

1.5 Regularity for the Euler–Lagrange Equation

In this section, we will see that the concept of harmonic majorants can be used to settle the regularity of a solution to the Euler–Lagrange equation.

Given a constant $\lambda \geq 0$. Consider the problem of solving the Euler–Lagrange equation

$$\Delta u(z) = \lambda u(z), \quad z \in \mathbf{D}$$

subject to the Dirichlet-type energy condition

$$\int_{\mathbf{D}} \left(|\nabla u(z)|^2 + \lambda\left(u(z)\right)^2\right) dm(z) < \infty,$$

where $\nabla = (2\partial/\partial z, 2\partial/\partial \bar{z})$ stands for the gradient vector. As well-known, this problem is solvable (cf. [BergSc, p. 258]), and each solution u is C^∞ on $\mathbf{D}$. If

$\lambda = 0$, then any solution u is harmonic on $\mathbf{D}$, and hence there is a holomorphic function f such that $u = \Re f$. Moreover, if this harmonic solution u satisfies

$$E_p(u) = \sup_{w \in \mathbf{D}} \int_{\mathbf{D}} |\nabla u(z)|^2 (1 - |\sigma_w(z)|^2)^p dm(z) < \infty, \quad p \in [0,1],$$

then $f \in \mathcal{Q}_p$ and thus $f \in \bigcap_{0 < q < \infty} \mathcal{H}^q$. Here and henceforth, $\mathcal{H}^q$, $q \in (0, \infty)$, is the q-Hardy space of all functions $f \in \mathcal{H}$ obeying

$$\|f\|_{\mathcal{H}^q} = \left(\sup_{r \in (0,1)} \int_{\mathbf{T}} |f(r\xi)|^q |d\xi| \right)^{\frac{1}{q}} < \infty,$$

and $\mathcal{H}^\infty$ is the set of all $f \in \mathcal{H}$ with

$$\|f\|_{\mathcal{H}^\infty} = \sup_{z \in \mathbf{D}} |f(z)| < \infty.$$

Thus, $|f|^q$ admits a harmonic majorant. Since $|u| \leq |f|$, we get that $|u|^q$ has a harmonic majorant for any $q > 0$. Even more is true.

Theorem 1.5.1. *Let $p \in (0,1)$ and $\lambda \geq 0$ be a constant. If u satisfies $\Delta u = \lambda u$ on $\mathbf{D}$ and $E_p(u) < \infty$, then $|u|^q$ admits a harmonic majorant on $\mathbf{D}$ for all $q > 0$.*

Proof. The argument is split into two cases.

Case 1: $q \in (0, 2/p]$. In this case, the following estimate for u is elementary:

$$|u|^q \leq (|u| + 1)^q \leq (|u| + 1)^{2/p} \preceq (|u|^{2/p} + 1).$$

So, if u is such that $\Delta u = \lambda u$ on $\mathbf{D}$ and $E_p(u) < \infty$, then

$$\int_{\mathbf{D}} |\nabla u(z)|^2 (1 - |z|^2)^p dm(z) \leq E_p(u) < \infty,$$

and hence $|u|^{2/p}$ admits a harmonic majorant owing to Theorem 1 in [Ya]. Therefore, $|u|^q$ admits a harmonic majorant.

Case 2: $q \in (2/p, \infty)$. Note that $p \in (0,1)$. So $q > 2/p$ implies $q > 2$. Without loss of generality, we may assume that u is nonconstant. A simple calculation with $q > 2$ gives

$$\Delta(|u|^q) = q(q-1)|u|^{q-2}|\nabla u|^2 + q\lambda|u|^q \geq 0$$

which indicates that $|u|^q$ is subharmonic on $\mathbf{D}$.

Suppose $r \in (1/4, 1)$. Applying the Green formula to $|u|^q$ and $\log r/|z|$ on the annulus $\{z \in \mathbf{D} : 1/4 < |z| < r\}$, and taking $\partial/\partial n$ as the radial derivative, we have

$$I(r) = \pi \int_{1/4 < |z| < r} \left(\log \frac{r}{|z|} \right) \Delta(|u(z)|^q) dm(z) = I_1(r) + I_2(r) + I_3(r),$$

where

$$I_1(r) = -\int_{|\zeta|=r} |u(\zeta)|^q \left(\frac{\partial}{\partial n}\log\frac{r}{|\zeta|}\right)|d\zeta|,$$

$$I_2(r) = -\int_{|\zeta|=1/4} \left(\log\frac{r}{|\zeta|}\right)\left(\frac{\partial}{\partial n}|u(\zeta)|^q\right)|d\zeta|,$$

and

$$I_3(r) = -4\int_{|\zeta|=1/4} |u(\zeta)|^q |d\zeta|.$$

Clearly, $I_1(r)$ is equal to $2\pi V_r(0)$, where V_r is the Poisson integral of $|u|^q$ on $\{z \in \mathbf{D} : |z| = r\}$. Observe that V_r is the least harmonic majorant of $|u|^q$ over $\{z \in \mathbf{D} : |z| < r\}$ and is nondecreasing with r. So, if we can prove $\sup_{1/4<r<1} I_1(r) < \infty$, then $|u|^q$ will admit a harmonic majorant $\lim_{r\to 1} V_r$ due to the Harnack Principle. Instead of verifying $\sup_{1/4<r<1} I_1(r) < \infty$, we show:

$$\sup_{1/4<r<1} I(r) < \infty, \quad \sup_{1/4<r<1} I_2(r) < \infty \quad \text{and} \quad \sup_{1/4<r<1} I_3(r) < \infty.$$

Since $I_3(r)$ is actually a constant (independent of r) and $\lim_{r\to 1} I_2(r)$ is finite due to the smoothness of $|u|^q$ in case of $q > 2$, it remains to prove $\sup_{1/4<r<1} I(r) < \infty$. This will be done by checking

$$II(r) = \int_{1/4<|z|<r} (1-|z|^2)\Delta\big(|u(z)|^q\big)dm(z) < \infty$$

thanks to

$$-\log|z| \le 4(1-|z|^2), \quad 1/4 < |z| < 1.$$

If $z = x + iy$, then

$$2^{-1}\Delta\big(|\nabla u|^2\big) = \left(\frac{\partial^2 u}{\partial x^2}\right)^2 + 2\left(\frac{\partial^2 u}{\partial x\partial y}\right)^2 + \left(\frac{\partial^2 u}{\partial y^2}\right)^2 + \lambda|\nabla u|^2 \ge 0,$$

and hence $|\nabla u|^2$ is subharmonic on $\mathbf{D}$, and so is $|\nabla u \circ \sigma_w|^2$ for any $w \in \mathbf{D}$. A simple calculation with this subharmonicity produces

$$\begin{aligned}
\big((1-|w|^2)|\nabla u(w)|\big)^2 &= |\nabla u \circ \sigma_w(0)|^2 \\
&\preceq \int_{|z|\le\frac{1}{2}} |\nabla u \circ \sigma_w(z)|^2(1-|z|^2)^p dm(z) \\
&\preceq E_p(u), \quad w \in \mathbf{D}.
\end{aligned}$$

As a result, we get

$$|u(z)| \preceq |u(0)| + \sqrt{E_p(u)}\log\frac{1}{1-|z|}, \quad z \in \mathbf{D}.$$

Note that both $p \in (0,1)$ and $q \in (2,\infty)$ ensure

$$(1-|z|)^{1-p} \left(\log \frac{1}{1-|z|} \right)^{q-2} \preceq 1, \quad z \in \mathbf{D}.$$

So

$$
\begin{aligned}
II(r) \ &\preceq\ \int_{1/4<|z|<1} |u(z)|^{q-2} |\nabla u(z)|^2 (1-|z|^2) dm(z) \\
&\quad + \int_{1/4<|z|<1} |u(z)|^q (1-|z|^2) dm(z) \\
&\preceq\ \int_{1/4<|z|<1} \left(|u(0)| + \sqrt{E_p(u)} \log \frac{1}{1-|z|} \right)^{q-2} |\nabla u(z)|^2 (1-|z|^2) dm(z) \\
&\quad + \int_{1/4<|z|<1} \left(|u(0)| + \sqrt{E_p(u)} \log \frac{1}{1-|z|} \right)^{q} (1-|z|^2) dm(z) \\
&\preceq\ |u(0)|^{q-2} \int_{1/4<|z|<1} |\nabla u(z)|^2 (1-|z|^2) dm(z) \\
&\quad + \left(E_p(u) \right)^{\frac{q-2}{2}} \int_{1/4<|z|<1} |\nabla u(z)|^2 (1-|z|^2)^p dm(z) \\
&\quad + |u(0)|^q \int_{1/4<|z|<1} (1-|z|^2) dm(z) \\
&\quad + \left(E_p(u) \right)^{\frac{q}{2}} \int_{1/4<|z|<1} \left(\log \frac{1}{1-|z|} \right)^q (1-|z|^2) dm(z) \\
&\preceq\ |u(0)|^{q-2} E_p(u) + |u(0)|^q + \left(E_p(u) \right)^{\frac{q}{2}} < \infty.
\end{aligned}
$$

The proof is complete. $\square$

1.6   Notes

Note 1.6.1. Section 1.1 summarizes the first four sections of [BanFlu] and the main function-theoretic characterizations of $\mathcal{Q}_p$ presented in [Xi3]. Example 1.1.1 is a special case of [Ban, (2.21)] which is verified via the isoperimetric inequality.

A broad exploration of the Q_p spaces naturally leads to a treatment of some other Q_p type function spaces and classes — see e.g., [AuMW], [EsWuXi], [Lis], [Lix], [PerR], [Rat], [Wul] and [Zhu3].

From 1999 to 2006, the scale of Q_p spaces was generalized in different ways to higher dimensions — see also [CnDe], [GuCDS], [GuMa], [CeKaSo], [ReKa], [Hu], [LiO], [ElGRT], [BernGRT] and [Sha]. Recently, the basic theory of the $\mathcal{Q}_p$-spaces was extended to bounded symmetric domains — see also [Eng].

Note 1.6.2. Section 1.2 is basically an adaptation of the sixth section of [DoGiVu]. Related to this topic, we can say something about the Schwarz derivative. If f is univalent on $\mathbf{D}$, then

$$\mathsf{SD}(f)(z) = \left(\frac{f''}{f'}\right)' - 2^{-1}\left(\frac{f''}{f'}\right)^2$$

is called the Schwarz derivative of f. It is well known that

$$\log f' \in \mathcal{B} \quad \text{with} \quad \|\log f'\|_{\mathcal{B}} \leq 6$$

and $\log f'$ belongs to $\mathcal{BMOA}$ if and only if

$$|\mathsf{SD}(f)(z)|^2(1 - |z|^2)^3 dm(z)$$

is a 1-Carleson measure on $\mathbf{D}$; see [BisJo]. Using Theorem 1.1.3 (v), we can readily find: if $p \in (0,1)$ then $\log f' \in \mathcal{Q}_p$ must imply that

$$|\mathsf{SD}(f)(z)|^2(1 - |z|^2)^{p+2} dm(z)$$

is a p-Carleson measure on $\mathbf{D}$. Although being unable to prove whether or not the converse holds too, we believe that it should be true since the Schwarz derivative can be regarded as a type of second derivative, measuring the rate of change of the best approximating Möbius mappings rather than linear transforms.

Note 1.6.3. Section 1.3 is a combination of [DoGiVu, Theorem 4.2] and [BuFeVu, Section 2]. Two further remarks are in order. The first one is about S_ϕ on the limiting Besov space $\mathcal{B}_1$ which consists of all functions $f \in \mathcal{H}$ with

$$\|f\|_{\mathcal{B}_1} = \int_{\mathbf{D}} |f''(z)|dm(z) < \infty.$$

Since $\mathcal{B}_1 \subset \mathcal{H}^\infty$, the chain rule gives that if ϕ is an entire function on $\mathbf{C}$, then

$$\big(\mathsf{S}_\phi(f)\big)'' = f''\mathsf{S}_{\phi'}(f) + (f')^2\mathsf{S}_{\phi''}(f), \quad f \in \mathcal{B}_1$$

and so that $\mathsf{S}_\phi(f) \in \mathcal{B}_1$. In other words, the nonlinear operator S_ϕ always maps $\mathcal{B}_1$ to itself; see also [BuFeVu, Proposition 5] and [AlvVMV] for the action of S_ϕ between $\mathcal{B}$ and the Bergman spaces. The second one is about S_ϕ from the Bloch-type spaces $\mathcal{B}^\alpha$ into $\mathcal{Q}_p$ and vice versa. Recall that $f \in \mathcal{B}^\alpha$, $\alpha \in (0,\infty)$ means:

$$f \in \mathcal{H} \quad \text{and} \quad \|f\|_{\mathcal{B}^\alpha} = \sup_{z \in \mathbf{D}}(1 - |z|^2)^\alpha |f'(z)| < \infty.$$

Using Theorem 1.3.4 and the existing result [Xi3, Theorem 2.1.1] that there are two $f_1, f_2 \in \mathcal{B}^\alpha$ such that

$$(1 - |z|^2)^\alpha\big(|f_1'(z)| + |f_2'(z)|\big) \approx 1, \quad z \in \mathbf{D},$$

one can readily get $\mathsf{S}_\phi : \mathcal{B}^\alpha \mapsto \mathcal{Q}_p$, $p \in [0, \infty)$, if and only if ϕ satisfies

$$\phi' = 0 \quad \text{whenever} \quad \alpha > 1$$

or

$$\phi(w) = aw + b \quad \text{whenever} \quad \alpha = 1$$

or

$$\sup_{w \in \mathbf{D}} \int_{\mathbf{D}} (1 - |z|^2)^{-2\alpha} (1 - |\sigma_w(z)|^2)^p dm(z) < \infty \quad \text{whenever} \quad \alpha \in (0, 1).$$

The case $\alpha \in (0, 1)$ can be found in [Xio, Theorem 2]. In addition, $\mathsf{S}_\phi : \mathcal{Q}_p \mapsto \mathcal{B}^\alpha$ holds for $p \in [0, \infty)$ and $\alpha \in (0, 1)$ (or $\alpha = 1$) if and only if ϕ is constant (or $\phi(w) = aw + b$); see also [Xio, Theorem 1] for $\alpha \in (0, 1)$. However, if $\alpha \in (1, \infty)$, then $f \in \mathcal{B}^\alpha$ is equivalent to

$$f \in \mathcal{H} \quad \text{with} \quad \sup_{z \in \mathbf{D}} (1 - |z|^2)^{\alpha - 1} |f(z)| < \infty,$$

and hence the inclusion $\mathcal{Q}_p \subseteq \mathcal{B}$ with growth

$$|f(z)| \preceq \left(\log \frac{2}{1 - |z|} \right) \|f\|_{\mathcal{Q}_p}, \quad f \in \mathcal{Q}_p,$$

the proof of [AlvVMV, Theorem 9] and its immediate remark yield that $\mathsf{S}_\phi : \mathcal{Q}_p \mapsto \mathcal{B}^\alpha$ when and only when ϕ is an entire function either of order $\rho = \limsup_{r \to \infty} (\log r)^{-1} \log \log \max\{|\phi(z)| : |z| = r\}$ less than 1 or of order $\rho = 1$ and type $\tau = \limsup_{r \to \infty} r^{-\rho} \log \max\{|\phi(z)| : |z| = r\} = 0$. A full description of the boundedness of S_ϕ acting between two Bloch-type spaces is given in [Xu].

Note 1.6.4. Section 1.4 is a direct outcome of reading [AuReTo] and its follow-up paper [RaReTo]. The key fact used in proving Theorem 1.4.4 is $F_p(f, \cdot)$ is continuous. This fact can be verified by the following basic estimate:

$$\frac{(1 - |w_1|)(1 - |w_2|)}{4} \leq \frac{1 - |\sigma_{w_1}(z)|^2}{1 - |\sigma_{w_2}(z)|^2} \leq \frac{4}{(1 - |w_1|)(1 - |w_2|)}, \quad z, \; w_1, \; w_2 \in \mathbf{D}.$$

For details, see also [WiXi2, Lemma 3.2].

Note 1.6.5. Section 1.5 is essentially [Xi4, Section 3.3]. However, the proof of Theorem 1.5.1 is motivated by the argument for [Ya, Theorem 1].

Chapter 2

Poisson versus Berezin with Generalizations

We are given $z \in \mathbf{D}$. Recall that

$$(2\pi)^{-1} \int_{\mathbf{T}} f \circ \sigma_z(\zeta)|d\zeta| \quad \text{and} \quad \pi^{-1} \int_{\mathbf{D}} f \circ \sigma_z(\zeta) dm(\zeta)$$

are the Poisson extensions of $f \in \mathcal{L}^1(\mathbf{T})$ (the Lebesgue 1-space on $\mathbf{T}$) and the Berezin transformation of $f \in \mathcal{L}^1(\mathbf{D})$ (the Lebesgue 1-space on $\mathbf{D}$), respectively. These two operators play an important role in the function theory of Hardy spaces and Bergman spaces. In this chapter, we use both to deduce directly or indirectly certain comparable derivative-free properties of each $\mathcal{Q}_p$ space via the following five aspects:

- Boundary Value and Brownian Motion;
- Derivative-free Module via Poisson Extension;
- Derivative-free Module via Berezin Transformation;
- Mixture of Derivative and Quotient;
- Dirichlet Double Integral without Derivative.

2.1 Boundary Value and Brownian Motion

The non-holomorphic Q classes on $\mathbf{T}$, which we denote $\mathcal{Q}_p(\mathbf{T})$, $-\infty < p < \infty$, are the classes of all functions $f \in \mathcal{L}^1(\mathbf{T})$ that satisfy

$$\|f\|_{\mathcal{Q}_p,3} = \sup_{I \subseteq \mathbf{T}} \left(|I|^{-p} \int_I \int_I \frac{|f(\zeta) - f(\eta)|^2}{|\zeta - \eta|^{2-p}} |d\zeta||d\eta| \right)^{\frac{1}{2}} < \infty.$$

The range of interest is $p \in (0,1)$. The forthcoming characterization involving Brownian motions is prepared for the characterization of the holomorphic Q classes in terms of the Poisson integrals of some absolute value functions.

Following the known $\mathcal{BMO}$-case; see also [Pet], we may assume that $(\Omega, \mathcal{B}, P)$ is a probability space. For $t \geq 0$ denote by $\gamma_{z,t}(\omega)$ the two-dimensional Brownian motion in $\mathbf{D}$ starting at the point $z \in \mathbf{D}$ at time $t = 0$. Given $z \in \mathbf{D}$, put

$$\tau_z(\omega) = \inf\{t \geq 0 : |\gamma_{z,t}(\omega)| = 1\},$$

which is the first hitting time of $\gamma_{z,t}$ on $\mathbf{T}$. Write

$$\tilde{\gamma}_{z,t}(\omega) = \gamma_{z,\min\{t,\tau_z(\omega)\}}(\omega)$$

for the Brownian motion starting at $z \in \mathbf{D}$ which is absorbed at its first point of impact on $\mathbf{T}$, at time $\tau_z(\omega)$.

For a sub-σ-algebra: $\mathcal{A} \subseteq \mathcal{B}$, and a function $f \in \mathcal{L}^1(\Omega, \mathcal{B}, P)$, it is known that $\nu(A) = \int_A f dP$ for $A \in \mathcal{A}$ is absolutely continuous with respect to $P|\mathcal{A}$, and so by the Radon–Nikodym Theorem there exists an $\mathcal{A}$-measurable $\mathcal{L}^1$ function g such that $\nu(A) = \int_A g dP$ for all $A \in \mathcal{A}$. Because this equation determines g uniquely a.e., $g = E(f|\mathcal{A})$ is called the conditional expectation of f with respect to $\mathcal{A}$. If one knows for each set of $\mathcal{A}$ whether or not it contains ω, then $E(f|\mathcal{A})(\omega)$ is the expected value of f. In particular, $E(f) = \int_\Omega f dP$.

The Poisson extension $\hat{f}$ of a function $f \in \mathcal{L}^1(\mathbf{T})$ from $\mathbf{T}$ to $\mathbf{D}$ is just the following operator acting on f:

$$\hat{f}(z) = \int_{\mathbf{T}} f(\zeta) \left(\frac{1 - |z|^2}{|\zeta - z|^2} \right) \frac{|d\zeta|}{2\pi}, \quad z \in \mathbf{D}.$$

Lemma 2.1.1. *Let $p \in (0,1)$ and $f \in \mathcal{L}^2(\mathbf{T})$. Then the following statements are equivalent:*

(i) $f \in \mathcal{Q}_p(\mathbf{T})$.

(ii) $|\nabla \hat{f}(z)|^2 (1 - |z|^2)^p dm(z)$ *is a p-Carleson measure on* $\mathbf{D}$.

(iii) $\displaystyle \sup_{w \in \mathbf{D}} \int_{\mathbf{D}} |\nabla \hat{f}(z)|^2 (1 - |\sigma_w(z)|^2)^p dm(z) < \infty.$

(iv) $\displaystyle \sup_{w \in \mathbf{D}} \int_{\mathbf{D}} \left(\int_{\mathbf{D}} |\nabla \hat{f}(u)|^2 \log \left| \frac{1 - \bar{u}z}{z - u} \right| dm(u) \right) (1 - |\sigma_w(z)|^2)^p \frac{dm(z)}{(1 - |z|^2)^2} < \infty.$

(v) $\displaystyle \sup_{w \in \mathbf{D}} \int_{\mathbf{D}} \left(\widehat{|f|^2}(z) - |\hat{f}(z)|^2 \right) (1 - |\sigma_w(z)|^2)^p \frac{dm(z)}{(1 - |z|^2)^2} < \infty.$

(vi) $\displaystyle \sup_{w \in \mathbf{D}} \int_{\mathbf{D}} \left(\int_\Omega |f(\gamma_{z,\tau_z}(\omega))|^2 dP(\omega) - |\hat{f}(z)|^2 \right) (1 - |\sigma_w(z)|^2)^p \frac{dm(z)}{(1 - |z|^2)^2} < \infty.$

(vii) $\displaystyle \sup_{w \in \mathbf{D}} \int_{\mathbf{D}} \left(E(|f(\gamma_{z,\tau_z})|^2) - |\hat{f}(z)|^2 \right) (1 - |\sigma_w(z)|^2)^p \frac{dm(z)}{(1 - |z|^2)^2} < \infty.$

(viii) $\displaystyle\sup_{w\in\mathbf{D}}\int_{\mathbf{D}}\left(\int_{\Omega}\left(\int_{0}^{\tau(\omega)}|\nabla\hat{f}(\tilde{\gamma}_{z,t})|^{2}dt\right)dP(\omega)\right)(1-|\sigma_{w}(z)|^{2})^{p}\frac{dm(z)}{(1-|z|^{2})^{2}}<\infty.$

(ix) $\displaystyle\sup_{w\in\mathbf{D}}\int_{\mathbf{D}}E\left(\int_{0}^{\tau}|\nabla\hat{f}(\tilde{\gamma}_{z,t})|^{2}dt\right)(1-|\sigma_{w}(z)|^{2})^{p}\frac{dm(z)}{(1-|z|^{2})^{2}}<\infty.$

Proof. Step 1: (i)⇔(ii)⇔(iii) This follows from [Xi3, Theorem 7.1.1 and Lemma 4.1.1] — see also Theorem 1.1.3 (viii).

Step 2: (iii)⇔(iv)⇔(v) The well-known Green formula is applied to infer that under $z\in\mathbf{D}$ and $p\in(0,1)$,

$$
\begin{aligned}
(1-|z|^{2})^{p} &= \frac{2}{\pi}\int_{\mathbf{D}}\left(\frac{\partial^{2}}{\partial u\partial\bar{u}}(1-|u|^{2})^{p}\right)\log|\sigma_{z}(u)|dm(u)\\
&= 2p\int_{\mathbf{D}}(1-p|u|^{2})(1-|u|^{2})^{p}\log\left|\frac{1-\bar{u}z}{u-z}\right|\frac{dm(u)}{(1-|u|^{2})^{2}}\\
&\approx \int_{\mathbf{D}}(1-|u|^{2})^{p}\log\left|\frac{1-\bar{u}z}{u-z}\right|\frac{dm(u)}{(1-|u|^{2})^{2}}.
\end{aligned}
$$

This estimate, together with Fubini's Theorem, Hardy–Littlewood's identity (cf. [Ga, p. 238]) and the conformal transform $\sigma_{w}(z)$, implies

$$
\begin{aligned}
&\int_{\mathbf{D}}|\nabla\hat{f}(z)|^{2}(1-|\sigma_{w}(z)|^{2})^{p}dm(z)\\
&= \int_{\mathbf{D}}|\nabla\hat{f}(\sigma_{w}(z))|^{2}(1-|z|^{2})^{p}dm(z)\\
&\approx \int_{\mathbf{D}}\left(\int_{\mathbf{D}}|\nabla\hat{f}(\sigma_{w}(z))|^{2}\log\left|\frac{1-\bar{u}z}{u-z}\right|dm(z)\right)(1-|u|^{2})^{p}\frac{dm(u)}{(1-|u|^{2})^{2}}\\
&\approx \int_{\mathbf{D}}\left(\int_{\mathbf{D}}|\nabla\hat{f}(z)|^{2}\log\left|\frac{1-\bar{u}z}{u-z}\right|dm(z)\right)(1-|\sigma_{w}(u)|^{2})^{p}\frac{dm(u)}{(1-|u|^{2})^{2}}\\
&\approx \int_{\mathbf{D}}\left(\widehat{|f|^{2}}(z)-|\hat{f}(z)|^{2}\right)(1-|\sigma_{w}(z)|^{2})^{p}\frac{dm(z)}{(1-|z|^{2})^{2}}.
\end{aligned}
$$

So, the desired equivalence follows.

Step 3: (i)⇔(vi)⇔(vii) Given a continuous function g on $\mathbf{T}$. Fix $z\in\mathbf{D}$ and let

$$
\nu_{z}(A)=P\left(\{\omega:\gamma_{z,\tau_{z}(\omega)}\in A\}\right)
$$

for every Lebesgue measurable subset A of $\mathbf{T}$. Changing variable yields

$$
\int_{\Omega}g\left(\gamma_{z,\tau_{z}(\omega)}(\omega)\right)dP(\omega)=\int_{\mathbf{T}}g(\zeta)d\nu_{z}(\zeta).
$$

Using Kakutani's Theorem in [Kak], we find that the right-hand-side function is harmonic on $\mathbf{D}$ with g being the boundary value function on $\mathbf{T}$, and hence it is just $\hat{g}(z)$. Note that this is true for all continuous functions on $\mathbf{T}$. Accordingly, we find

$$d\nu_z(\zeta) = \left(\frac{1-|z|^2}{|\zeta-z|^2}\right)\frac{|d\zeta|}{2\pi}, \quad \zeta \in \mathbf{T},$$

which is a conformally invariant probability measure on $\mathbf{T}$ in the sense of

$$d\nu_z(\zeta) = d\nu_{\sigma_w(z)}(\sigma_w(\zeta)), \quad w,z \in \mathbf{D}, \, \zeta \in \mathbf{T}.$$

In other words, if $S \subseteq \mathbf{T}$ is Lebesgue measurable, then the hitting probability of $\gamma_{z,t}$ in S is the harmonic measure of S at z:

$$P\left(\{\omega : \gamma_{z,\tau_z(\omega)}(\omega) \in S\}\right) = \nu_z(S) = \frac{1}{2\pi}\int_S \frac{1-|z|^2}{|\zeta-z|^2}|d\zeta|.$$

Accordingly, this equation, together with the previous Steps 1-2, implies the required equivalence.

Step 4: (i)$\Leftrightarrow$(viii)$\Leftrightarrow$(ix) For $z \in \mathbf{D}$, $t \geq 0$, and a Lebesgue measurable set $S \subseteq \mathbf{D}$, let

$$\mu_{z,t}(S) = P(\{\omega : \tilde{\gamma}_{z,t}(\omega) \in S\}).$$

From the density formula associated with the two-dimensional Brownian motions (cf. [Pet, p. 20]), it follows that

$$\mu_{z,t}(S) \leq P(\{\omega : \gamma_{z,t}(\omega) \in S\}) = \frac{1}{2\pi t}\int_S \exp\left(-\frac{|z|^2}{2t}\right)dm(z),$$

and so that $\mu_{z,t}$ is absolutely continuous with respect to dm. Accordingly, there is a density function $\delta(z,t,w)$ such that

$$\delta(z,t,w) \leq \frac{1}{2t}\exp\left(-\frac{|z-w|^2}{2t}\right) \quad \text{and} \quad \mu_{z,t}(S) = \pi^{-1}\int_S \delta(z,t,w)dm(w).$$

Using Hunt's main result in [Hun], we obtain that if $\delta(z,t,w)$ is determined via

$$P\left(\{\omega : \tilde{\gamma}_{z,t}(\omega) \in S\}\right) = \pi^{-1}\int_S \delta(z,t,w)dm(w)$$

for each Lebesgue measurable set $S \subseteq \mathbf{D}$, then

$$\int_0^\infty \delta(z,t,w)dt = \log\left|\frac{1-\bar{w}z}{z-w}\right| \quad \text{a.e. } w,z \in \mathbf{D}.$$

Hence

$$\int_{\mathbf{D}} |\nabla \hat{f}(z)|^2 \log \left| \frac{1 - \bar{w}z}{w - z} \right| \frac{dm(z)}{\pi} \approx \int_{\mathbf{D}} |\nabla \hat{f}(z)|^2 \left(\int_0^\infty \delta(w,t,z)dt \right) dm(z)$$

$$\approx \int_0^\infty \left(\int_{\mathbf{D}} |\nabla \hat{f}(z)|^2 \delta(w,t,z)dm(z) \right) dt$$

$$\approx \int_0^\infty \left(\int_{\mathbf{D}} |\nabla \hat{f}(z)|^2 d\mu_{w,t}(z) \right) dt$$

$$\approx \int_0^\infty \left(\int_{t<\tau(\omega)} |\nabla \hat{f}(\tilde{\gamma}_{w,t})|^2 dP(\omega) \right) dt$$

$$\approx \int_\Omega \left(\int_0^{\tau(\omega)} |\nabla \hat{f}(\tilde{\gamma}_{w,t}(\omega))|^2 dt \right) dP(\omega)$$

$$\approx E \left(\int_0^\tau |\nabla \hat{f}(\tilde{\gamma}_{w,t})|^2 dt \right),$$

establishing the desired equivalence due to Steps 1–2 above. $\qquad \square$

2.2 Derivative-free Module via Poisson Extension

As with $\mathcal{Q}_p$, we shall apply Lemma 2.1.1 to discuss the effect of absolute values on the behavior of $f \in \mathcal{Q}_p$ and at the same time, the value distribution of $f \in \mathcal{Q}_p$.

Theorem 2.2.1. *Let $p \in (0,1)$ and $f \in \mathcal{H}^2$. Then the following statements are equivalent:*

(i) $f \in \mathcal{Q}_p$.

(ii) $\displaystyle \sup_{w \in \mathbf{D}} \int_{\mathbf{T}} \int_{\mathbf{T}} \frac{|f(\zeta) - f(\eta)|^2}{|\zeta - \eta|^2} \left(\int_{\mathbf{D}} \left| \left(\log \frac{\zeta - z}{\eta - z} \right)' \right|^2 \frac{dm(z)}{(1 - |\sigma_w(z)|^2)^{-p}} \right) |d\zeta||d\eta| < \infty.$

(iii) $\displaystyle \sup_{w \in \mathbf{D}} \int_{\mathbf{D}} |f - \widehat{f(z)}|^2(z)(1 - |\sigma_w(z)|^2)^p \frac{dm(z)}{(1 - |z|^2)^2} < \infty.$

(iv) $|f| \in \mathcal{Q}_p(\mathbf{T})$ *and*

$$\sup_{w \in \mathbf{D}} \int_{\mathbf{D}} \left((\widehat{|f|}(z))^2 - |f(z)|^2 \right)(1 - |\sigma_w(z)|^2)^p \frac{dm(z)}{(1 - |z|^2)^2} < \infty.$$

(v) $|f| \in \mathcal{Q}_p(\mathbf{T})$ *and there is a nonnegative harmonic function h such that $|f| \leq h$ on $\mathbf{D}$ and*

$$\sup_{w \in \mathbf{D}} \int_{\mathbf{D}} \left((h(z))^2 - |f(z)|^2 \right)(1 - |\sigma_w(z)|^2)^p \frac{dm(z)}{(1 - |z|^2)^2} < \infty.$$

(vi) *If*

$$N_f(w, z) = \begin{cases} -\sum_{f(z_j)=w} \log |\sigma_{z_j}(z)| & , \quad w \in f(\mathbf{D}), \\ 0 & , \quad w \in \mathbf{C} \setminus f(\mathbf{D}), \end{cases}$$

then

$$\sup_{w \in \mathbf{D}} \int_{\mathbf{D}} \left(\int_{\mathbf{C}} N_f(v, z) dm(v) \right) (1 - |\sigma_w(z)|^2)^p \frac{dm(z)}{(1 - |z|^2)^2} < \infty.$$

Proof. Step 1: (i)⇔(ii)⇔(iii) Note that $\hat{f} = f$ for $f \in \mathcal{H}^2$. So, from Lemma 2.1.1 with $f \in \mathcal{H}^2$ it follows that $f \in \mathcal{Q}_p$ if and only if

$$\sup_{w \in \mathbf{D}} \int_{\mathbf{D}} \left(\widehat{|f|^2}(z) - |f(z)|^2 \right) (1 - |\sigma_w(z)|^2)^p \frac{dm(z)}{(1 - |z|^2)^2} < \infty.$$

This equivalence, along with

$$
\begin{aligned}
2(\widehat{|f - f(z)|^2})(z) &= 2(\widehat{|f|^2}(z) - |f(z)|^2) \\
&= \int_{\mathbf{T}} \int_{\mathbf{T}} |f(\zeta) - f(\eta)|^2 \left(\frac{1 - |z|^2}{|1 - z\bar{\zeta}||1 - z\bar{\eta}|} \right)^2 |d\zeta||d\eta| \\
&= \int_{\mathbf{T}} \int_{\mathbf{T}} \left| \frac{f(\zeta) - f(\eta)}{\zeta - \eta} \right|^2 \left| \frac{1}{\zeta - z} - \frac{1}{\eta - z} \right|^2 (1 - |z|^2)^2 |d\zeta||d\eta| \\
&= \int_{\mathbf{T}} \int_{\mathbf{T}} \left| \frac{f(\zeta) - f(\eta)}{\zeta - \eta} \right|^2 \left| \left(\log \frac{\zeta - z}{\eta - z} \right)' \right|^2 (1 - |z|^2)^2 |d\zeta||d\eta|,
\end{aligned}
$$

implies the desired equivalence right away.

Step 2: (i)⇒(iv) If $f \in \mathcal{Q}_p$, then $f \in \mathcal{Q}_p(\mathbf{T})$ and hence $|f| \in \mathcal{Q}_p(\mathbf{T})$ thanks to

$$\big||f(\zeta)| - |f(\eta)|\big| \leq |f(\zeta) - f(\eta)|, \quad \zeta, \eta \in \mathbf{T}.$$

Clearly,

$$\left(\widehat{|f|}(z) \right)^2 \leq \widehat{|f|^2}(z), \quad z \in \mathbf{D}.$$

Thus,

$$
\begin{aligned}
&\sup_{w \in \mathbf{D}} \int_{\mathbf{D}} \left(\left(\widehat{|f|}(z) \right)^2 - |f(z)|^2 \right) (1 - |\sigma_w(z)|^2)^p \frac{dm(z)}{(1 - |z|^2)^2} \\
&\leq \sup_{w \in \mathbf{D}} \int_{\mathbf{D}} \left(\widehat{|f|^2}(z) - |f(z)|^2 \right) (1 - |\sigma_w(z)|^2)^p \frac{dm(z)}{(1 - |z|^2)^2},
\end{aligned}
$$

and consequently, (iv) follows from Lemma 2.1.1.

Step 3: (iv)⇒(v) This follows from $|f| \leq \widehat{|f|} = h$ on $\mathbf{D}$.

Step 4: (v)⇒(i) Let $|f| \in \mathcal{Q}_p(\mathbf{T})$ and let there be a nonnegative harmonic function h such that $|f| \leq h$ on $\mathbf{D}$ and

$$\sup_{w \in \mathbf{D}} \int_{\mathbf{D}} \left((h(z))^2 - |f(z)|^2 \right) (1 - |\sigma_w(z)|^2)^p \frac{dm(z)}{(1 - |z|^2)^2} < \infty.$$

An application of Lemma 2.1.1 with $|f| \in \mathcal{Q}_p(\mathbf{T})$ yields

$$\sup_{w \in \mathbf{D}} \int_{\mathbf{D}} \left(\widehat{|f|^2}(z) - \left(\widehat{(|f|}(z))^2 \right) (1 - |\sigma_w(z)|^2)^p \frac{dm(z)}{(1 - |z|^2)^2} < \infty.$$

Note that $\widehat{|f|}$ is the least nonnegative harmonic function greater than or equal to $|f|$ on $\mathbf{D}$ (see e.g., [Du, p. 28]). So $\widehat{|f|} \leq h$ on $\mathbf{D}$ and

$$\widehat{|f|^2}(z) - |f(z)|^2 \leq \widehat{|f|^2}(z) - \left(\widehat{|f|}(z) \right)^2 + \left(h(z) \right)^2 - |f(z)|^2, \quad z \in \mathbf{D}.$$

These inequalities imply that Lemma 2.1.1 (v) holds for f, and thus $f \in \mathcal{Q}_p$.

Step 5: (i)⇔(vi) Note that if $f \in \mathcal{H}^2$, then

$$\int_{\mathbf{D}} |f'(u)|^2 \log \left| \frac{1 - \bar{u}z}{u - z} \right| dm(u) = \int_{\mathbf{C}} N_f(v, z) dm(v), \quad z \in \mathbf{D}.$$

This, together with Lemma 2.1.1 (iv), implies the desired equivalence. □

Corollary 2.2.2. *Let $p \in (0, 1)$. Then:*

(i) *For a function ψ obeying $\psi > 0$ a.e. on $\mathbf{T}$, $\log \psi \in \mathcal{L}^1(\mathbf{T})$ and $\psi \in \mathcal{L}^2(\mathbf{T})$, there exists a function $f \in \mathcal{Q}_p$ such that $|f| = \psi$ a.e. on $\mathbf{T}$ if and only if $\psi \in \mathcal{Q}_p(\mathbf{T})$ and*

$$\sup_{w \in \mathbf{D}} \int_{\mathbf{D}} \left((\widehat{\psi}(z))^2 - \exp \left(\widehat{\log \psi^2}(z) \right) \right) (1 - |\sigma_w(z)|^2)^p \frac{dm(z)}{(1 - |z|^2)^2} < \infty.$$

(ii) *A universal covering map f of a domain $\Omega \subseteq \mathbf{C}$ is in $\mathcal{Q}_p$ if and only if*

$$\sup_{w \in \mathbf{D}} \int_{\mathbf{D}} \left(\int_{\Omega} g_{\Omega}(v, f(z)) dm(v) \right) (1 - |\sigma_w(z)|^2)^p \frac{dm(z)}{(1 - |z|^2)^2} < \infty.$$

Proof. (i) If $\psi \in \mathcal{Q}_p(\mathbf{T})$ satisfies the required supremum condition, then we define the outer function

$$\mathcal{O}_\psi(z) = \exp \left(\int_{\mathbf{D}} \frac{\zeta + z}{\zeta - z} \log \psi(\zeta) \frac{|d\zeta|}{2\pi} \right), \quad z \in \mathbf{D}.$$

Since $|\mathcal{O}_\psi(\zeta)| = \psi(\zeta)$ for a.e. $\zeta \in \mathbf{T}$ and $|\mathcal{O}_\psi(z)| = \exp \left(\widehat{\log \psi}(z) \right)$ for any $z \in \mathbf{D}$, we conclude from Theorem 2.2.1 that $\mathcal{O}_\psi \in \mathcal{Q}_p$ which is the desired function.

On the other hand, if there exists a function $f \in \mathcal{Q}_p$ such that $|f| = \psi$ a.e. on $\mathbf{T}$, then $\|f\|_{\mathcal{H}^2} < \infty$ and hence f can be written as IO — product of the inner function I and the outer function O. Since $|I| \leq 1$ on $\mathbf{D}$ and $|I| = 1$ a.e. on $\mathbf{T}$ as well as

$$O(z) = \exp \left(\int_{\mathbf{D}} \frac{\zeta + z}{\zeta - z} \log \psi(\zeta) \frac{|d\zeta|}{2\pi} \right), \quad z \in \mathbf{D},$$

we have $|f| \leq |O|$ on $\mathbf{D}$ and $|f| = \psi = |O|$ a.e. on $\mathbf{T}$. Accordingly,

$$\sup_{w \in \mathbf{D}} \int_{\mathbf{D}} \left((\widehat{\psi}(z))^2 - \exp\left(\widehat{\log \psi^2}(z)\right) \right) (1 - |\sigma_w(z)|^2)^p \frac{dm(z)}{(1 - |z|^2)^2}$$

$$\leq \sup_{w \in \mathbf{D}} \int_{\mathbf{D}} \left(\left(\widehat{|f|}(z)\right)^2 - |f(z)|^2 \right) (1 - |\sigma_w(z)|^2)^p \frac{dm(z)}{(1 - |z|^2)^2} < \infty.$$

(ii) If f is a universal covering map of $\mathbf{\Omega}$, it can be seen from [Ne, p. 209] that

$$N_f(v, z) = g_{\mathbf{\Omega}}(v, f(z)), \quad v \in \mathbf{\Omega}, \ z \in \mathbf{D},$$

and hence the required equivalence follows from Theorem 2.2.1 (vi). $\square$

2.3   Derivative-free Module via Berezin Transformation

The hyperbolic distance between two points in $\mathbf{D}$ is given by

$$d_{\mathbf{D}}(z_1, z_2) = \log\left(\frac{1 + |\sigma_{z_1}(z_2)|}{1 - |\sigma_{z_1}(z_2)|}\right)^{\frac{1}{2}}, \quad z_1, z_2 \in \mathbf{D}.$$

This distance determines the hyperbolic or Bergman metric on $\mathbf{D}$, is invariant under $Aut(\mathbf{D})$, and produces the hyperbolic open ball

$$B(z, r) = \{w \in \mathbf{D} : d_{\mathbf{D}}(z, w) < r\}, \quad z \in \mathbf{D}, \ r > 0.$$

Meanwhile, for $p \in (0, \infty)$ and $\beta \in (-1, \infty)$, the weighted Bergman space $\mathcal{A}^{q,\beta}$ is the collection of all $f \in \mathcal{H}$ satisfying

$$\|f\|_{\mathcal{A}^{q,\beta}} = \left(\int_{\mathbf{D}} |f(z)|^q (1 - |z|^2)^\beta dm(z) \right)^{\frac{1}{q}} < \infty.$$

It is known that (see e.g. [Zhu2, Chapter 2])

$$\|f\|_{\mathcal{A}^{q,\beta}} \approx \left(\int_{\mathbf{D}} |f'(z)|^q (1 - |z|^2)^{q+\beta} dm(z) \right)^{\frac{1}{q}}.$$

The right-hand-side integral is of the Dirichlet type. The next result indicates that in some cases these integrals have equivalent bidisk forms without derivatives.

Lemma 2.3.1. *Let $p \in (0, 2)$ and $f \in \mathcal{H}$. Then*

$$I_1(f) = \int_{\mathbf{D}} |f'(z)|^2 (1 - |z|^2)^p dm(z) < \infty$$

if and only if

$$I_2(f) = \int_{\mathbf{D}} \int_{\mathbf{D}} \left(\frac{|f(z) - f(w)|^2}{|1 - z\bar{w}|^4} \right) (1 - |z|^2)^p dm(z) dm(w) < \infty.$$

In this case, $I_1(f) \approx I_2(f)$.

Proof. First of all, note that

$$\int_{\mathbf{D}} |F(w) - F(0)|^2 dm(w) \approx \int_{\mathbf{D}} |F'(w)|^2 (1 - |w|^2)^2 dm(w)$$

holds for any $F \in \mathcal{H}$; see e.g. [Zhu1, Theorem 4.27]. So, this comparability result with $F(w) = f \circ \sigma_z(w)$ implies

$$
\begin{aligned}
I_2(f) &= \int_{\mathbf{D}} (1 - |z|^2)^{p-2} \left(\int_{\mathbf{D}} |f \circ \sigma_z(w) - f(z)|^2 dm(w) \right) dm(z) \\
&\approx \int_{\mathbf{D}} (1 - |z|^2)^{p-2} \left(\int_{\mathbf{D}} |(f \circ \sigma_z)'(w)|^2 (1 - |w|^2)^2 dm(w) \right) dm(z) \\
&\approx \int_{\mathbf{D}} (1 - |z|^2)^{p} \left(\int_{\mathbf{D}} |f'(w)|^2 \left(\frac{1 - |w|^2}{|1 - z\bar{w}|^2} \right)^2 dm(w) \right) dm(z).
\end{aligned}
$$

If $I_1(f) < \infty$, then by Fubini's Theorem and $p \in (0, 2)$, we have

$$I_2(f) \approx \int_{\mathbf{D}} |f'(w)|^2 \left(\int_{\mathbf{D}} \frac{(1 - |w|^2)^2 (1 - |z|^2)^p}{|1 - z\bar{w}|^4} dm(z) \right) dm(w) \preceq I_1(f).$$

Conversely, if $I_2(f) < \infty$, then for $r \in (0, 1]$ and $z \in \mathbf{D}$, we apply the following estimates (cf. [Zhu1, §4.3]):

$$|f'(z)|^2 \preceq \frac{1}{m(B(z,r))} \int_{B(z,r)} |f'(w)|^2 dm(w)$$

and

$$\frac{(1 - |w|^2)^2}{|1 - z\bar{w}|^4} \approx \frac{1}{(1 - |z|^2)^2} \approx \frac{1}{m(B(z,r))}, \quad w \in B(z,r)$$

to get

$$I_2(f) \succeq \int_{\mathbf{D}} (1 - |z|^2)^p \left(\frac{1}{m(B(z,r))} \int_{B(z,r)} |f'(w)|^2 dm(w) \right) dm(z) \succeq I_1(f).$$

We are done. $\qquad\qquad\qquad\qquad\qquad\qquad\qquad\qquad\qquad\qquad\qquad\qquad\qquad\qquad\square$

Now we bring the Berezin transformation into play. In so doing, put

$$\tilde{f}(z) = \pi^{-1} \int_{\mathbf{D}} f(\zeta) |\sigma_z'(\zeta)|^2 dm(\zeta), \quad z \in \mathbf{D},$$

and

$$A_r(f)(z) = \frac{1}{m(B(z,r))} \int_{B(z,r)} f(\zeta) dm(\zeta), \quad z \in \mathbf{D}, \ r > 0.$$

Theorem 2.3.2. *Let* $p \in (0, 2)$, $r \in (0, \infty)$ *and* $f \in \mathcal{A}^{2,0}$. *Then the following statements are equivalent:*

(i) $f \in \mathcal{Q}_p$.

(ii) $\displaystyle \sup_{w \in \mathbf{D}} \int_{\mathbf{D}} \int_{\mathbf{D}} \left(\frac{|f(z) - f(\zeta)|^2}{|1 - z\bar{\zeta}|^4} \right) \left(1 - |\sigma_w(z)|^2 \right)^p dm(z) dm(\zeta) < \infty$.

(iii) $\displaystyle \sup_{w \in \mathbf{D}} \int_{\mathbf{D}} \left(\widetilde{|f|^2}(z) - |f(z)|^2 \right) \left(1 - |\sigma_w(z)|^2 \right)^p \frac{dm(z)}{(1 - |z|^2)^2} < \infty$.

(iv) $\displaystyle \sup_{w \in \mathbf{D}} \int_{\mathbf{D}} \left(A_r(|f|^2)(z) - |A_r(f)(z)|^2 \right) \left(1 - |\sigma_w(z)|^2 \right)^p \frac{dm(z)}{(1 - |z|^2)^2} < \infty$.

(v) $\displaystyle \sup_{w \in \mathbf{D}} \int_{\mathbf{D}} A_r \left(|f - A_r(f)(z)|^2 \right)(z) \left(1 - |\sigma_w(z)|^2 \right)^p \frac{dm(z)}{(1 - |z|^2)^2} < \infty$.

Proof. Step 1: (i)$\Leftrightarrow$(ii) We have that $f \in \mathcal{Q}_p$ if and only if $\sup_{w \in \mathbf{D}} I_1(f \circ \sigma_w) < \infty$ which is by Lemma 2.3.1 equivalent to $\sup_{w \in \mathbf{D}} I_2(f \circ \sigma_w) < \infty$. So the desired equivalence follows from a simple calculation with the conformal map σ_w.

Step 2: (ii)$\Leftrightarrow$(iii) A routine computation with $f \in \mathcal{A}^{2,0}$ gives

$$
\begin{aligned}
\widetilde{|f|^2}(z) - |f(z)|^2 &= \pi^{-1} \int_{\mathbf{D}} |f \circ \phi_z(\zeta) - f(z)|^2 dm(\zeta) \\
&= \pi^{-1} \int_{\mathbf{D}} |f(\zeta) - f(z)|^2 \frac{(1 - |z|^2)^2}{|1 - z\bar{\zeta}|^4} dm(\zeta).
\end{aligned}
$$

Accordingly,

$$
\begin{aligned}
&I_2(f \circ \sigma_w) \\
&= \int_{\mathbf{D}} \left(1 - |\sigma_w(z)|^2 \right)^p \left(\int_{\mathbf{D}} |f(\zeta) - f(z)|^2 \frac{(1 - |z|^2)^2}{|1 - z\bar{\zeta}|^4} dm(\zeta) \right) \frac{dm(z)}{(1 - |z|^2)^2} \\
&= \pi \int_{\mathbf{D}} \left(\widetilde{|f|^2}(z) - |f(z)|^2 \right) \left(1 - |\sigma_w(z)|^2 \right)^p \frac{dm(z)}{(1 - |z|^2)^2}.
\end{aligned}
$$

This implies the required equivalence.

Step 3: (iii)$\Rightarrow$(iv)$\Rightarrow$(v)$\Rightarrow$(i) From Step 1 and

$$
\frac{(1 - |w|^2)^2}{|1 - z\bar{w}|^4} \approx \frac{1}{(1 - |z|^2)^2}, \quad w \in \mathbf{D}
$$

it turns out that

$$
\widetilde{|f|^2}(z) - |f(z)|^2 \approx \int_D \int_D \frac{|f(\zeta) - f(\eta)|^2 (1 - |z|^2)^4}{|1 - z\bar{\zeta}|^4 |1 - z\bar{\eta}|^4} dm(\zeta) dm(\eta)
$$

$$
\succeq \int_{B(z,2r)} \int_{B(z,2r)} \frac{|f(\zeta) - f(\eta)|^2 (1 - |z|^2)^4}{|1 - z\bar{\zeta}|^4 |1 - z\bar{\eta}|^4} dm(\zeta) dm(\eta)
$$

$$
\succeq \int_{B(z,2r)} \int_{B(z,2r)} \left(\frac{|f(\zeta) - f(\eta)|}{m(B(z,2r))} \right)^2 dm(\zeta) dm(\eta)
$$

$$
\succeq A_r \big(|f - A_r(f)(z)|^2 \big)(z).
$$

Namely, (iii) implies (iv).

In the meantime, the elementary estimates

$$
|f'(0)|^2 \preceq \int_{B(0,r)} |f(\zeta) - c|^2 dm(\zeta), \quad c \in \mathbf{C}
$$

and

$$
\frac{(1 - |z|^2)^2}{|1 - z\bar{\zeta}|^4} \approx \frac{1}{m(B(z,r))}, \quad \zeta \in B(z,r)
$$

yield

$$
(1 - |z|^2)^2 |f'(z)|^2 \preceq \int_{B(0,r)} |f \circ \sigma_z(\zeta) - A_r(f)(z)|^2 dm(\zeta)
$$

$$
\approx \int_{B(z,r)} |f(\zeta) - A_r(f)(z)|^2 \left(\frac{1 - |z|^2}{|1 - z\bar{\zeta}|^2} \right)^2 dm(\zeta)
$$

$$
\approx A_r \big(|f - A_r(f)(z)|^2 \big)(z)
$$

$$
\approx A_r(|f|^2)(z) - |A_r(f)(z)|^2.
$$

Consequently,

$$
I_1(f \circ \sigma_w) = \int_D |f'(z)|^2 (1 - |\sigma_w(z)|^2)^p dm(z)
$$

$$
\preceq \int_D A_r \big(|f - A_r(f)(z)|^2 \big)(z) (1 - |\sigma_w(z)|^2)^p \frac{dm(z)}{(1 - |z|^2)^2}
$$

$$
\approx \int_D \big(A_r(|f|^2)(z) - |A_r(f)(z)|^2 \big) (1 - |\sigma_w(z)|^2)^p \frac{dm(z)}{(1 - |z|^2)^2}.
$$

In other words, (iv) implies (v) which implies (i). □

2.4 Mixture of Derivative and Quotient

In this section, we extend the characterization presented in the last section to the mixed situation involving derivative and quotient.

Lemma 2.4.1. *Let $p \in (0, \infty)$ and $q \in [0, \infty)$.*

(i) *Suppose $F \in \mathcal{H}^p$ and $F(0) = 0$. If $q \in (0, \min\{2, p\}]$, then*

$$\int_{\mathbf{T}} |F(\zeta)|^p |d\zeta| \preceq \int_{\mathbf{D}} |F(z)|^{p-q} |F'(z)|^q (1 - |z|^2)^{q-1} dm(z).$$

Moreover, if $q \in [2, p+2)$, then

$$\int_{\mathbf{D}} |F(z)|^{p-q} |F'(z)|^q (1 - |z|^2)^{q-1} dm(z) \preceq \int_{\mathbf{T}} |F(\zeta)|^p |d\zeta|.$$

(ii) *Suppose $F \in \mathcal{A}^{p,0}$ and $F(0) = 0$. If $q \in [0, p+2)$, then*

$$\int_{\mathbf{D}} |F(z)|^p dm(z) \approx \int_{\mathbf{D}} |F(z)|^{p-q} |F'(z)|^q (1 - |z|^2)^q dm(z).$$

Proof. For any $F \in \mathcal{A}^{p,0}$, let

$$I(F; p, q) = \int_{\mathbf{D}} |F(z)|^{p-q} |F'(z)|^q (1 - |z|^2)^q dm(z).$$

First of all, we recall two basic facts for $f \in \mathcal{H}$, $f_r(z) = f(rz)$, $r \in (0, 1)$ and $p \in (0, \infty)$. The first one is the Hardy–Stein identity which reads as:

$$\|f_r\|_{\mathcal{H}^p}^p = 2\pi |f(0)|^p + p^2 \int_{\mathbf{D}} |f_r(z)|^{p-2} |f_r'(z)|^2 (-\log |z|) dm(z).$$

The second one is the following Littlewood–Paley inequalities:

$$\|f_r\|_{\mathcal{H}^p}^p \preceq |f(0)|^p + I(f_r; p, p), \quad p \in (0, 2]$$

and

$$|f(0)|^p + I(f_r; p, p) \preceq \|f_r\|_{\mathcal{H}^p}^p, \quad p \in [2, \infty).$$

Next, we check (i) and (ii).

(i) For the first inequality, let $q \in (0, \min\{2, p\}]$. Clearly, the inequality is valid for $p = q = 2$. So we just consider following two cases.

Case 1: $0 < q < 2 < p$. Using the Hardy–Stein identity, Hölder's inequality and the second Littlewood–Paley inequality above for $F_r(z) = F(rz)$, we derive

$$\begin{aligned}
\|F_r\|_{\mathcal{H}^p}^p &\preceq \int_{\mathbf{D}} |F_r(z)|^{p-2} |F_r'(z)|^2 (-\log |z|) dm(z) \\
&\preceq \left(I(F_r; p, q) \right)^{\frac{p-2}{p-q}} \left(I(F_r; p, p) \right)^{\frac{2-q}{p-q}} \\
&\preceq \left(I(F; p, q) \right)^{\frac{p-2}{p-q}} \left(\|F_r\|_{\mathcal{H}^p} \right)^{\frac{p(2-q)}{p-q}},
\end{aligned}$$

thereupon producing $\|F\|^p_{\mathcal{H}^p} \preceq I(F; p, q)$.

Case 2: $0 < q < p < 2$. By the first Littlewood–Paley inequality above, Hölder's inequality and the Hardy–Stein identity for F_r, we find

$$\begin{aligned} \|F_r\|^p_{\mathcal{H}^p} &\preceq I(F_r; p, p) \\ &\preceq \left(I(F_r; p, q)\right)^{\frac{2-p}{2-q}} \left(I(F_r; p, 2)\right)^{\frac{p-q}{2-q}} \\ &\preceq \left(I(F; p, q)\right)^{\frac{2-p}{2-q}} \|F_r\|^{\frac{p(p-q)}{2-q}}_{\mathcal{H}^p}, \end{aligned}$$

thereupon implying $\|F\|^p_{\mathcal{H}^p} \preceq I(F; p, q)$.

As with the second estimate, suppose $q \in [2, p + 2)$. Since $F \in \mathcal{H}^p$, this function can be written as $F = BG$ where G has no zeros with $\|G\|_{\mathcal{H}^p} = \|F\|_{\mathcal{H}^p}$ and B is a Blaschke product. Accordingly,

$$|F|^{p-q}|F'|^q \le 2^{q-1}(|G|^p|B|^{p-q}|B'|^q + |B|^p|G|^{p-q}|G'|^q).$$

Since $G \ne 0$, letting $h = G^{\frac{p}{q}}$ yields $h \in \mathcal{H}^q$, $|h'|^q = pq^{-1}|G|^{p-q}|G'|^q$ and

$$\begin{aligned} &\int_{\mathbf{D}} |B(z)|^p |G(z)|^{p-q} |G'(z)|^q (1 - |z|^2)^{q-1} dm(z) \\ &\preceq \int_{\mathbf{D}} |h'(z)|^q (1 - |z|^2)^{q-1} dm(z) \\ &\preceq \|h\|^q_{\mathcal{H}^q} \approx \|F\|^p_{\mathcal{H}^p}. \end{aligned}$$

For the other estimate we use the Carleson embedding for $\mathcal{H}^p$ — see [Ga, pp. 238-239] to get

$$\int_{\mathbf{D}} |G(z)|^p |B(z)|^{p-q} |B'(z)|^q (1 - |z|^2)^{q-1} dm(z) \preceq \sup_{a \in \mathbf{D}} C(a) \|G\|^p_{\mathcal{H}^p},$$

where

$$C(a) = \int_{\mathbf{D}} \frac{1 - |a|^2}{|1 - \bar{a}z|^2} |B(z)|^{p-q} |B'(z)|^q (1 - |z|^2)^{q-1} dm(z).$$

The desired inequality will be established by proving $\sup_{a \in \mathbf{D}} C(a) < \infty$. In so doing, we are making use of $z \mapsto \sigma_a(z)$, $(1 - |z|^2)|B'(z)| \le 1 - |B(z)|^2$ and the Hardy–Stein identity to derive

$$\begin{aligned} C(a) &= \int_{\mathbf{D}} |B \circ \sigma_a(z)|^{p-q} |(B \circ \sigma_a)'(z)|^q (1 - |z|^2)^{q-1} dm(z) \\ &\preceq \int_{\mathbf{D}} |B \circ \sigma_a(z)|^{q-2} |(B \circ \sigma_a)'(z)|^2 (1 - |z|^2) dm(z) \\ &\preceq \|B \circ \sigma_a\|^q_{\mathcal{H}^q} \preceq 1. \end{aligned}$$

(ii) Proving that estimate amounts to verifying $I(F; p, 0) \approx I(F; p, q)$. To do so, let us consider two cases as follows.

Case 1: $q \in (0, 2]$. If $p \in (0, q)$, then by the Hölder inequality,

$$I(F; p, 0) \preceq I(F; p, p) \preceq \left(I(F; p, q)\right)^{\frac{p}{q}} \left(I(F; p, 0)\right)^{1 - \frac{p}{q}},$$

and hence $I(F; p, 0) \preceq I(F; p, q)$. At the same time, if $p \in [q, \infty)$, then by the first inequality in (i) we have that for $r \in (0, 1)$,

$$\int_{\mathbf{T}} |F(r\zeta)|^p |d\zeta| \preceq \int_{\mathbf{D}} |F(rz)|^{p-q} |rF'(rz)|^q (-\log|z|)^{q-1} dm(z)$$

and consequently, integrating both sides with respect to rdr and changing variable twice ($z \mapsto rz$ and $t = |z|/r$), we get

$$
\begin{aligned}
I(F; p, 0) \quad &\preceq \quad \int_0^1 \left(\int_{|z| < r} |F(z)|^{p-q} |F'(z)|^q r^{q-2} \left(\log \frac{r}{|z|} \right)^{q-1} dm(z) \right) rdr \\
&\preceq \quad \int_{\mathbf{D}} |F(z)|^{p-q} |F'(z)|^q \left(\int_{|z|}^1 r^{p-1} \left(\log \frac{r}{|z|} \right)^{q-1} dr \right) dm(z) \\
&\preceq \quad \int_{\mathbf{D}} |F(z)|^{p-q} |F'(z)|^q (-\log|z|)^q dm(z) \\
&\preceq \quad I(F; p, q).
\end{aligned}
$$

In the last estimate we have used the fact that $|F(z)|^{p-q}|F'(z)|^q$ is subharmonic and the inequality $-|z| \log|z| \le 1 - |z|^2$.

To prove $I(F; p, q) \preceq I(F; p, 0)$, applying the Hardy–Stein identity to F_r, integrating the induced identity with respect to rdr and using the inequality

$$\int_{|z|}^1 r \log \frac{r}{|z|} dr \ge (1 - |z|^2)/4,$$

we further get

$$
\begin{aligned}
I(F; p, 2) \quad &\preceq \quad \int_{\mathbf{D}} |F(z)|^{p-2} |F'(z)|^2 \left(\int_{|z|}^1 r \log \frac{r}{|z|} dr \right) dm(z) \\
&\approx \quad \int_0^1 \left(\int_{|z| < r} |F(z)|^{p-2} |F'(z)|^2 \log \frac{r}{|z|} dm(z) \right) rdr \\
&\approx \quad I(F; p, 0).
\end{aligned}
$$

Accordingly, we can use the Hölder inequality with $q \in (0, 2]$ to obtain

$$I(F; q, q) \le \left(I(F; p, 2)\right)^{\frac{q}{2}} \left(I(F; p, 0)\right)^{1 - \frac{q}{2}} \preceq I(F; p, 0),$$

as desired.

Case 2: $q \in (2, \infty)$. Note that if $p - q + 2 \in (0, \infty)$ and $r \in (0, 1)$, then the second inequality in (i) holds, namely,

$$\int_{\mathbf{D}} |F(rz)|^{p-q} |rF'(rz)|^q (1 - |z|^2)^{q-1} dm(z) \preceq \int_{\mathbf{T}} |F(r\zeta)|^p |d\zeta|.$$

Thus, by integration from r to 1 with respect to $r\,dr$, we obtain $I(F; p, q) \preceq I(F; p, 0)$. To establish the opposite inequality, we observe

$$\int_{|z|}^1 r \log \frac{r}{|z|} \, dr \approx (1 - |z|^2)\left(1 + \log \frac{1}{|z|^2}\right).$$

This comparability result, along with the Hölder inequality and the estimate

$$|z|^{p - \frac{2p}{q}} \left(1 + \log \frac{1}{|z|^2}\right) \preceq 1, \quad z \in \mathbf{D},$$

produces

$$
\begin{aligned}
I(F; p, 0) &\preceq \int_{\mathbf{D}} |F(z)|^{p-2} |F'(z)|^2 \left(\int_{|z|}^1 r \log \frac{r}{|z|} dr\right) dm(z) \\
&\preceq \left(\int_{\mathbf{D}} \left|\frac{F(z)}{z}\right|^p dm(z)\right)^{1 - \frac{2}{q}} \left(I(F; p, q)\right)^{\frac{2}{q}} \\
&\preceq \left(I(F; p, 0)\right)^{1 - \frac{2}{q}} \left(I(F; p, q)\right)^{\frac{2}{q}},
\end{aligned}
$$

thereupon reaching $I(F; p, 0) \preceq I(F; p, q)$. $\qquad\square$

The following characterization extends Theorem 2.3.2 from $q = 0$ to $q > 0$.

Theorem 2.4.2. *Given $p \in (0, 2)$ and $q \in [0, 4)$. For $f \in \mathcal{H}$ let*

$$\Xi(f; \zeta, z) = \frac{(1 - |\zeta|^2)|f'(\zeta)|}{|f(\zeta) - f(z)|}, \quad \zeta, z \in \mathbf{D}.$$

Then $f \in \mathcal{Q}_p$ if and only if.

$$\sup_{w \in \mathbf{D}} \int_{\mathbf{D}} \int_{\mathbf{D}} \left(\frac{|f(z) - f(\zeta)|^2}{|1 - z\bar{\zeta}|^4}\right) \left(\Xi(f; \zeta, z)\right)^q (1 - |\sigma_w(z)|^2)^p dm(z) dm(\zeta) < \infty.$$

Proof. Suppose the last condition is valid. We then employ Lemma 2.4.1 (ii) with $F = f \circ \sigma_w - f(w)$ to obtain

$$
\begin{aligned}
\left((1 - |w|^2)|f'(w)|\right)^2 & \\
\preceq \int_{\mathbf{D}} & |f \circ \sigma_w(z) - f(w)|^2 dm(z) \\
\preceq \int_{\mathbf{D}} & |f \circ \sigma_w(z) - f(w)|^{2-q} |(f \circ \sigma_w)'(z)|^q (1 - |z|^2)^q dm(z).
\end{aligned}
$$

Integrating the last estimate against $(1 - |\sigma_a(w)|^2)^p dm(w)$, $a \in \mathbf{D}$, and changing the variable $z \mapsto \sigma_w(z)$, we find

$$\left(F_p(f,a)\right)^2 \preceq \int_{\mathbf{D}} \int_{\mathbf{D}} \left(\frac{|f(z) - f(\zeta)|^2}{|1 - z\bar{\zeta}|^4}\right) \left(\Xi(f; \zeta, z)\right)^q (1 - |\sigma_a(z)|^2)^p dm(z) dm(\zeta),$$

thereupon deriving $f \in \mathcal{Q}_p$.

Conversely, suppose $f \in \mathcal{Q}_p$. Note that by the change of variable $\lambda = \sigma_a(w)$, $a \in \mathbf{D}$,

$$\begin{aligned}
\int_{\mathbf{D}} \frac{(1 - |\sigma_a(w)|^2)^p}{|1 - \bar{w}u|^4} dm(w) &= (1 - |a|^2)^2 \int_{\mathbf{D}} \frac{(1 - |\lambda|^2)^p}{|1 - \bar{a}\lambda - a\bar{u} + \lambda\bar{u}|^4} dm(\lambda) \\
&= \frac{(1 - |a|^2)^2}{|1 - a\bar{u}|^4} \int_{\mathbf{D}} \frac{(1 - |\lambda|^2)^p}{|1 - \bar{\lambda}\sigma_a(u)|^4} dm(\lambda) \\
&\preceq \frac{(1 - |a|^2)^p (1 - |u|^2)^{p-2}}{1 - \bar{a}u|^{2p}}, \quad u \in \mathbf{D}.
\end{aligned}$$

So, using the change of variable: $\lambda = \sigma_w(z)$ and Lemma 2.4.1 (ii) again, we find

$$\begin{aligned}
&\int_{\mathbf{D}} \left(\frac{|f(z) - f(\zeta)|^2}{|1 - z\bar{\zeta}|^4}\right) \left(\Xi(f; \zeta, z)\right)^q (1 - |\sigma_a(z)|^2)^p dm(z) dm(\zeta) \\
&= \int_{\mathbf{D}} I(f \circ \sigma_w - f(w); 2, q)(1 - |w|^2)^{-2}(1 - |\sigma_a(w)|^2)^p dm(w) \\
&\preceq \int_{\mathbf{D}} \left(\int_{\mathbf{D}} |f \circ \sigma_w(z) - f(w)|^2 dm(z)\right) (1 - |w|^2)^{-2}(1 - |\sigma_a(w)|^2)^p dm(w) \\
&\preceq \int_{\mathbf{D}} \left(\int_{\mathbf{D}} |(f \circ \sigma_w)'(z)|^2(1 - |z|^2)^2 dm(z)\right) \frac{(1 - |\sigma_a(w)|^2)^p}{(1 - |w|^2)^2} dm(w) \\
&\preceq \int_{\mathbf{D}} |f'(u)|^2(1 - |u|^2)^2 \left(\int_{\mathbf{D}} \frac{(1 - |\sigma_a(w)|^2)^p}{|1 - \bar{w}u|^4} dm(w)\right) dm(u) \\
&\preceq \left(F_p(f,a)\right)^2 \preceq \|f\|_{\mathcal{Q}_p,1}^2,
\end{aligned}$$

whence implying the desired assertion. $\qquad\square$

2.5　Dirichlet Double Integral without Derivative

In this section, we use the derivative-free description of the weighted Dirichlet integral over the bidisk to characterize $\mathcal{Q}_p$ functions.

Lemma 2.5.1. *Let $f \in \mathcal{H}$. If $\alpha \in (-\infty, 1/2]$, $\beta, \gamma \in (-1, \infty)$ and $\min\{\beta, \gamma\} + 2\alpha > -1$, then*

$$\int_{\mathbf{D}} |f'(z)|^2(1 - |z|^2)^{1-2\alpha} dm(z)$$

$$\approx \int_{\mathbf{D}} \int_{\mathbf{D}} \left(\frac{|f(z) - f(w)|^2}{|1 - \bar{z}w|^{3+2\alpha+\beta+\gamma}}\right) (1 - |z|^2)^\beta (1 - |w|^2)^\gamma dm(z) dm(w).$$

Proof. It suffices to consider the case $\beta = \gamma$ since $\beta \geq \gamma$ implies

$$\frac{2^{\gamma-\beta}(1-|z|^2)^\beta(1-|w|^2)^\beta}{|1-\bar{z}w|^{3+2\alpha+2\beta}} \leq \frac{(1-|z|^2)^\beta(1-|w|^2)^\gamma}{|1-\bar{z}w|^{3+2\alpha+\beta+\gamma}}$$

$$\leq \frac{2^{\beta-\gamma}(1-|z|^2)^\gamma(1-|w|^2)^\gamma}{|1-\bar{z}w|^{3+2\alpha+2\gamma}}.$$

If $\alpha = 1/2$, then a direct computation yields

$$\int_{\mathbf{D}} |f(z) - f(0)|^2(1-|z|^2)^\beta dm(z) \precsim \int_{\mathbf{D}} |f'(z)|^2(1-|z|^2)^{\beta+2} dm(z).$$

Applying this inequality to $f \circ \sigma_w$, we derive

$$
\begin{aligned}
I(f;w) &= \int_{\mathbf{D}} |f(z)-f(w)|^2 |1-\bar{z}w|^{-4-2\beta}(1-|z|^2)^\beta dm(z) \\
&= (1-|w|^2)^{-(2+\beta)} \int_{\mathbf{D}} |f\circ\sigma_w(z) - f\circ\sigma_w(0)|^2(1-|z|^2)^\beta dm(z) \\
&\precsim (1-|w|^2)^{-(2+\beta)} \int_{\mathbf{D}} |(f\circ\sigma_w)'(z)|^2(1-|z|^2)^{\beta+2} dm(z) \\
&\precsim \int_{\mathbf{D}} |f'(z)|^2 |1-\bar{z}w|^{-2\beta-4}(1-|z|^2)^{\beta+2} dm(z).
\end{aligned}
$$

Integrating $I(f;w)$ over $\mathbf{D}$ with respect to $(1-|w|^2)^\beta dm(w)$ and changing the order of integration, we further get

$$
\begin{aligned}
\|f\|_{\mathbf{D}}^2 &\succsim \int_{\mathbf{D}} |f'(z)|^2 \left(\int_{\mathbf{D}} \frac{(1-|w|^2)^\beta}{|1-\bar{w}z|^{4+2\beta}} dm(w) \right) (1-|z|^2)^{\beta+2} dm(z) \\
&\succsim \int_{\mathbf{D}} I(f;w)(1-|w|^2)^\beta dm(w)
\end{aligned}
$$

thanks to $\beta > -1$.

As for the opposite inequality, we make the following estimates:

$$
\begin{aligned}
(1-|z|^2)^2 |f'(z)|^2 &= |(f\circ\sigma_z)'(0)|^2 \\
&\precsim \int_{|u|<1/2} |f\circ\sigma_z(u) - f\circ\sigma_z(0)|^2 dm(u) \\
&\precsim \int_{\mathbf{D}} |f\circ\sigma_z(u) - f\circ\sigma_z(0)|^2(1-|u|^2)^\beta dm(u) \\
&\approx (1-|z|^2)^{\beta+2} \int_{\mathbf{D}} \frac{|f(z)-f(w)|^2(1-|w|^2)^\beta}{|1-\bar{w}z|^{4+2\beta}} dm(w),
\end{aligned}
$$

and then integrate both sides of the last inequalities against $(1-|z|^2)^{-2} dm(z)$ to produce

$$\|f\|_{\mathbf{D}}^2 \precsim \int_{\mathbf{D}} I(f;w)(1-|w|^2)^\beta dm(w),$$

as desired.

Now suppose $\alpha < 1/2$. Setting $\delta = 3/2 + \alpha + \beta$, $f(z) = \sum_{k=0}^{\infty} a_k z^k$, and for any integer m,

$$1_m = \begin{cases} 1 & , \quad m \geq 0, \\ 0 & , \quad m < 0, \end{cases}$$

plus making substitutions: $z = re^{i\theta}$, $w = se^{i\phi}$, $t = \phi - \theta$ and $\zeta = se^{it}$, we have

$$I(f; \beta, \delta)$$

$$= \int_{\mathbf{D}} \int_{\mathbf{D}} \left(\frac{|f(z) - f(w)|^2}{|1 - \bar{z}w|^{2\delta}} \right) (1 - |z|^2)^\beta (1 - |w|^2)^\beta dm(z) dm(w)$$

$$\approx \int_0^1 \int_{-\pi}^{\pi} \int_0^1 \int_{-\pi}^{\pi} \left(\frac{|f(re^{i\theta}) - f(se^{i(\theta+t)})|^2}{|1 - rse^{it}|^{2\delta}} \right) (1 - r^2)^\beta (1 - s^2)^\beta rs\,d\theta\,dr\,dt\,ds$$

$$\approx \int_0^1 \int_{-\pi}^{\pi} \int_0^1 \sum_{k=1}^{\infty} |a_k|^2 \frac{|r^k - s^k e^{ikt}|^2}{|1 - rse^{it}|^{2\delta}} (1 - r^2)^\beta (1 - s^2)^\beta rs\,dr\,dt\,ds$$

$$\approx \sum_{k=1}^{\infty} |a_k|^2 \int_0^1 \int_{\mathbf{D}} \frac{|r^k - \zeta^k|^2}{|1 - r\zeta|^{2\delta}} (1 - r^2)^\beta (1 - s^2)^\beta r\,dm(\zeta)\,dr$$

$$\approx \sum_{k=1}^{\infty} |a_k|^2 I(k),$$

where

$$I(k) = \int_0^1 \int_{\mathbf{D}} |q_r(\zeta)|^2 (1 - r^2)^\beta (1 - s^2)^\beta r\,dm(\zeta)\,dr \quad \text{with} \quad q_r(\zeta) = \frac{r^k - \zeta^k}{(1 - r\zeta)^\delta}.$$

If we can prove $I(k) \approx k^{2\alpha}$ for $k \in \mathbf{N}$, we will then have

$$I(f; \delta, \beta) \approx \sum_{k=1}^{\infty} |a_k|^2 k^{2\alpha} \approx \int_{\mathbf{D}} |f'(z)|^2 (1 - |z|^2)^{1-2\alpha} dm(z),$$

thereby establishing the desired comparability.

In so doing, let $B(x, y) = \int_0^1 t^{x-1}(1 - t)^{y-1} dt$. Then for $x, y \in (0, \infty)$ and $j, k \in \mathbf{N}$,

$$B(j + 1 + x, y) - B(j + k + 1 + x, y)$$

$$= \int_0^1 t^{j+x} \left(\sum_{n=0}^{k-1} t^n \right) (1 - t)^y dt$$

$$\approx \sum_{n=0}^{k-1} (n + j + 1)^{-y-1} \approx (j + 1)^{-y} - (k + j + 1)^{-y}.$$

Since $q_r(\zeta)$ is a member of $\mathcal{A}^{2,\beta}$, it can be represented by the reproducing formula for this space. Consequently,

$$
I(k)
$$
$$
= \int_0^1 \int_{\mathbf{D}} |q_r(\zeta)|^2 (1 - |\zeta|^2)^\beta dm(\zeta)(1 - r^2)^\beta r dr
$$
$$
= \int_0^1 \int_{\mathbf{D}} \left| \frac{1+\beta}{\pi} \int_{\mathbf{D}} \frac{q_r(\eta)(1 - |\eta|^2)^\beta}{(1 - \bar{\eta}\zeta)^{2+\beta}} dm(\eta) \right|^2 \left(\frac{1 - |\zeta|^2}{(1 - r^2)^{-1}} \right)^\beta dm(\zeta) r dr
$$
$$
\approx \int_0^1 \int_{\mathbf{D}} \left| \sum_{j=0}^\infty \frac{\int_{\mathbf{D}} q_r(\eta)\bar{\eta}^j \zeta^j (1 - |\eta|^2)^\beta dm(\eta)}{B(j+1,\beta+1)} \right|^2 \left(\frac{1 - |\zeta|^2}{(1 - r^2)^{-1}} \right)^\beta dm(\zeta) r dr
$$
$$
\approx \int_0^1 \left(\sum_{j=0}^\infty \frac{\left| \int_{\mathbf{D}} q_r(\eta)\bar{\eta}^j ((1 - |\eta|^2)(1 - r^2))^\beta dm(\eta) \right|^2}{B(j+1,\beta+1)} \right) r dr.
$$

Note also that

$$
q_r(\eta) = (r^k - \eta^k) \sum_{n=0}^\infty \frac{\Gamma(n+\delta) r^n \eta^n}{\Gamma(\delta)\Gamma(n+1)}.
$$

Thus, we can conclude

$$
\int_{\mathbf{D}} q_r(\eta)\bar{\eta}^j (1 - |\eta|^2)^\beta dm(\eta)
$$
$$
\approx \frac{\Gamma(j+\delta)B(j+1,\beta+1)r^{k+j}}{\Gamma(j+1)} - \frac{1_{j-k}\Gamma(j-k+\delta)B(j+1,\beta+1)r^{j-k}}{\Gamma(j-k+1)},
$$

so that

$$
I(k) \approx \lim_{m \to \infty} \left(\sum_{j=m-k+1}^\infty I_1(j,k) + 2 \sum_{j=0}^{m-k} (I_1(j,k) - I_2(j,k)) \right),
$$

where

$$
I_1(j,k) = \frac{B(j+1,\beta+1)B(j+k+1,\beta+1)(\Gamma(j+\delta))^2}{(\Gamma(j+1))^2}
$$

and

$$
I_2(j,k) = \frac{(B(j+k+1,\beta+1))^2 \Gamma(j+k+\delta)\Gamma(j+\delta)}{\Gamma(j+k+1)\Gamma(j+1)}.
$$

By the above estimate on $B(\cdot,\cdot)$, it follows that

$$
I_1(j,k) \approx j^{-1-\beta}(k+j)^{-1-\beta} j^{2\delta-2} \preceq j^{2\alpha-1}, \quad j \in \mathbf{N},
$$

and so that $\lim_{m\to\infty} \sum_{j=m-k+1}^{m} I_1(j,k) = 0$ due to $\alpha \in (-\infty, 1/2)$. Using the estimates

$$\frac{\Gamma(j+\delta-1)}{\Gamma(j+1/2-\alpha)} \approx (j+1)^{\alpha+\delta-3/2} \quad \text{and} \quad B(k, \delta-1) \approx k^{1-\delta},$$

we further get that $I_1(j,k) - I_2(j,k)$ is equal to the product of both

$$B(k+j+1, \beta+1)\Gamma(j+\delta)\Gamma(\beta+1)$$
$$\Gamma(j+1)\Gamma(1/2-\alpha)$$

and $B(j+\delta, 1/2-\alpha) - B(j+k+\delta, 1/2-\alpha)$, whence producing

$$I_1(j,k) - I_2(j,k) \approx (j+1)^{\delta-1}(K+j+1)^{-\beta-1}\big((j+1)^{\alpha-1/2} - (j+k+1)^{\alpha-1/2}\big).$$

Combining the previous estimates we finally obtain

$$I(k)$$
$$\approx \lim_{m\to\infty} \sum_{j=0}^{m-k} (j+1)^{\delta-1}(K+j+1)^{-\beta-1}\big((j+1)^{\alpha-1/2} - (j+k+1)^{\alpha-1/2}\big)$$
$$\approx \int_0^\infty x^{\delta-1}(k+x+^{-\beta-1}\big(x^{\alpha-1/2} - (k+x)^{\alpha-1/2}\big)dx$$
$$\approx k^{2\alpha}\int_0^\infty t^{2\alpha+\beta}(1+t)^{\alpha-\beta-3/2}\big((1+t)^{1/2-\alpha} - t^{1/2-\alpha}\big)dt \approx k^{2\alpha},$$

as required. $\qquad\qquad\qquad\qquad\qquad\qquad\qquad\qquad\qquad\qquad\qquad\qquad\square$

Clearly, Lemma 2.5.1 extends Lemma 2.3.1 from one parameter to three. As a direct application of Lemma 2.5.1 we derive the following result which is a natural generalization of Theorem 2.3.2.

Theorem 2.5.2. *Given $p \in (0, \infty)$, $\beta, \gamma \in (-1, \infty)$ and $\min\{\beta, \gamma\} - p > -2$. Let $f \in \mathcal{H}$. Then $f \in \mathcal{Q}_p$ if and only if*

$$\sup_{a\in\mathbf{D}} \int_{\mathbf{D}} \int_{\mathbf{D}} \Big(\frac{|f(z)-f(w)|^2}{|1-\bar{z}w|^{4-p+\beta+\gamma}}\Big)\frac{(1-|a|^2)^p(1-|z|^2)^\beta(1-|w|^2)^\gamma}{|1-\bar{a}z|^{p+\beta-\gamma}|1-\bar{a}w|^{p+\gamma-\beta}}dm(z)dm(w)$$

is finite.

Proof. Applying Lemma 2.5.1 to $f \circ \sigma_a$ with $p = 1 - 2\alpha$, and making use of the identities

$$1 - |\sigma_a(z)|^2 = \frac{(1-|a|^2)(1-|z|^2)}{|1-\bar{a}z|^2}, \quad \sigma_a \circ \sigma_a(z) = z, \quad \text{and} \quad |\sigma'_a(z)| = \frac{1-|a|^2}{|1-\bar{a}z|^2},$$

we reach the assertion. $\qquad\qquad\qquad\qquad\qquad\qquad\qquad\qquad\qquad\qquad\qquad\square$

2.6 Notes

Note 2.6.1. Section 2.1 is taken from in [Xi4, 3.1]. A good source for more on the Brownian motions associated with $\mathcal{BMO}$ is [Pet].

Note 2.6.2. Section 2.2 comes from [Xi4, 3.3]. For the extremal case $p = 1$ of Theorem 2.2.1 (iii), see the main result in [Kob1] of which a special form is: $f \in \mathcal{BMOA}$ if and only if $|f - f(w)|^2$ has the least harmonic majorant for any $w \in \mathbf{D}$. And in this case, Theorem 2.2.1 (iii) is replaced by

$$
\sup_{z \in \mathbf{D}} (|f \widehat{- f(z)}|^2)(z) = \sup_{z \in \mathbf{D}} (\widehat{|f|^2}(z)) - |f(z)|^2)
$$

$$
\approx \sup_{z \in \mathbf{D}} \left((\widehat{|f|}(z))^2 - |f(z)|^2 \right) < \infty.
$$

See [Leu, Section 3]. As with the limiting case $p \to 0$, we have however noted that $f \in \mathcal{D}$ if and only if $f \in \mathcal{H}^2$ and

$$
\int_{\mathbf{T}} \int_{\mathbf{T}} \frac{|f(\zeta) - f(\eta)|^2}{|\zeta - \eta|^2} |d\zeta||d\eta| < \infty.
$$

See e.g. [Pee]. This means that Theorem 2.2.1 (ii) is natural. Moreover, regarding Corollary 2.2.2 (i), we may ask an interesting question: is there a function $\psi > 0$ a.e. on $\mathbf{T}$ with $\psi \in \mathcal{Q}_p(\mathbf{T})$, $\log \psi \in \mathcal{L}^1(\mathbf{T})$ but $\mathcal{O}_\psi \notin \mathcal{Q}_p$? This question has a positive answer for $p = 1$ — see also [Dy], although the answer is unknown for $p \in (0, 1)$.

Note 2.6.3. Section 2.3 is produced by considering a special case of the weight functions considered in the main results of [WulZh2]. It is well known that $f \in \mathcal{B} = \mathcal{Q}_2$ if and only if $f \in \mathcal{A}^{2,0}$ and

$$
\sup_{z \in \mathbf{D}} (|f \widetilde{- f(z)}|^2)(z) = \sup_{z \in \mathbf{D}} (\widetilde{|f|^2}(z) - |f(z)|^2)
$$

$$
\approx \sup_{z \in \mathbf{D}} (A_r(|f|^2)(z) - |A_r(f)(z)|^2)
$$

$$
\approx \sup_{z \in \mathbf{D}} A_r(|f - A_r(f)(z)|^2)(z) < \infty.
$$

See also [Zhu1, Theorem 7.1.9]. In other words, when extending Theorem 2.3.2 to the end point case $p = 2$, we should obtain the corresponding characterization of the Bloch space. Meanwhile, if considering the other endpoint $p = 0$, then we find that $f \in \mathcal{D}$ is described by $f \in \mathcal{H}$ obeying

$$
\int_{\mathbf{D}} \int_{\mathbf{D}} \frac{|f(z) - f(w)|^2}{|1 - z\bar{w}|^4} dm(z)dm(w) < \infty.
$$

See e.g., [Zhu1, Theorem 5.3.4].

Note 2.6.4. Section 2.4 is taken from [Xi5] whose Berezin-type characterization is a by-product of [Kw] that describes the holomorphic diagonal Besov space $\mathcal{B}_p = \{f \in \mathcal{H} : f' \in \mathcal{A}^{p,p-2}\}$, $p \in (1,\infty)$ respectively the Bloch space $\mathcal{B}$ by means of

$$\int_{\mathbf{D}} \int_{\mathbf{D}} \frac{|f(z) - f(w)|^p}{|1 - \bar{w}z|^4} \Big(\Xi(f; z, w)\Big)^\alpha dm(z)dm(w) < \infty,$$

where $\alpha \in [0, 2 + p)$, respectively

$$\sup_{a \in \mathbf{D}} \int_{\mathbf{D}} \int_{\mathbf{D}} \frac{|f(z) - f(w)|^p}{|1 - \bar{w}z|^4} \Big(\Xi(f; z, w)\Big)^\alpha (1 - |\sigma_a(w)|^2)^2 dm(z)dm(w) < \infty,$$

where $p \in (0, \infty)$, $\alpha \in [0, 2 + p)$. In addition, Lemma 2.4.1 (ii) is from [Kw], and Lemma 2.4.1 (i) due to [KimKw] and [Lue].

Note 2.6.5. Section 2.5 comprises [RocWu, Theorem 1] and its proof and conformally invariant version (cf. [Xi5]). The argument for the case $\alpha = 1/2$ of Lemma 2.5.1 is motivated by the proofs of [ArFiJaPe, Lemmas 3.2–3.3–3.4] — on the other hand, this special case can be found in [ArFiPe2, Proposition 3.6].

Chapter 3

Isomorphism, Decomposition and Discreteness

When studying a space of holomorphic functions, it is very useful to be able to establish an isomorphism of the space to a well-understood space, or to express every member in the space as a linear combination of some appropriate elementary functions, or to obtain a discrete presentation of the space. In this chapter, we will see that each $\mathcal{Q}_p$ function has such representations through a consideration of the following five aspects:

- Carleson Measures under an Integral Operator;
- Isomorphism to a Holomorphic Morrey Space;
- Decomposition via Bergman Style Kernels;
- Discreteness by Derivatives;
- Characterization in Terms of a Conjugate Pair.

3.1 Carleson Measures under an Integral Operator

For $p > 0$, we write $\mathcal{CM}_p$ for the class of all p-Carleson measures — nonnegative Radon measures μ on $\mathbf{D}$ with

$$\|\mu\|_{\mathcal{CM}_p} = \sup_{I \subseteq \mathbf{T}} \frac{\mu(S(I))}{|I|^p} < \infty.$$

Lemma 3.1.1. *Let $p, q \in (0, \infty)$ and μ be nonnegative Radon measures μ on $\mathbf{D}$. Then $\mu \in \mathcal{CM}_p$ if and only if*

$$\|\mu\|_{\mathcal{CM}_{p,q}} = \sup_{w \in \mathbf{D}} \int_{\mathbf{D}} \frac{(1 - |w|^2)^q}{|1 - \bar{w}z|^{p+q}} d\mu(z) < \infty.$$

Proof. Suppose $\|\mu\|_{\mathcal{CM}_{p,q}} < \infty$. Given a Carleson box $S(I) = \{z \in \mathbf{D} : 1 - h \leq |z| < 1,\ |\theta - \arg z| \leq h\}$ with $h = |I|$. Letting $w = (1 - h)e^{i(\theta + h/2)}$, we achieve

$$\|\mu\|_{\mathcal{CM}_{p,q}} \geq \mu(S(I)) \inf_{z \in S(I)} \frac{(1 - |w|^2)^q}{|1 - \bar{w}z|^{p+q}} \succeq \frac{\mu(S(I))}{|I|^p},$$

which implies $\|\mu\|_{\mathcal{CM}_p} \preceq \|\mu\|_{\mathcal{CM}_{p,q}} < \infty$.

On the other hand, if $\mu \in \mathcal{CM}_p$, then $\|\mu\|_{\mathcal{CM}_p} < \infty$.

When $|w| < 3/4$, we have

$$\int_{\mathbf{D}} \frac{(1 - |w|^2)^q}{|1 - \bar{w}z|^{p+q}} d\mu(z) \preceq \mu(\mathbf{D}) \preceq \|\mu\|_{\mathcal{CM}_p}.$$

When $3/4 \leq |w| < 1$, we put $\mathbf{E}_n = \{z \in \mathbf{D} : |z - w/|w|| < 2^n(1 - |w|)\}$ and then obtain $\mu(\mathbf{E}_n) \preceq \|\mu\|_{\mathcal{CM}_p} 2^{np}(1 - |w|^2)^p$ for $n \in \mathbf{N}$. Note that

$$\frac{(1 - |w|^2)^q}{|1 - \bar{w}z|^{p+q}} \preceq \frac{1}{(1 - |w|^2)^p}, \quad z \in \mathbf{E}_1,$$

and for $n \geq 1$ and $\mathbf{E}_0 = \emptyset$,

$$\frac{(1 - |w|^2)^q}{|1 - \bar{w}z|^{p+q}} \preceq \frac{1}{2^{2(p+q)n}(1 - |w|^2)^p}, \quad z \in \mathbf{E}_n \setminus \mathbf{E}_{n-1}.$$

Thus

$$
\begin{aligned}
\|\mu\|_{\mathcal{CM}_{p,q}} &\leq \sup_{w \in \mathbf{D}} \sum_{n=1}^{\infty} \int_{\mathbf{E}_n \setminus \mathbf{E}_{n-1}} \frac{(1 - |w|^2)^q}{|1 - \bar{w}z|^{p+q}} d\mu(z) \\
&\preceq \sup_{w \in \mathbf{D}} \sum_{n=1}^{\infty} \frac{\mu(\mathbf{E}_n)}{2^{2n(p+q)}(1 - |w|^2)^p} \\
&\preceq \|\mu\|_{\mathcal{CM}_p} \sum_{n=1}^{\infty} 2^{-nq},
\end{aligned}
$$

as desired. $\square$

For $a, b > 0$ we define formally a $\mathsf{T}_{a,b}$ operator; that is, the following integral operator:

$$\mathsf{T}_{a,b}f(z) = \int_{\mathbf{D}} \frac{(1 - |w|^2)^{b-1}}{|1 - \bar{w}z|^{a+b}} f(w)dm(w), \quad z \in \mathbf{D}.$$

In the next result we show that $\mathcal{CM}_p$ is invariant under $\mathsf{T}_{a,b}$.

Lemma 3.1.2. *Let* $p \in (0, 2)$, $a > \frac{2-p}{2}$, $b > \frac{1+p}{2}$ *and* f *be Lebesgue measurable on* **D**. *If* $|f(z)|^2(1-|z|^2)^p dm(z)$ *belongs to* $\mathcal{CM}_p$ *then* $|\mathsf{T}_{a,b}f(z)|^2(1-|z|^2)^{p+2a-2}dm(z)$ *also belongs to* $\mathcal{CM}_p$.

Proof. Assuming
$$d\mu_{f,p}(z) = |f(z)|^2 (1 - |z|^2)^p dm(z)$$
and
$$d\mu_{T_{a,b}f,p}(z) = |T_{a,b}f(z)|^2 (1 - |z|^2)^{p+2a-2} dm(z),$$
we verify that $\|\mu_{f,p}\|_{\mathcal{CM}_p} < \infty$ yields $\|\mu_{T_{a,b}f,p}\|_{\mathcal{CM}_p} < \infty$. To that end, we use $2^k I$ as the subarc of **T** with the same center as I and the length $2^k|I|$ for $k \in \mathbf{N}$ with $k \le -\log_2 |I|$. Given any Carleson box $S(I)$. Under $\|\mu_{f,p}\|_{\mathcal{CM}_p} < \infty$ we make the following estimates:

$$\mu_{T_{a,b}f,p}\big(S(I)\big)$$
$$= \int_{S(I)} |T_{a,b}f(z)|^2 (1 - |z|^2)^{p+2a-2} dm(z)$$
$$\le \int_{S(I)} \left(\left(\int_{S(2I)} + \int_{\mathbf{D}\backslash S(2I)} \right) \frac{|f(w)|(1 - |w|^2)^{b-1}}{|1 - \bar{w}z|^{a+b}} dm(w) \right)^2 \frac{dm(z)}{(1 - |z|^2)^{2-p-2a}}$$
$$\preceq \int_{S(I)} \left(\int_{S(2I)} \frac{|f(w)|(1 - |w|^2)^{b-1}}{|1 - \bar{w}z|^{a+b}} dm(w) \right)^2 \frac{dm(z)}{(1 - |z|^2)^{2-p-2a}}$$
$$+ \int_{S(I)} \left(\int_{\mathbf{D}\backslash S(2I)} \frac{|f(w)|(1 - |w|^2)^{b-1}}{|1 - \bar{w}z|^{a+b}} dm(w) \right)^2 \frac{dm(z)}{(1 - |z|^2)^{2-p-2a}}$$
$$= \mathrm{Int}_1 + \mathrm{Int}_2.$$

For Int_1, we shall apply the classical Schur's Lemma for bounded operators on $\mathcal{L}^2(\mathbf{D})$. Indeed, we consider

$$k(z, w) = \frac{(1 - |z|^2)^{b-1+p/2}(1 - |w|^2)^{b-1-p/2}}{|1 - \bar{w}z|^{a+b}}$$

and its induced integral operator

$$(\mathsf{T}f)(z) = \int_{\mathbf{D}} f(w)k(z, w)\, dm(w).$$

Using [Zhu1, Lemma 4.2.2] (or [Xi3, Lemma 1.4.1]), we derive that if

$$\max\left\{ -a - \frac{p}{2}, \frac{p}{2} - b \right\} < \gamma < \min\left\{ a - 1 + \frac{p}{2}, b - 1 - \frac{p}{2} \right\},$$

then

$$\int_{\mathbf{D}} k(z, w)(1 - |w|^2)^\gamma \, dm(w) \preceq (1 - |z|^2)^\gamma$$

and

$$\int_{\mathbf{D}} k(z, w)(1 - |z|^2)^\gamma \, dm(z) \preceq (1 - |w|^2)^\gamma.$$

Accordingly, the operator T is bounded from $\mathcal{L}^2(\mathbf{D})$ to $\mathcal{L}^2(\mathbf{D})$. Once the function f in $\mathsf{T}f$ is replaced by $g(w) = (1 - |w|^2)^{p/2}|f(w)|1_{S(2I)}(w)$, where 1_E stands for the characteristic function of E, we find

$$
\begin{aligned}
\text{Int}_1 \; &\preceq \; \int_{\mathbf{D}} \left(\int_{\mathbf{D}} g(w)k(z,w)\,dm(w) \right)^2 dm(z) \\
&\preceq \; \int_{\mathbf{D}} |g(z)|^2 dm(z) \preceq |I|^p \|\mu_{f,p}\|_{\mathcal{CM}_p}.
\end{aligned}
$$

To handle Int_2, we note that both

$$
2^n|I| \preceq |1 - \bar{w}z|, \quad z \in S(I), \;\; w \in S(2^{n+1}I) \setminus S(2^nI),
$$

and

$$
\int_{S(2^nI)} (1 - |z|^2)^{c-2} dm(z) \preceq (2^n|I|)^c, \quad c > 1,
$$

hold for all $n \in \mathbf{N} \cup \{0\}$. When writing

$$
\mathbf{D} \setminus S(2I) = \bigcup_{n=1}^{\infty} S(2^{n+1}I) \setminus S(2^nI),
$$

and using $a > (2 - p)/2$ as well as $p \in (0, 2)$, we then get by the Hölder inequality

$$
\begin{aligned}
\text{Int}_2 \\
\preceq \; & \int_{S(I)} \left(\sum_{n=1}^{\infty} \int_{S(2^{n+1}I) \setminus S(2^nI)} \frac{|f(w)|(1 - |w|^2)^{b-1}}{|1 - \bar{w}z|^{a+b}}\,dm(w) \right)^2 \frac{dm(z)}{(1 - |z|^2)^{2-p-2a}} \\
\preceq \; & \int_{S(I)} \left(\sum_{n=1}^{\infty} (2^n|I|)^{-(a+b)} \int_{S(2^{n+1}I)} \frac{|f(w)|}{(1 - |w|^2)^{1-b}}\,dm(w) \right)^2 \frac{dm(z)}{(1 - |z|^2)^{2-p-2a}} \\
\preceq \; & |I|^{2a+p} \left(\sum_{n=1}^{\infty} (2^n|I|)^{-(a+b)} \int_{S(2^{n+1}I)} \frac{|f(w)|}{(1 - |w|^2)^{1-b}}\,dm(w) \right)^2 \\
\preceq \; & |I|^{2a+p} \left(\sum_{n=1}^{\infty} (2^n|I|)^{-a-\frac{p}{2}} \left(\mu_{f,p}(S(2^{n+1}I)) \right)^{\frac{1}{2}} \right)^2 \\
\preceq \; & |I|^p \left(\sum_{n=1}^{\infty} 2^{-na} \right)^2 \|\mu_{f,p}\|_{\mathcal{CM}_p}.
\end{aligned}
$$

The foregoing estimates on Int_1 and Int_2 give $\mu_{\mathsf{T}_{a,b}f,p} \in \mathcal{CM}_p$. $\square$

As an immediate application of Lemma 3.1.2, we can estimate the distance of a Bloch function to the Q_p space.

Theorem 3.1.3. *For $p \in (0,2)$ and $f \in \mathcal{B}$ let $dist_{\mathcal{B}}(f, \mathcal{Q}_p) = \inf\{\|f - g\|_{\mathcal{B}} : g \in \mathcal{Q}_p\}$. Then*

$$dist_{\mathcal{B}}(f, \mathcal{Q}_p) \approx \inf\left\{\epsilon > 0 : 1_{\Omega_\epsilon(f)}(z)(1 - |z|^2)^{p-2}dm(z) \in \mathcal{CM}_p\right\},$$

where $\Omega_\epsilon(f) = \{z \in \mathbf{D} : (1 - |z|^2)|f'(z)| \geq \epsilon\}$ and $1_{\mathbf{E}}$ stands for the characteristic function of a set $\mathbf{E}$.

Proof. Because of $f \in \mathcal{B}$, this function has the following integral representation:

$$f(z) = f(0) + \frac{1}{\pi}\int_{\mathbf{D}}\frac{(1 - |w|^2)f'(w)}{\bar{w}(1 - \bar{w}z)^2}dm(w) = f_1(z) + f_2(z),$$

where

$$f_1(z) = f(0) + \frac{1}{\pi}\int_{\Omega_\epsilon(f)}\frac{(1 - |w|^2)f'(w)}{\bar{w}(1 - \bar{w}z)^2}dm(w)$$

and

$$f_2(z) = \frac{1}{\pi}\int_{\mathbf{D}\setminus\Omega_\epsilon(f)}\frac{(1 - |w|^2)f'(w)}{\bar{w}(1 - \bar{w}z)^2}dm(w).$$

Note that

$$|f_1'(z)| \leq \|f\|_{\mathcal{B}}\int_{\mathbf{D}}\frac{1 - |w|^2}{|1 - z\bar{w}|^3}\left(\frac{1_{\Omega_\epsilon(f)}(w)}{1 - |w|^2}\right)dm(w).$$

So, if $1_{\Omega_\epsilon(f)}(z)(1 - |z|^2)^{p-2}dm(z) \in \mathcal{CM}_p$ then from Lemma 3.1.2 it follows that $(1 - |z|^2)^p|f_1'(z)|^2dm(z) \in \mathcal{CM}_p$ and consequently $f_1 \in \mathcal{Q}_p$. At the same time, we have

$$|f_2'(z)| \leq \epsilon\int_{\mathbf{D}}|1 - z\bar{w}|^{-3}dm(w) \preceq \epsilon(1 - |z|^2)^{-1}, \quad z \in \mathbf{D},$$

hence getting

$$dist_{\mathcal{B}}(f, \mathcal{Q}_p) \leq \|f - f_1\|_{\mathcal{B}} \preceq \epsilon.$$

This gives

$$dist_{\mathcal{B}}(f, \mathcal{Q}_p) \preceq \inf\left\{\epsilon > 0 : 1_{\Omega_\epsilon(f)}(z)(1 - |z|^2)^{p-2}dm(z) \in \mathcal{CM}_p\right\}.$$

In order to verify the reversed version of the above-established inequality, we suppose

$$dist_{\mathcal{B}}(f, \mathcal{Q}_p) < \inf\left\{\epsilon > 0 : 1_{\Omega_\epsilon(f)}(z)(1 - |z|^2)^{p-2}dm(z) \in \mathcal{CM}_p\right\} = \epsilon_0.$$

Of course, we only consider the case $\epsilon_0 > 0$ — otherwise there is nothing to argue. Then, there exists an ϵ_1 such that $0 < \epsilon_1 < \epsilon_0$ and $dist_{\mathcal{B}}(f, \mathcal{Q}_p) \leq \epsilon_1$. Hence, by definition, we can find a function $g \in \mathcal{Q}_p$ such that $\|f - g\|_{\mathcal{B}} < \epsilon_1$. Now for any $\epsilon \in (\epsilon_1, \epsilon_0)$ we have

$$1_{\Omega_\epsilon(f)}(z)(1 - |z|^2)^{p-2}dm(z) \notin \mathcal{CM}_p.$$

But, $\|f - g\|_{\mathcal{B}} < \epsilon_1$ yields

$$(1 - |z|^2)|g'(z)| > (1 - |z|^2)|f'(z)| - \epsilon_1, \quad z \in \mathbf{D},$$

and so

$$1_{\Omega_\epsilon(f)}(z) \leq 1_{\Omega_{\epsilon - \epsilon_1}(g)}(z), \quad z \in \mathbf{D}.$$

The last inequality implies

$$1_{\Omega_{\epsilon - \epsilon_1}(g)}(z)(1 - |z|^2)^{p-2}dm(z) \notin \mathcal{CM}_p.$$

But

$$1_{\Omega_{\epsilon - \epsilon_1}(g)}(z)(1 - |z|^2)^{p-2}dm(z) \leq (\epsilon - \epsilon_1)^{-2}|g'(z)|^2(1 - |z|^2)^p 1_{\Omega_{\epsilon - \epsilon_1}(g)}(z)dm(z)$$

gives

$$1_{\Omega_{\epsilon - \epsilon_1}(g)}(z)(1 - |z|^2)^{p-2}dm(z) \in \mathcal{CM}_p,$$

a contradiction. Therefore, we must have $\epsilon_0 \leq dist_{\mathcal{B}}(f, \mathcal{Q}_p)$, as required. $\square$

3.2 Isomorphism to a Holomorphic Morrey Space

For $p \in (-\infty, \infty)$, denote by $\mathcal{L}^{2,p}(\mathbf{T})$ the Morrey space of all Lebesgue measurable functions f on $\mathbf{T}$ that satisfy

$$\|f\|_{\mathcal{L}^{2,p}(\mathbf{T})} = \sup_{I \subseteq \mathbf{T}} \left(|I|^{-p} \int_I |f(\zeta) - f_I|^2|d\zeta|\right)^{\frac{1}{2}} < \infty,$$

where the supremum as before ranges over all subarcs I of $\mathbf{T}$. Clearly, $\mathcal{L}^{2,0}(\mathbf{T})$ and $\mathcal{L}^{2,1}(\mathbf{T})$ coincide with $\mathcal{L}^2(\mathbf{T})$ and $\mathcal{BMO}(\mathbf{T})$ respectively. Moreover,

$$\mathcal{BMO}(\mathbf{T}) \subseteq \mathcal{L}^{2,p_1}(\mathbf{T}) \subseteq \mathcal{L}^{2,p_2}(\mathbf{T}) \subseteq \mathcal{L}^2(\mathbf{T}), \quad 0 < p_2 < p_1 < 1.$$

Even more surprisingly, there is a closed relationship between $\mathcal{L}^{2,p}(\mathbf{T})$ and $\mathcal{Q}_p(\mathbf{T})$. To see this, we need a characterization of the Morrey space.

Theorem 3.2.1. *Let $p \in (0, 2)$ and f be Lebesgue measurable on $\mathbf{T}$. Then the following statements are equivalent:*

(i) $f \in \mathcal{L}^{2,p}(\mathbf{T})$.

(ii) *If $d\mu_f(z) = |\nabla \hat{f}(z)|^2(1 - |z|^2)dm(z)$ then $\mu_f \in \mathcal{CM}_p$.*

(iii) $\displaystyle\sup_{w \in \mathbf{D}} (1 - |w|^2)^{1-p} \int_{\mathbf{D}} |\nabla \hat{f}(z)|^2 (1 - |\sigma_w(z)|^2)dm(z) < \infty.$

(iv) $\displaystyle\sup_{w \in \mathbf{D}} (1 - |w|^2)^{1-p} \int_{\mathbf{D}} |\nabla \hat{f}(z)|^2 \big(-\log|\sigma_w(z)|\big)dm(z) < \infty.$

(v) $\|f\|_{\mathcal{L}^{2,p}(\mathbf{T}),1} = \sup_{z \in \mathbf{D}} \left((1-|z|^2)^{1-p}(\widehat{|f - \hat{f}(z)|^2})(z)\right)^{\frac{1}{2}} < \infty.$

(vi) $\|f\|_{\mathcal{L}^{2,p}(\mathbf{T}),2} = \sup_{z \in \mathbf{D}} \left((1-|z|^2)^{1-p}(\widehat{|f|^2}(z) - |\hat{f}(z)|^2)\right)^{\frac{1}{2}} < \infty.$

Proof. By Lemma 3.1.1, Hardy–Littlewood's identity

$$\int_{\mathbf{D}} |\nabla \hat{f}(z)|^2 (1-|z|^2) dm(z) \;\approx\; \int_{\mathbf{D}} |\nabla \hat{f}(z)|^2 \log \frac{1}{|z|} dm(z)$$

$$\approx\; \int_{\mathbf{T}} |f(\zeta) - \hat{f}(0)|^2 |d\zeta|,$$

and some calculations with σ_w, we can figure out that (ii)$\Leftrightarrow$(iii)$\Leftrightarrow$(iv)$\Leftrightarrow$(v)$\Leftrightarrow$(vi). Thus, it suffices to prove (i)$\Leftrightarrow$(v).

Suppose (i) holds. For each nonzero $z \in \mathbf{D}$ let I_z be the subarc of $\mathbf{T}$ with center $z/|z|$ and length $1 - |z|$, and for $z = 0$ let $I_z = \mathbf{T}$. Moreover, for $n = 0, 1, \ldots, N-1$ let $J_n = 2^n I_z$ where N is the smallest natural number such that $2^N |I_z| \geq 1$. Finally, let $J_N = \mathbf{T}$. With these, we make the following estimates for a given point $z \in \mathbf{D}$:

$$(\widehat{|f - \hat{f}(z)|^2})(z)$$

$$\preceq\; \int_{\mathbf{T}} |f(\zeta) - f_{I_z}|^2 \frac{1-|z|^2}{|1 - \bar{z}\zeta|^2} |d\zeta|$$

$$\preceq\; \left(\int_{J_0} + \sum_{n=0}^{N-1} \int_{J_{n+1} \backslash J_n}\right) |f(\zeta) - f_{J_0}|^2 \frac{1-|z|^2}{|1 - \bar{z}\zeta|^2} |d\zeta|$$

$$\preceq\; (1-|z|^2)^{-1} \left(\int_{J_0} + \sum_{n=0}^{N-1} 2^{-2n} \int_{J_{n+1} \backslash J_n}\right) |f(\zeta) - f_{J_0}|^2 |d\zeta|$$

$$\preceq\; |J_0|^{-1} \int_{J_0} |f(\zeta) - f_{J_0}|^2 |d\zeta| + \sum_{n=0}^{N-1} \frac{2^{-n}}{|J_{n+1}|} \int_{J_{n+1} \backslash J_n} |f(\zeta) - f_{J_0}|^2 |d\zeta|.$$

Note that the Cauchy–Schwarz inequality and $f \in \mathcal{L}^{2,p}(\mathbf{T})$ imply

$$|f_{J_{n+1}} - f_{J_n}| \;\leq\; \frac{2}{|J_{n+1}|} \int_{J_{n+1}} |f(\zeta) - f_{J_{n+1}}| |d\zeta|$$

$$\preceq\; \left(\frac{2}{|J_{n+1}|} \int_{J_{n+1}} |f(\zeta) - f_{J_{n+1}}|^2 |d\zeta|\right)^{\frac{1}{2}}$$

$$\preceq\; 2|J_{n+1}|^{\frac{p-1}{2}} \|f\|_{\mathcal{L}^{2,p}(\mathbf{T})}$$

$$\preceq\; 2^{1+(p-1)(n+1)/2} (1-|z|)^{(p-1)/2} \|f\|_{\mathcal{L}^{2,p}(\mathbf{T})}.$$

So

$$|f_{J_{n+1}} - f_{J_0}| \leq |f_{J_{n+1}} - f_{J_n}| + \cdots + |f_{J_1} - f_{J_0}|$$
$$\preceq \max\{1, 2^{n(p-1)/2}\} n (1 - |z|)^{\frac{p-1}{2}} \|f\|_{\mathcal{L}^{2,p}(\mathbf{T})}.$$

Accordingly,

$$\frac{1}{|J_{n+1}|} \int_{J_{n+1}} |f(\zeta) - f_{J_0}|^2 |d\zeta|$$

$$\leq \left(\left(\frac{1}{|J_{n+1}|} \int_{J_{n+1}} |f(\zeta) - f_{J_{n+1}}|^2 |d\zeta| \right)^{\frac{1}{2}} + |f_{J_{n+1}} - f_{J_0}| \right)^2$$

$$\preceq n^2 \max\{1, 2^{n(p-1)}\} (1 - |z|)^{p-1} \|f\|^2_{\mathcal{L}^{2,p}(\mathbf{T})}.$$

Finally, we obtain

$$(\widehat{|f - \hat{f}(z)|^2})(z) \preceq (1 - |z|^2)^{p-1} \|f\|^2_{\mathcal{L}^{2,p}(\mathbf{T})} \sum_{n=0}^{\infty} \frac{n^2 \max\{1, 2^{n(p-1)}\}}{2^n}.$$

In other words, (v) is valid because of $p \in (0, 2)$.

Conversely, if (v) holds, then for any subarc $I \subset \mathbf{T}$, we take such a $z \in \mathbf{D}$ that $z/|z|$ is the center of I and $|z| = 1 - |I|$, to get

$$|1 - \bar{\zeta} z| \preceq 1 - |z|^2, \quad \zeta \in I,$$

and then

$$|I|^{-p} \int_I |f(\zeta) - f_I|^2 |d\zeta| \leq 4|I|^{-p} \int_I |f(\zeta) - \hat{f}(z)|^2 |d\zeta|$$

$$\preceq (1 - |z|^2)^{1-p} \int_I |f(\zeta) - \hat{f}(z)|^2 \frac{1 - |z|^2}{|1 - \bar{\zeta} z|^2} |d\zeta|$$

$$\preceq (1 - |z|^2)^{1-p} (\widehat{|f - \hat{f}(z)|^2})(z)$$

$$\preceq \|f\|^2_{\mathcal{L}^{2,p}(\mathbf{T}),1},$$

which gives (i) right away. $\square$

Corollary 3.2.2. *For $p \in (0, 2)$ let $\mathcal{H}^{2,p} = \mathcal{L}^{2,p} \cap \mathcal{H}^2$ be the holomorphic Morrey space on $\mathbf{D}$. If $f \in \mathcal{H}^2$, then the following statements are equivalent:*

(i) $f \in \mathcal{H}^{2,p}(\mathbf{T})$.

(ii) $|f'(z)|^2 (1 - |z|^2) dm(z) \in \mathcal{CM}_p$.

(iii) $\sup_{w \in \mathbf{D}} (1 - |w|^2)^{1-p} \int_{\mathbf{D}} |f'(z)|^2 (1 - |\sigma_w(z)|) dm(z) < \infty.$

(iv) $\sup_{w \in \mathbf{D}} (1 - |w|^2)^{1-p} \int_{\mathbf{D}} |f'(z)|^2 \big(- \log |\sigma_w(z)| \big) dm(z) < \infty.$

(v) $\sup_{z \in \mathbf{D}} \big((1 - |z|^2)^{1-p} (\widehat{|f - f(z)|^2})(z) \big)^{\frac{1}{2}} < \infty.$

(vi) $\sup_{z \in \mathbf{D}} \big((1 - |z|^2)^{1-p} (\widehat{|f|^2}(z) - |f(z)|^2) \big)^{\frac{1}{2}} < \infty.$

Proof. Note that $\hat{f} = f$ for $f \in \mathcal{H}^2$. So the desired result follows from Theorem 3.2.1. $\qquad\square$

Now, it is natural to compare $\mathcal{Q}_p$ with $\mathcal{H}^{2,p}$. In so doing, we consider the following formal α-order derivative of $f \in \mathcal{H}$ at $z \in \mathbf{D}$:

$$f^{(\alpha)}(z) = \frac{\Gamma(b + \alpha)}{\pi \Gamma(b)} \int_{\mathbf{D}} \frac{\bar{w}^{\lceil \alpha - 1 \rceil} (1 - |w|^2)^{b-1} f'(w)}{(1 - \bar{w}z)^{b+\alpha}} dm(w), \quad b + \alpha > 0, \quad b > 1.$$

Here, $\lceil x \rceil$ stands for the smallest integer greater than or equal to $x \in \mathbf{R}$, and $\Gamma(\cdot)$ is still the Gamma function.

A straightforward calculation gives

$$(z^n)^{(\alpha)} = \begin{cases} 0 & , \quad n < \lceil \alpha - 1 \rceil + 1, \\ \left(\frac{\Gamma(b+n+\alpha-1-\lceil \alpha-1 \rceil)\Gamma(n+1)}{\Gamma(b+n)\Gamma(n-\lceil \alpha-1 \rceil)} \right) z^{n-1-\lceil \alpha-1 \rceil} & , \quad n \geq \lceil \alpha - 1 \rceil + 1. \end{cases}$$

and thus illustrates that if $\alpha = n \in \mathbf{N}$, then $f^{(\alpha)}$ is just the usual n-th derivative of f.

The next result tells us that each holomorphic Q class is isomorphic to a holomorphic Morrey space.

Theorem 3.2.3. *Let* $p \in (0,2)$, $\alpha \in (\frac{2-p}{2}, \infty)$ *and* $f \in \mathcal{H}$. *Then the following statements are equivalent:*

(i) $f \in \mathcal{Q}_p$.

(ii) $|f^{(\alpha)}(z)|^2 (1 - |z|^2)^{2\alpha - 2 + p} dm(z) \in \mathcal{CM}_p$.

(iii) $f^{(\frac{1-p}{2})} \in \mathcal{H}^{2,p}$.

Consequently, $f \in \mathcal{H}^{2,p}$ *if and only if* $f^{(\frac{p-1}{2})} \in \mathcal{Q}_p$.

Proof. Step 1: (i)$\Leftrightarrow$(ii) If $f \in \mathcal{Q}_p$, then by Theorem 1.1.3 we see that $|f'(z)|^2 (1 - |z|^2)^p dm(z)$ is a p-Carleson measure, and so is $|f^{(\alpha)}(z)|^2 (1 - |z|^2)^{2\alpha - 2 + p} dm(z)$ thanks to an application of Lemma 3.1.2 to

$$|f^{(\alpha)}(z)| \leq \frac{\Gamma(b + \alpha)}{\pi \Gamma(b)} \int_{\mathbf{D}} \frac{(1 - |w|^2)^{b-1} |f'(w)|}{|1 - \bar{w}z|^{b+\alpha}} dm(w), \quad z \in \mathbf{D}.$$

Conversely, suppose $|f^{(\alpha)}(z)|^2(1-|z|^2)^{2\alpha-2+p}dm(z) \in \mathcal{CM}_p$. Then we consider the following function $g(z)$ generated by $f^{(\alpha)}$:

$$g(z) = \frac{\Gamma(b+1)z^{\lceil\alpha-1\rceil}}{\pi\Gamma(b+\alpha-1)} \int_{\mathbf{D}} \left(\frac{(1-|\zeta|^2)^{b+\alpha-2}}{(1-\bar{\zeta}z)^{b+1}}\right) f^{(\alpha)}(\zeta)dm(\zeta).$$

From Lemma 3.1.2 it follows that $|g(z)|^2(1-|z|^2)^p dm(z) \in \mathcal{CM}_p$. To establish the relationship between g and f', we express f and $f^{(\alpha)}$ as the Taylor series below:

$$f(z) = \sum_{j=0}^{\infty} a_j z^j \quad\text{and}\quad f^{(\alpha)}(z) = \sum_{j=0}^{\infty} a_{j,\alpha} z^j, \quad z \in \mathbf{D}.$$

If $m = \lceil\alpha-1\rceil$, then $m \geq 0$ (if otherwise $m \leq -1$, then $\alpha - 1 \leq -1$ and hence $\alpha \leq 0$ but $\alpha > (2-p)/2 > 0$, a contradiction). Consequently, the formula regarding $(z^j)^{(\alpha)}$ implies

$$a_{j,\alpha} = a_{j+m+1}\left(\frac{\Gamma(j+b+\alpha)\Gamma(j+m+2)}{\Gamma(j+1)\Gamma(j+m+1+b)}\right), \quad j \in \mathbf{N} \cup \{0\}.$$

This, along with a calculation with the integral determining g, implies

$$\begin{aligned}
g(z) &= \sum_{j=0}^{\infty}\left(\frac{\Gamma(j+b+1)}{\Gamma(j+b+\alpha)}\right)a_{j,\alpha}z^{j+m}\\
&= \sum_{j=0}^{\infty}\left(\frac{\Gamma(j+b+1)\Gamma(j+m+2)}{\Gamma(j+m+b+1)\Gamma(j+1)}\right)a_{j+m+1}z^{j+m}.
\end{aligned}$$

Case 1: $m = 0$. This forces $\alpha = 1$ due to $\alpha > (2-p)/2 > 0$ and thus $g = f'$. Of course, $f \in \mathcal{Q}_p$.

Case 2: $m > 0$. In this case, we use $\beta(x,y) = \frac{\Gamma(x)\Gamma(y)}{\Gamma(x+y)}$ as the Beta function, and then obtain

$$g(z) = \sum_{j=0}^{\infty}\left(\frac{\beta(j+b+1,m)}{\beta(j+1,m)}\right)(j+m+1)a_{j+m+1}z^{j+m}.$$

Writing

$$s_m(z) = \sum_{j=1}^{m} j a_j z^{j-1} \quad\text{and}\quad h(z) = \sum_{j=0}^{\infty}\left(1-\frac{\beta(j+b+1,m)}{\beta(j+1,m)}\right)a_{j+m+1}z^{j+m+1},$$

we find $f'(z) = g(z) + s_m(z) + h'(z)$. Note that s_m is a polynomial and $|g(z)|^2(1-|z|^2)^p$ is a p-Carleson measure. Thus, in order to prove $f \in \mathcal{Q}_p$ it is enough to verify $h \in \mathcal{D}$. To this end, we observe that the Beta function enjoys

$$0 < 1 - \frac{\beta(j+b+1,m)}{\beta(j+1,m)} \leq \frac{(b+1)m}{j+m+1}, \quad j \in \mathbf{N} \cup \{0\}.$$

Accordingly,

$$\|h\|_{\mathcal{D}}^2 = \pi \sum_{j=0}^{\infty} \left(1 - \frac{\beta(j+b+1,m)}{\beta(j+1,m)} \right)^2 (j+m+1)|a_{j+m+1}|^2$$

$$\preceq \sum_{j=0}^{\infty} \frac{|a_{j+m+1}|^2}{j+m+1}.$$

Notice also that the assumption that $|f^{(\alpha)}(z)|^2(1-|z|^2)^{2\alpha-2+p}dm(z) \in \mathcal{CM}_p$ gives

$$\sum_{j=0}^{\infty} \frac{|a_{j,\alpha}|^2}{(j+1)^{2\alpha-1+p}} \approx \int_{\mathbf{D}} |f^{(\alpha)}(z)|^2(1-|z|^2)^{2\alpha-2+p}dm(z) < \infty.$$

So

$$|a_{j,\alpha}| \approx \frac{|a_{j,\alpha}|}{(j+1)^{\alpha}} \Rightarrow \sum_{j=0}^{\infty} \frac{|a_{j+m+1}|^2}{(j+m+1)^{p-1}} < \infty.$$

Finally, we reach

$$\|h\|_{\mathcal{D}}^2 \preceq \sum_{j=0}^{\infty} \frac{|a_{j+m+1}|^2}{(j+m+1)^{p-1}} < \infty,$$

as required.

Step 2: (i)⇔(iii) Since the first step reveals that $f \in \mathcal{Q}_p$ if and only if $|f^{(\frac{3-p}{2})}(z)|^2(1-|z|^2)dm(z)$ is an element of $\mathcal{CM}_p$, we conclude from Corollary 3.2.2 that $f \in \mathcal{Q}_p$ is equivalent to $f^{(\frac{1-p}{2})} \in \mathcal{H}^{2,p}$.

Taking (i)⇔(ii) into account, we realize that $g = f^{(\frac{p-1}{2})} \in \mathcal{Q}_p$ is equivalent to $|g^{(\frac{3-p}{2})}|^2(1-|z|^2)dm(z) \in \mathcal{CM}_p$; that is, $|f'(z)|^2(1-|z|^2)dm(z) \in \mathcal{CM}_p$, i.e., $f \in \mathcal{H}^{2,p}$ thanks to Corollary 3.2.2. We are done. □

3.3 Decomposition via Bergman Style Kernels

As a special case of the foregoing α-derivatives, we find that if $f \in \mathcal{H}$ satisfies $F_p(f,0) < \infty$, then by the following integral reproducing formula:

$$f'(w) = \frac{b}{\pi} \int_{\mathbf{D}} f'(z) \left(\frac{(1-|z|^2)^{b-1}}{(1-\bar{z}w)^{b+1}} \right) dm(z), \quad b > \frac{p+1}{2}.$$

When approximating the above integral by a Riemann sum, we are naturally led to sketching a decomposition result for functions in $\mathcal{Q}_p$.

As with this aspect, we will still use $B(z,r) = \{w \in \mathbf{D} : d_{\mathbf{D}}(z,w) < r\}$ as the hyperbolic $0 < r$-ball centered at $z \in \mathbf{D}$. Moreover, for $\tau > 0$ we say that a sequence of points $\{z_j\}$ in $\mathbf{D}$ is τ-separated, respectively τ-dense, provided

$$\inf_{j \neq k} d_{\mathbf{D}}(z_j, z_k) \geq \tau, \quad \text{respectively} \quad \mathbf{D} = \bigcup_{j=1}^{\infty} B(z_j, \tau).$$

We need two basic lemmas as follows.

Lemma 3.3.1. *For $b > 0$, $z, w \in \mathbf{D}$ let $k_w(z) = \frac{(1-|z|^2)^{b-1}}{(1-\bar{z}w)^{b+1}}$. If $d_{\mathbf{D}}(z, z_0) \leq 1$, then*

$$|k_w(z) - k_w(z_0)| \preceq d_{\mathbf{D}}(z, z_0)|k_w(z_0)|, \quad w \in \mathbf{D}.$$

Proof. Noting the estimate

$$1 - |z|^2 \approx 1 - |z_0|^2 \approx |1 - \bar{z_0}z|, \quad \text{as} \quad d_{\mathbf{D}}(z, z_0) \leq 1,$$

and using the mean-value theorem for derivatives, we obtain that if $d_{\mathbf{D}}(z, z_0) \leq 1/2$, then

$$
\begin{aligned}
\left|1 - \frac{k_w(z)}{k_w(z_0)}\right| &\leq \left|1 - \left(\frac{1-|z|^2}{1-|z_0|^2}\right)^{b-1}\right| + \left|\left(\frac{1-|z|^2}{1-|z_0|^2}\right)^{b-1}\right|\left|1 - \left(\frac{1-\bar{z_0}w}{1-\bar{z}w}\right)^{b+1}\right| \\
&\preceq \frac{|z - z_0|}{1-|z_0|^2} + \frac{|z - z_0|}{|1-\bar{z}w|} \\
&\preceq |\sigma_{z_0}(z)| \\
&\preceq d_{\mathbf{D}}(z, z_0),
\end{aligned}
$$

as required. $\square$

Lemma 3.3.2. *For any $\tau \in (0, 1]$, there is a $\tau/2$-separated and τ-dense sequence $\{z_j\}$ in $\mathbf{D}$ such that any $z \in \mathbf{D}$ lies in at most finite many of $\{B(z_j, 2\tau)\}$.*

Proof. Given $\tau > 0$, we can choose a sequence of points $\{z_j\}$ in $\mathbf{D}$ that is $\tau/2$-separated and τ-dense. Let $B(z_0, 2\tau)$ be fixed and covered by $\{B(z_j, \tau/4)\}_{j=1}^{n}$ with

$$\inf_{j \neq k} d_{\mathbf{D}}(z_j, z_k) \geq \tau/8 \quad \text{and} \quad \bigcup_{j=1}^{n} B(z_j, \tau/16) \subseteq B(z_0, 2\tau).$$

Clearly, $\{B(z_j, \tau/16)\}_{j=1}^{n}$ are disjoint, and so

$$\sum_{j=1}^{n} m\big(B(z_j, \tau/16)\big) \leq m\big(B(z_0, 2\tau)\big).$$

By a direct calculation, there is a constant $\kappa > 0$ independent of τ and $\{z_j\}_{j=0}^{n}$ such that

$$m\big(B(z_0, 2\tau)\big) \leq \kappa \min_{1 \leq j \leq n} m\big(B(z_j, \tau/16)\big).$$

Thus $n \leq \kappa$ and therefore, there exists a natural number N such that any hyperbolic 2τ-ball can be covered by N hyperbolic $\tau/4$-balls.

Suppose now that there is a point $z \in \mathbf{D}$ that belongs to $\cap_{k=1}^{N+1} B(z_{j_k}, 2\tau)$. Then $z_{j_k} \in B(z, 2\tau)$ for $k = 1, 2, \ldots, N + 1$. According to the existence of N, we may assume that $B(z, 2\tau)$ is covered by $\{B(w_n, \tau/4)\}_{n=1}^{N}$. Thus there is some $B(w_n, \tau/4)$ that contains two points of $\{z_{j_k}\}_{k=1}^{N+1}$. From this we in turn derive that the hyperbolic distance between these two points is less than $\tau/2$, which contradicts the fact that $\{z_j\}$ is $\tau/2$-separated. $\square$

For simplicity, we use δ_a as the Dirac measure with mass 1 at a. Below is the expected decomposition theorem for $\mathcal{Q}_p$, $p \in (0,2)$, in terms of the normalized Bergman-type kernels $\left((1-|z_j|^2)/(1-\overline{z_j}z)\right)^b$, $b > (1+p)/2$.

Theorem 3.3.3. *Let $p \in (0,2)$ and $b > \frac{1+p}{2}$. Then there is a sequence $\{z_j\}$ in $\mathbf{D}$ such that:*

(i) *For any complex-valued sequence $\{\lambda\}_{j=1}^{\infty}$ satisfying $\sum_{j=1}^{\infty} |\lambda_j|^2 (1-|z_j|^2)^p \delta_{z_j} \in \mathcal{CM}_p$, the function*

$$f(z) = \sum_{j=1}^{\infty} \lambda_j \left(\frac{1-|z_j|^2}{1-\overline{z_j}z}\right)^b$$

belongs to $\mathcal{Q}_p$ with

$$\|f\|_{\mathcal{Q}_p} \preceq \left\| \sum_{j=1}^{\infty} |\lambda_j|^2 (1-|z_j|^2)^p \delta_{z_j} \right\|_{\mathcal{CM}_p}.$$

(ii) *If $f \in \mathcal{Q}_p$, then*

$$f(z) = \sum_{j=1}^{\infty} \lambda_j \left(\frac{1-|z_j|^2}{1-\overline{z_j}z}\right)^b$$

for some sequence of complex numbers $\{\lambda\}_{j=1}^{\infty}$ obeying

$$\left\| \sum_{j=1}^{\infty} |\lambda_j|^2 (1-|z_j|^2)^p \delta_{z_j} \right\|_{\mathcal{CM}_p} \preceq \|f\|_{\mathcal{Q}_p}.$$

Proof. For $\tau > 0$, suppose that $\{z_j\}$ is a sequence constructed in Lemma 3.3.2.

(i) Suppose $d\nu_p = \sum_{j=1}^{\infty} |\lambda_j|^2 (1-|z_j|^2)^p \delta_{z_j}$ is in $\mathcal{CM}_p$. Then

$$\sum_{j=1}^{\infty} |\lambda_j|^2 (1-|z_j|^2)^p < \infty.$$

Using $2b > 1 + p$ and the Cauchy–Schwarz inequality, we have that for z in any compacta of $\mathbf{D}$,

$$\sum_{j=1}^{\infty} |\lambda_j| \left|\frac{1-|z_j|^2}{1-\overline{z_j}z}\right|^b \leq \left(\sum_{j=1}^{\infty} |\lambda_j|^2 (1-|z_j|^2)^p\right)^{\frac{1}{2}} \left(\sum_{j=1}^{\infty} \frac{(1-|z_j|^2)^{2b-p}}{(1-|z|)^{2b}}\right)^{\frac{1}{2}} < \infty.$$

Accordingly, f defined in (ii) belongs to $\mathcal{H}$, but also obeys

$$f'(w) = b \sum_{j=1}^{\infty} \lambda_j \overline{z_j} \left(\frac{(1-|z_j|^2)^b}{(1-\overline{z_j}w)^{b+1}}\right), \quad w \in \mathbf{D}.$$

A direct calculation with $z = \sigma_{z_j}(\zeta)$ gives

$$\int_{B(z_j, \tau/4)} \frac{(1 - |z|^2)^{b-1}}{(1 - \bar{z}w)^{b+1}} dm(z) = \left(\frac{m(B(z_j, \tau/4))(1 - |z_j|^2)^{b-1}}{C_j(1 - \overline{z_j}w)^{b+1}} \right),$$

where

$$C_j = \frac{2b\pi(e^{5\tau} - 1)^2(e^{5\tau} + 1)^{-2}}{\left(1 - (e^{5\tau} - 1)^2(e^{5\tau} + 1)^{-2}|z_j|^2\right)\left(1 - (4e^{5\tau})^b(1 + e^{5\tau})^{-2b}\right)}$$

is bounded. Consequently, we obtain

$$
\begin{aligned}
f'(w) &= b\sum_{j=1}^{\infty} \lambda_j \overline{z_j} \frac{C_j(1 - |z_j|^2)}{m((B(z_j, \tau/4))} \int_{\mathbf{D}} 1_{B(z_j, \tau/4)}(z) \frac{(1 - |z|^2)^{b-1}}{(1 - \bar{z}w)^{b+1}} dm(z) \\
&= b\int_{\mathbf{D}} \frac{(1 - |z|^2)^{b-1}}{(1 - \bar{z}w)^{b+1}} \left(\sum_{j=1}^{\infty} \lambda_j \overline{z_j} \frac{C_j(1 - |z_j|^2)}{m(B(z_j, \tau/4))} 1_{B(z_j, \tau/4)}(z) \right) dm(z).
\end{aligned}
$$

To prove $f \in \mathcal{Q}_p$, let

$$F(z) = \sum_{j=1}^{\infty} \frac{\lambda_j C_j \overline{z_j}(1 - |z_j|^2)1_{B(z_j, \tau/4)}(z)}{m((B(z_j, \tau/4))}.$$

According to Lemma 3.1.2, it suffices to verify that the measure $d\mu_{F,p}(z) = |F(z)|^2(1 - |z|^2)^p dm(z)$ belongs to $\mathcal{CM}_p$. To this end, we employ the boundedness of C_j and the disjointness of $\{B(z_j, \tau/4)\}$ to establish the following estimates for any $w \in \mathbf{D}$:

$$
\begin{aligned}
&\int_{\mathbf{D}} \left(\frac{1 - |w|^2}{|1 - \bar{w}z|^2} \right)^p d\mu_{F,p}(z) \\
&= \int_{\mathbf{D}} (1 - |\sigma_w(z)|^2)^p |F(z)|^2 dm(z) \\
&\preceq \int_{\mathbf{D}} \sum_{j=1}^{\infty} |\lambda_j|^2 \frac{1 - |z_j|^2}{m(B(z_j, \tau/4))} 1_{B(z_j, \tau/4)}(z)(1 - |\sigma_w(z)|^2)^p dm(z) \\
&\preceq \sum_{j=1}^{\infty} \int_{\cup B(z_j, \tau/4)} \left(\frac{|\lambda_j|(1 - |z_j|^2)}{m(B(z_j, \tau/4))} \right)^2 (1 - |\sigma_w(z)|^2)^p dm(z) \\
&\preceq \sum_{\mathbf{D}} |\lambda_j|^2(1 - |\sigma_w(z_j)|^2)^p \\
&\preceq \|\nu_p\|_{\mathcal{CM}_p},
\end{aligned}
$$

which in turn implies $\mu_{F,p} \in \mathcal{CM}_p$ thanks to Lemma 3.1.1.

(ii) To handle this part, for those points $\{z_j\}$ fixed at the beginning of the argument, pick a sequence of Lebesgue measurable sets $\{D_j\}$ (cf. [CoiRoc, p.18]) such that

$$B(z_j, \frac{\tau}{4}) \subseteq D_j \subseteq B(z_j, \tau); \quad \mathbf{D} = \bigcup_{j=1}^{\infty} D_j.$$

Without loss of generality, we may also assume that $|z_j| > 0$ for $j \in \mathbf{N}$. If $f \in \mathcal{Q}_p$, then $f' \in A^{2,p}$ and hence by Lemma 3.3.2 (i),

$$\begin{aligned}
f'(w) &= \frac{b}{\pi} \int_{\mathbf{D}} f'(z) \left(\frac{(1-|z|^2)^{b-1}}{(1-\bar{z}w)^{b+1}} \right) dm(z) \\
&= \frac{b}{\pi} \sum_{j=1}^{\infty} \int_{D_j} f'(z) \left(\frac{(1-|z|^2)^{b-1}}{(1-\bar{z}w)^{b+1}} \right) dm(z).
\end{aligned}$$

Putting

$$\mathsf{S}(f)(w) = \frac{1}{\pi} \sum_{j=1}^{\infty} f'(z_j) m(D_j) \left(\frac{(1-|z_j|^2)^{b-1}}{\bar{z}_j(1-\bar{z}_jw)^b} \right),$$

we get by the triangle inequality and the definition of $k_w(z)$,

$$\begin{aligned}
&|f'(w) - \mathsf{S}(f)'(w)| \\
&= \frac{b}{\pi} \sum_{j=1}^{\infty} \int_{D_j} \left(f'(z) \frac{(1-|z|^2)^{b-1}}{(1-\bar{z}w)^{b+1}} - \frac{f'(z_j)(1-|z_j|^2)^{b-1}}{(1-\bar{z}_jw)^{b+1}} \right) dm(z) \\
&\leq \sum_1 + \sum_2,
\end{aligned}$$

where

$$\sum_1 = \frac{b}{\pi} \sum_{j=1}^{\infty} \int_{D_j} |f'(z)||k_w(z) - k_w(z_j)| dm(z)$$

and

$$\sum_2 = \frac{b}{\pi} \sum_{j=1}^{\infty} \int_{D_j} (|f'(z) - f'(z_j)|) |k_w(z_j)| dm(z).$$

Using Lemma 3.3.1 we have

$$\sum_1 \preceq \tau \int_{\mathbf{D}} |f'(z)||k_w(z)| dm(z).$$

To control $\sum_2$, we change variables and make simple supremum estimates to get

$$\sum_2 \leq \sum_{j=1}^{\infty} \int_{B(z_j, \tau)} |f'(z) - f'(z_j)||k_w(z_j)|dm(z)$$

$$= \sum_{j=1}^{\infty} \int_{B(0, \tau)} |f'(\sigma_{z_j}(z)) - f'(\sigma_{z_j}(0))| \left(\frac{1 - |z_j|^2}{|1 - \overline{z_j}z|^2}\right)^2 |k_w(z_j)|dm(z)$$

$$\leq \sum_{j=1}^{\infty} \int_{B(0, \tau)} \frac{s(1 - |z_j|^2)}{(1 - s_0)^2} \sup_{\zeta \in B(z_j, \tau)} |f''(\zeta)| \left(\frac{1 - |z_j|^2}{|1 - \overline{z_j}z|^2}\right)^2 |k_w(z_j)|dm(z)$$

$$\leq \sum_{j=1}^{\infty} \frac{s(1 - |z_j|^2)}{(1 - s_0)^2} \sup_{\zeta \in B(z_j, \tau)} |f''(\zeta)| m\big(B(z_j, \tau)\big)|k_w(z_j)|,$$

where

$$s = \frac{e^\tau - e^{-\tau}}{e^\tau + e^{-\tau}} \quad \text{and} \quad s_0 = \frac{e - e^{-1}}{e + e^{-1}}.$$

To continue our argument, we note four basic facts as follows:

First, there is constant $c_1 > 0$ independent of $\tau \in (0, 1]$ and $z \in \mathbf{D}$ such that

$$|f''(\zeta)| \leq \frac{c_1}{m(B(\zeta, \tau))} \int_{B(\zeta, \tau)} |f''(\xi)|dm(\xi);$$

Second, there is a constant $c_2 > 0$ independent of $\tau \in (0, 1]$, $\zeta \in \mathbf{D}$ and z_j such that

$$\zeta \in B(z_j, r) \Rightarrow B(\zeta, \tau) \subseteq B(z_j, 2\tau) \quad \text{and} \quad m\big(B(z_j, \tau)\big) \leq c_2 m\big(B(\zeta, \tau)\big);$$

Third, there is a constant $c_3 > 0$ independent of $\tau \in (0, 1]$, $z \in \mathbf{D}$ and z_j such that

$$1 - |z_j|^2 \leq c_3(1 - |z|^2), \quad z \in B(z_j, 2\tau);$$

Fourth, Lemma 3.3.1 and its symmetric form imply

$$d_{\mathbf{D}}(z, z_0) \leq 1 \Rightarrow k_w(z) \approx k_w(z_0), \quad w \in \mathbf{D}.$$

With the help of the above estimates and Lemma 3.3.2 we then have

$$\sum_2 \preceq s \sum_{j=1}^{\infty} \int_{D_j} |f''(z)|(1 - |z|^2)|k_w(z)|dm(z)$$

$$\preceq s \int_{\mathbf{D}} |f''(z)|(1 - |z|^2)|k_w(z)|dm(z).$$

Putting the estimates on $\sum_1$ and $\sum_2$ together, we obtain

$$|f'(w) - \mathsf{S}(f)'(w)| \preceq \tau \int_{\mathbf{D}} |f'(z)||k_w(z)|dm(z)$$

$$+ s \int_{\mathbf{D}} |f''(z)|(1 - |z|^2)|k_w(z)|dm(z), \quad w \in \mathbf{D}.$$

Since $f \in \mathcal{Q}_p$, we conclude that

$$|f'(z)|^2(1 - |z|^2)^p dm(z) \quad \text{and} \quad \big(|f''(z)|(1 - |z|^2)\big)^2(1 - |z|^2)^p dm(z)$$

are in $\mathcal{CM}_p$ (cf. Theorem 1.1.3 or 3.2.3). Applying this to Lemma 3.1.2, we find

$$\|f - \mathsf{S}(f)\|_{\mathcal{Q}_p, 1} \precsim s\|f\|_{\mathcal{Q}_p, 1}.$$

If τ is small enough then so is s, and hence the operator $\mathsf{S}(f)$ is invertible with the bounded inverse being

$$\mathsf{S}^{-1} = (\mathsf{Id} - (\mathsf{Id} - \mathsf{S}))^{-1} = \sum_{n=0}^{\infty}(\mathsf{Id} - \mathsf{S})^n,$$

where Id stands for the identity operator acting on $\mathcal{Q}_p$. With this, we obtain

$$f(z) = \mathsf{S}\mathsf{S}^{-1}(f)(z) = \frac{1}{\pi}\sum_{j=1}^{\infty}\big(\mathsf{S}^{-1}(f)'(z_j)\big)m(D_j)\Big(\frac{(1 - |z_j|^2)^{b-1}}{\overline{z_j}(1 - \overline{z_j}z)^b}\Big)$$

$$= \sum_{j=1}^{\infty}\lambda_j\Big(\frac{1 - |z_j|^2}{1 - \overline{z_j}z}\Big)^b,$$

where

$$\lambda_j = \frac{\mathsf{S}^{-1}(f)'(z_j)m(D_j)}{\pi\overline{z_j}(1 - |z_j|^2)}.$$

Of course, we have to verify

$$d\mu_{\{z_j\}, p} = \sum_{j=1}^{\infty}|\lambda_j|^2(1 - |z_j|^2)^p\delta_{z_j} \in \mathcal{CM}_p.$$

To do so, we observe that the mean-value-inequality

$$|F(z_j)|^2(1 - |z_j|^2)^p \precsim \frac{1}{m\big(B(z_j, \tau/20)\big)}\int_{B(z_j, \tau/20)}|F(z)|^2(1 - |z|^2)^p dm(z)$$

holds for any $F \in \mathcal{H}$. Thus, for any $w \in \mathbf{D}$ we apply the disjointness of $\{D_j\}$ to obtain

$$\int_{\mathbf{D}}\Big(\frac{1 - |w|^2}{|1 - \overline{w}z|^2}\Big)^p d\mu_{\{z_j\}, p}(z)$$

$$\leq \sum_{j=1}^{\infty}\Big(\frac{1 - |w|^2}{|1 - \overline{w}z|^2}\Big)^p|\mathsf{S}^{-1}(f)'(z_j)|^2 m(D_j)(1 - |z_j|^2)^{p-2}$$

$$\precsim \sum_{j=1}^{\infty}\frac{\big(m(D_j)(1 - |z_j|^2)^{-1}\big)^2}{m\big(B(z_j, \tau/20)\big)}\int_{B(z_j, \tau/20)}|\mathsf{S}^{-1}(f)'(z)|^2(1 - |\sigma_w(z)|^2)^p dm(z)$$

$$\precsim \int_{\mathbf{D}}|\mathsf{S}^{-1}(f)'(z)|^2(1 - |\sigma_w(z)|^2)^p dm(z)$$

$$\precsim \|\mathsf{S}^{-1}(f)\|_{\mathcal{Q}_p, 1}^2,$$

which implies by Lemma 3.1.1 the desired result. $\qquad\qquad\qquad\qquad$ $\square$

3.4 Discreteness by Derivatives

Given a sequence of points $S = \{z_j\}$ in $\mathbf{D}$. We are interested in the relation between the geometry of S and the values which derivatives of functions in $\mathcal{Q}_p$ can take on S. Once done, a discrete form of $\mathcal{Q}_p$ will be established.

Using the fact that for $p \in (0, \infty)$, $\mathcal{Q}_p$ is a subspace of $\mathcal{B}$ we find that

$$\sup_{j \in \mathbf{N}} (1 - |z_j|^2)|f'(z_j)| \leq \|f\|_{\mathcal{B}} \preceq \|f\|_{\mathcal{Q}_p}, \quad f \in \mathcal{Q}_p.$$

This means that the sequence $\mathsf{T}f = \{(1 - |z_j|^2)f'(z_j)\}_{j=1}^{\infty}$ is in $\ell^{\infty}(S)$. On the basis of Theorem 3.3.3 we then ask for the relationship between $\{\mathsf{T}f : f \in \mathcal{Q}_p\}$ and the sequential space $\mathcal{SCM}_p(S)$ consisting of all complex numbers $\{c_j\}$ such that $\sum_{j=1}^{\infty}(1-|z_j|^2)^p|c_j|^2\delta_{z_j}$ belongs to $\mathcal{CM}_p$. In so doing, we introduce a variation of Schur's Lemma for the sequences.

Lemma 3.4.1. *Given $p \in [0, \infty)$ and a sequence $\{z_j\}$ in $\mathbf{D}$. Let $A = (a_{n,k})$ be an infinite matrix with nonnegative entries. If there are a sequence $\{h_j\}$ of positive numbers and two positive constants c_1 and c_2 independent of $w \in \mathbf{D}$ such that*

$$\sum_{n \neq k} a_{n,k}h_n \leq c_1 h_k, \quad k \in \mathbf{N}$$

and

$$\sum_{k \neq n} a_{n,k}h_k(1 - |\sigma_w(z_k)|^2)^p \leq c_2 h_n(1 - |\sigma_w(z_n)|^2)^p, \quad n \in \mathbf{N},$$

then

$$\sum_k \Big| \sum_{n \neq k} a_{n,k}s_n \Big|^2 (1 - |\sigma_w(z_k)|^2)^p \leq c_1 c_2 \sum_n |s_n|^2 (1 - |\sigma_w(z_n)|^2)^p.$$

Proof. Using the hypothesis, Cauchy–Schwarz's inequality and Fubini's Theorem, we have

$$\sum_k \Big| \sum_{n \neq k} a_{n,k}s_n \Big|^2 (1 - |\sigma_w(z_k)|^2)^p$$

$$\leq \sum_k \Big(\sum_{n \neq k} a_{n,k}h_n \Big) \Big(\sum_{n \neq k} a_{n,k}|s_n|^2 h_n^{-1} \Big) (1 - |\sigma_w(z_k)|^2)^p$$

$$\leq c_1 \sum_n |s_n|^2 h_n^{-1} \Big(\sum_{k \neq n} h_k a_{n,k}(1 - |\sigma_w(z_k)|^2)^p \Big)$$

$$\leq c_1 c_2 \sum_n |s_n|^2 (1 - |\sigma_w(z_n)|^2)^p,$$

as required. $\qquad\qquad\qquad\qquad\qquad\qquad\qquad\qquad\qquad\qquad\qquad\qquad$ $\square$

This technique will be used in the argument for the following result.

Theorem 3.4.2. *Given $p \in (0,2)$. Let $\{z_j\}$ be a sequence of points in $\mathbf{D}$ with $\tau = \inf_{j \neq k} d_{\mathbf{D}}(z_j, z_k)$.*

(i) *If $\tau > 0$, then $\mathsf{T} : \mathcal{Q}_p \mapsto \mathcal{SCM}_p$ is injective.*

(ii) *If $\mathsf{T} : \mathcal{Q}_p \mapsto \mathcal{SCM}_p$ is surjective, then $\tau > 0$. Conversely, there is a $\tau_0 > 0$ such that if $\tau > \tau_0$, then $\mathsf{T} : \mathcal{Q}_p \mapsto \mathcal{SCM}_p$ is surjective.*

Proof. (i) Suppose $\tau > 0$. Then $\{B(z_j, \tau/2)\}$ are disjoint. If $f \in \mathcal{Q}_p$, then

$$d\mu_{f,p}(z) = (1 - |z|^2)^p |f'(z)|^2 dm(z) \in \mathcal{CM}_p,$$

and hence for any $w \in \mathbf{D}$ we use

$$1 - |z|^2 \approx 1 - |z_j|^2 \quad \text{and} \quad |1 - \overline{z_j} w| \approx |1 - \overline{z} w| \quad \text{as} \quad d_{\mathbf{D}}(z, z_j) < \frac{\tau}{2}$$

and the mean-value-inequality for f' to get

$$\sum_{j=1}^{\infty} \left((1 - |z_j|^2) |f'(z_j)| \right)^2 \left(1 - |\sigma_{z_j}(w)|^2 \right)^p$$

$$\preceq \sum_{j=1}^{\infty} \frac{(1 - |z_j|^2)^2 \left(1 - |\sigma_{z_j}(w)|^2 \right)^p}{m\big(B(z_j, \tau/2)\big)} \int_{B(z_j, \tau/2)} |f'(z)|^2 dm(z)$$

$$\preceq \sum_{j=1}^{\infty} \int_{B(z_j, \tau/2)} |f'(z)|^2 \left(1 - |\sigma_{z_j}(w)|^2 \right)^p dm(z)$$

$$\preceq \big(F_p(f, w) \big)^2$$

$$\preceq \|f\|_{\mathcal{Q}_p, 1}^2.$$

That is to say, $\mathsf{T} f \in \mathcal{SCM}_p$, as desired.

(ii) Again, noting $\mathcal{Q}_p \subseteq \mathcal{B}$, we have that if $f \in \mathcal{Q}_p$, then the Bergman reproducing formula for f' gives

$$f'(w) = \frac{2}{\pi} \int_{\mathbf{D}} f'(z) \frac{1 - |z|^2}{(1 - w\overline{z})^3} dm(z), \quad w \in \mathbf{D}.$$

By the proof of Lemma 3.3.1 with $b = 2$ we find that

$$\left| f'(w) - \left(\frac{1 - |z_j|^2}{1 - w\overline{z_j}} \right)^3 f'(z_j) \right| \preceq \frac{d_{\mathbf{D}}(w, z_j)}{1 - |w|^2}.$$

Consequently,

$$|(1 - |z_j|^2) f'(z_j) - (1 - |z_k|^2) f'(z_k)| \preceq \|f\|_{\mathcal{B}} d_{\mathbf{D}}(z_j, z_k) \preceq \|f\|_{\mathcal{Q}_p} d_{\mathbf{D}}(z_j, z_k).$$

Of course, if T maps $\mathcal{Q}_p$ onto $\mathcal{SCM}_p$, then there are functions $f_j \in \mathcal{Q}_p$ that have $\|f_j\|_{\mathcal{Q}_p,1}$ uniformly bounded and that satisfy

$$(1 - |z_k|^2) f'(z_j) = \begin{cases} 1 & , \quad j = k, \\ 0 & , \quad j \neq k. \end{cases}$$

Hence $\tau > 0$.

Conversely, let τ be sufficiently large. To prove that T is onto, it is enough to find the right inverse of T. As a good approximation of the inverse, we are motivated by the decomposition theorem for $\mathcal{Q}_p$ to consider the following R operator acting on $\{s_j\} \in \mathcal{SCM}_p$:

$$\mathsf{R}(\{s_j\})(z) = \sum_{n=1}^{\infty} \frac{s_n}{b \overline{z_n}} \left(\frac{1 - |z_n|^2}{1 - \overline{z_n} z} \right)^b, \quad b > \frac{1 + p}{2}.$$

The proof of Theorem 3.3.3 tells us that $\mathsf{R}(\{s_j\}) \in \mathcal{Q}_p$ whenever $\{s_j\} \in \mathcal{SCM}_p$. Now a simple calculation reveals that for the identity operator Id on $\mathcal{SCM}_p$,

$$(\mathsf{Id} - \mathsf{TR})(\{s_j\}) = \left\{ -\sum_{n \neq k} s_n \frac{(1 - |z_k|^2)(1 - |z_n|^2)^b}{(1 - \overline{z_n} z_k)^{b+1}} \right\}.$$

So to prove that the operator norm of $\mathsf{Id} - \mathsf{TR}$ on $\mathcal{SCM}_p$ can be small enough, let

$$a_{n,k} = \frac{(1 - |z_k|^2)(1 - |z_n|^2)^b}{(1 - \overline{z_n} z_k)^{b+1}}.$$

According to Lemma 3.1.1 we are going to show that there is a constant $c > 0$ independent of $w \in \mathbf{D}$ such that

$$\sum_{k=1}^{\infty} \Big| \sum_{n \neq k} s_n a_{n,k} \Big|^2 (1 - |\sigma_w(z_k)|^2)^p \leq c \sum_{n=1}^{\infty} |s_n|^2 (1 - |\sigma_w(z_n)|^2)^p.$$

By Lemma 3.4.1, it suffices to demonstrate that there exists sufficiently small constant $c > 0$ independent of $w \in \mathbf{D}$ such that

$$J_k = \sum_{n \neq k} a_{n,k} (1 - |z_n|^2)^\gamma \leq c(1 - |z_k|^2)^\gamma$$

and

$$K_n = \sum_{k \neq n} a_{n,k} (1 - |z_k|^2)^\gamma \left(1 - |\sigma_w(z_k)|^2 \right)^p \leq c(1 - |z_n|^2)^\gamma \left(1 - |\sigma_w(z_k)|^2 \right)^p,$$

where γ is determined by

$$\max\{p, 1 - b\} < \gamma < \min\{b + 2, 1 + p + 2b\}.$$

When $\gamma \in (1-b, b+2)$, we can easily work out the following estimate:

$$J_k \preceq (1-|z_k|^2) \int_{d_{\mathbf{D}}(z,z_k)>\tau/2} \frac{(1-|z|^2)^{\gamma+b-2}}{|1-\overline{z_k}z|^{b+1}} dm(z)$$

$$\preceq (1-|z_k|^2)^\gamma \int_{d_{\mathbf{D}}(z,0)>\tau/2} \frac{(1-|z|^2)^{\gamma+b-2}}{|1-z|^{2\gamma-2}} dm(z).$$

Similarly, when $\gamma \in (p, 1+p+2b)$ we can achieve

$$K_n = (1-|w|^2)^p (1-|z_n|^2)^b \sum_{k\neq n} \frac{(1-|z_k|^2)^{1+p+\gamma}}{|1-\overline{z_n}z_k|^{b+1}|1-\bar{w}z_k|^{2p}}$$

$$\preceq (1-|w|^2)^p (1-|z_n|^2)^b \sum_{k\neq n} \int_{B(z_k,\tau/2)} \frac{(1-|z|^2)^{p+\gamma-1}}{|1-\overline{z_n}z|^{b+1}|1-\bar{w}z|^{2p}} dm(z)$$

$$\preceq (1-|w|^2)^p (1-|z_n|^2)^b \int_{d_{\mathbf{D}}(z,z_n)>\tau/2} \frac{(1-|z|^2)^{p+\gamma-1}}{|1-\overline{z_n}z|^{b+1}|1-\bar{w}z|^{2p}} dm(z)$$

$$\preceq \frac{(1-|w|^2)^p}{(1-|z_n|^2)^{-(p+\gamma)}} \int_{d_{\mathbf{D}}(z,0)>\tau/2} \frac{(1-|z|^2)^{p+\gamma-1}|1-\overline{z_n}z|^{2b-2\gamma}}{|1-\bar{w}z_n+(\bar{w}-\overline{z_n})z|^{2p}} dm(z)$$

$$\preceq (1-|z_n|^2)^\gamma (1-|\sigma_w(z_n)|^2)^p \int_{d_{\mathbf{D}}(z,0)>\tau/2} \frac{(1-|z|^2)^{\gamma-1-p}}{|1-z|^{2\gamma-2b}} dm(z).$$

Owing to

$$\int_{\mathbf{D}} \frac{(1-|z|^2)^{\gamma+b-2}}{|1-z|^{2\gamma-2}} dm(z) < \infty \quad \text{and} \quad \int_{\mathbf{D}} \frac{(1-|z|^2)^{\gamma-1-p}}{|1-z|^{2\gamma-2b}} dm(z) < \infty,$$

for any γ between $\max\{p, 1-b\}$ and $\min\{b+2, 1+p+2b\}$, we can insure

$$\lim_{\tau\to\infty} \int_{d_{\mathbf{D}}(z,0)>\tau/2} \frac{(1-|z|^2)^{\gamma+b-2}}{|1-z|^{2\gamma-2}} dm(z) = 0$$

and

$$\lim_{\tau\to\infty} \int_{d_{\mathbf{D}}(z,0)>\tau/2} \frac{(1-|z|^2)^{\gamma-1-p}}{|1-z|^{2\gamma-2b}} dm(z) = 0.$$

In other words, the operator norm of $\mathsf{Id} - \mathsf{TR}$, thanks to Lemma 3.4.1, will be as small as we wish as $\tau \to \infty$. It follows that $\mathsf{TR} = \mathsf{Id} - (\mathsf{Id} - \mathsf{TR})$ is invertible. Consequently, $\mathsf{R(TR)}^{-1}$ is the operator that maps $\mathcal{SCM}_p$ onto $\mathcal{Q}_p$, and the proof is finished. $\square$

3.5 Characterization in Terms of a Conjugate Pair

We now utilize the covering lemma — Lemma 3.3.2 to study the representation of $\mathcal{Q}_p$ functions based on a conjugate pair. To this end, we need the following result.

Lemma 3.5.1. *Let $p \in (0, \infty)$ and $f \in \mathcal{H}$. Then:*

(i) $\left(F_p(f, 0)\right)^2 = p \int_{\mathbf{D}} f(z)\overline{zf'(z)}(1 - |z|^2)^{p-1}dm(z).$

(ii) $\left(F_p(f, 0)\right)^2 \preceq \int_{\mathbf{D}} |f(z)f'(z)|(1 - |z|^2)^{p-1}dm(z).$

The reversed inequality holds for $f \in \mathcal{D}^{2,p}$ if and only if $p \in (1, \infty)$.

Proof. (i) Assuming $f(z) = \sum_{j=0}^{\infty} a_j z^j$, and integrating by parts, we get

$$\left(F_p(f, 0)\right)^2$$

$$= \int_0^1 \left(\int_{\mathbf{T}} \left(\sum_{j=1}^{\infty} j a_j r^{j-1} \zeta^{j-1} \right) \left(\sum_{j=1}^{\infty} j \overline{a_j} r^{j-1} \overline{\zeta}^{j-1} \right) |d\zeta| \right) (1 - r^2)^p r \, dr$$

$$= \pi \sum_{j=1}^{\infty} j^2 |a_j|^2 \int_0^1 r^{j-1}(1 - r)^p dr$$

$$= p\pi \sum_{j=1}^{\infty} j |a_j|^2 \int_0^1 r^j (1 - r)^{p-1} dr.$$

Meanwhile, we also have

$$\int_{\mathbf{D}} f(z)\overline{zf'(z)}(1 - |z|^2)^{p-1}dm(z)$$

$$= \int_0^1 \left(\int_{\mathbf{T}} \left(\sum_{j=0}^{\infty} a_j r^j \zeta^j \right) \left(\sum_{j=0}^{\infty} j \overline{a_j} r^j \overline{\zeta}^j \right) |d\zeta| \right) r(1 - r^2)^{p-1} dr$$

$$= \pi \sum_{j=1}^{\infty} j |a_j|^2 \int_0^1 r^j (1 - r)^{p-1} dr,$$

hence reaching the required formula.

(ii) The inequality follows from (i) right away. Now if $p \in (1, \infty)$ then by the Cauchy–Schwarz inequality,

$$\int_{\mathbf{D}} |f(z)\overline{zf'(z)}|(1 - |z|^2)^{p-1}dm(z)$$

$$\leq F_p(f, 0) \left(\int_{\mathbf{D}} |f(z)|^2 (1 - |z|^2)^{p-2} dm(z) \right)^{\frac{1}{2}}$$

$$\preceq \left(F_p(f, 0)\right)\left(|f(0)| + F_p(f, 0)\right).$$

Next we prove that if $p \in (0, 1)$, then the last inequality cannot be true in general. To do so, set

$$G_p(f) = \left(\int_{\mathbf{D}} |f(z)f'(z)|(1 - |z|^2)^{p-1}dm(z) \right)^{\frac{1}{2}}.$$

Suppose otherwise $G_p(f) \preceq F_p(f,0)$ for any $f \in \mathcal{D}^{2,p}$. Then for any $f_1, f_2 \in \mathcal{D}^{2,p}$ with $f_2 \not\equiv 0$ we have

$$G_p(f_1 \pm f_2, 0) \preceq F_p(f_1 \pm f_2, 0) \preceq F_p(f_1, 0) + F_p(f_2, 0).$$

Using the identity

$$f_1' f_2 + f_1 f_2' = \frac{(f_1 + f_2)(f_1' + f_2') - (f_1 - f_2)(f_1' - f_2')}{2}$$

and the triangle inequality, we further get

$$
\begin{aligned}
& H_p(f_1, f_2) \\
= {} & \int_{|f_1(z)| \le |f_2(z)|} |f_1'(z) f_2(z)| (1 - |z|^2)^{p-1} dm(z) \\
\preceq {} & \left(F_p(f_1, 0) \right)^2 + \left(F_p(f_2, 0) \right)^2 + \int_{|f_1(z)| \le |f_2(z)|} |f_1(z) f_2'(z)| (1 - |z|^2)^{p-1} dm(z) \\
\preceq {} & \left(F_p(f_1, 0) \right)^2 + \left(F_p(f_2, 0) \right)^2.
\end{aligned}
$$

Consequently,

$$
\begin{aligned}
I_p(f_1, f_2) &= \int_{\mathbf{D}} |f_1'(z) f_2(z)| (1 - |z|^2)^{p-1} dm(z) \\
&= H_p(f_1, f_2) + \int_{|f_1(z)| \ge |f_2(z)|} |f_1'(z) f_2(z)| (1 - |z|^2)^{p-1} dm(z) \\
&\preceq \left(F_p(f_1, 0) \right)^2 + \left(F_p(f_2, 0) \right)^2.
\end{aligned}
$$

Nevertheless, the last estimate would yield $f_2 \equiv 0$, a contradiction. As a matter of fact, the techniques used in the last two sections allow us to find a positive number τ_0 close to 1 with the following property: If $\tau \in (\tau_0, 1)$ and $\{z_j\}$ satisfies $\inf_{m \ne n} |\sigma_{z_m}(z_n)| \ge \tau$, then $T_{\{z_j\}, p}(f) = \{(1 - |z_j|^2)^{1+p/2} f'(z_j)\}$ is a surjective operator from $\mathcal{D}^{2,p}$ to ℓ^2. Keeping this in mind, we now take a sequence $\{z_j\}$ on $\mathbf{D}$ such that $\inf_{m \ne n} |\sigma_{z_m}(z_n)| \ge \tau$ and $\mathbf{D}$ is covered by the pseudo-hyperbolic disks $\Delta(z_j, r) = \{z \in \mathbf{D} : |\sigma_{z_j}(z)| < r\}$ where $r \in (3^{-1/2}, 1)$. Then, for each $j \in \mathbf{N}$ there is a point w_j in the closure $\overline{\Delta(z_j, r)}$ of $\Delta(z_j, r)$ satisfying

$$|f_2(w_j)| = \max\{|f_2(z)| : z \in \overline{\Delta(z_j, r)}\}.$$

From the covering of $\mathbf{D}$ it follows that

$$\int_{\mathbf{D}} |f_2(z)|^2 (1 - |z|^2)^{p-2} dm(z) \preceq \sum_{j=1}^{\infty} (1 - |w_j|^2)^p |f_2(w_j)|^2.$$

Note also that $|\sigma_{w_m}(z_n)| \le r$, $s = 2r(1 + r^2)^{-1}$ and $m \ne n$ imply

$$|\sigma_{w_m}(w_n)| \ge \frac{|\sigma_{z_m}(z_n)| - s}{1 - s|\sigma_{z_m}(z_n)|}.$$

Hence, $r \in (3^{-1/2}, 1)$ yields

$$|\sigma_{w_m}(w_n)| \geq \delta = (\tau - s)(1 - s\tau)^{-1} > \tau_0 \quad \text{as} \quad \tau > (\tau_0 + s)(1 + s\tau_0)^{-1},$$

and consequently $\mathsf{T}_{\{w_j\},p} : \mathcal{D}^{2,p} \mapsto \ell^2$ is onto. Furthermore, from the Open Mapping Theorem it turns out that for any $\{c_j\} \in \ell^2$ there exists an $f_1 \in \mathcal{D}^{2,p}$ such that

$$\|f_1\|_{\mathcal{D}^{2,p}} \preceq \|\{c_j\}\|_{\ell^2} \quad \text{and} \quad c_j = (1 - |w_j|^2)^{1+p/2} f_1'(w_j).$$

With the preceding constructions, we derive

$$\left(\sum_{j=1}^{\infty} (1 - |w_j|^2)^p |f_2(w_j)|^2 \right)^{\frac{1}{2}}$$

$$\preceq \sup_{\|\{c_j\}\|_{\ell^2} \leq 1} \sum_{j=1}^{\infty} |f_2(w_j)| (1 - |w_j|^2)^{p/2} |c_j|$$

$$\preceq \sup_{\|f\|_{\mathcal{D}^{2,p}} \leq 1} \sum_{j=1}^{\infty} |f_2(w_j) f_1'(w_j)| (1 - |w_j|^2)^{1+p}$$

$$\preceq \sup_{\|f_1\|_{\mathcal{D}^{2,p}} \leq 1} \sum_{j=1}^{\infty} \int_{\Delta(w_j, \delta/2)} |f_1'(z) f_2(z)| (1 - |z|^2)^{p-1} dm(z)$$

$$\preceq \sup_{\|f_1\|_{\mathcal{D}^{2,p}} \leq 1} I_p(f_1, f_2)$$

$$\preceq 1 + \big(F_p(f_2, 0) \big)^2,$$

hence producing

$$\int_{\mathbf{D}} |f_2(z)|^2 (1 - |z|^2)^{p-2} dm(z) < \infty, \quad p \in (0, 1).$$

This finiteness is not valid unless $f_2 \equiv 0$. We are done. $\qquad \square$

The above lemma leads to the following description involving the conjugate pair.

Theorem 3.5.2. *Let $p \in (0, \infty)$ and $f \in \mathcal{H}$. Then:*

(i) $\|f\|_{\mathcal{Q}_{p,1}}^2 = p \sup_{w \in \mathbf{D}} \int_{\mathbf{D}} f(z) \overline{f'(z)} \sigma_w'(z) \overline{\sigma_w(z)} (1 - |\sigma_w(z)|^2)^{p-1} dm(z).$

(ii) $\|f\|_{\mathcal{Q}_{p,1}}^2 \preceq \sup_{w \in \mathbf{D}} \int_{\mathbf{D}} |f(z) f'(z)| |\sigma_w(z) \sigma_w'(z)| (1 - |\sigma_w(z)|^2)^{p-1} dm(z).$

The reversed inequality holds for all $f \in \mathcal{Q}_p$ if and only if $p \in (1, \infty)$.

Proof. (i) Applying Lemma 3.5.1 (i) to $f \circ \sigma_w$ and doing some simple calculations with the conformal mapping σ_w, we can verify the assertion.

(ii) The forward inequality trivially follows from (i). Regarding the opposite estimate, we choose $f(z) = \log(1 - z)$ to find $\|f\|_{\mathcal{Q}_p}^2 < \infty$ but

$$
\sup_{w \in \mathbf{D}} \int_{\mathbf{D}} |f(z) f'(z)| |\sigma_w(z) \sigma_w'(z)| \left(1 - |\sigma_w(z)|^2\right)^{p-1} dm(z)
$$

$$
\geq \int_{\mathbf{D}} |\log(1 - z)| |z| |1 - z|^{-1} (1 - |z|^2)^{p-1} dm(z)
$$

$$
\geq \int_{|1-z|<1-|z|^2} \left(-\log(1 - |z|^2)\right) |z| (1 - |z|^2)^{p-2} dm(z)
$$

$$
\succeq \int_0^1 t^2 \left(-\log(1 - t^2)\right) (1 - t^2)^{p-2} dt
$$

$$
= \infty \quad \text{for} \quad p \in (0, 1). \qquad \square
$$

3.6 Notes

Note 3.6.1. Section 3.1 is more or less a by-product of [WuXie1], [WuXie2] and [Zha2]. Lemma 3.1.1 has a higher dimensional version — see also [KotLR]. The operator in Lemma 3.1.2 is a generalized combination of Lemmas 4.1.1 and 7.2.2 in [Xi1]. Theorem 3.1.3 for $p \in (0, 1]$ is a special case of [Zha2, Theorem 2] — in particuar, the case $p = 1$ goes back to Jones' result (cf. [An] and [GhZ]). Furthermore, using the second-order derivatives of f_1 and f_2, Fubini's Theorem and the limiting case $p = 0$ of Lemma 3.1.1 (cf. Lemma 6.4.1 (i) of this book), we can readily extend Theorem 3.1.3 to $p = 0$ — more precisely, if $f \in \mathcal{B}$ and $\mathcal{B}_1 \subseteq \mathcal{X} \subseteq \mathcal{D}$, then

$$
dist_{\mathcal{B}}(f, \mathcal{X}) = \inf\{\|f - g\|_{\mathcal{B}} : g \in \mathcal{X}\} \approx \inf\left\{\epsilon > 0 : \int_{\mathbf{D}} \frac{1_{\Omega_\epsilon(f)}(z)}{(1 - |z|^2)^2} dm(z) < \infty\right\}.
$$

Here, we should also observe

$$
dist_{\mathcal{B}}(f, \mathcal{D}) \leq dist_{\mathcal{B}}(f, \mathcal{X}) \leq dist_{\mathcal{B}}(f, \mathcal{B}_1).
$$

But, it is still an open problem to give an equivalent estimate (similar to the above description) for $dist_{\mathcal{B}}(f, \mathcal{H}^\infty)$ – see [An] and [AnClPo]. Since $\mathcal{B}_1 \subset \mathcal{H}^\infty \subset \mathcal{BMOA}$, we conjecture that if $f \in \mathcal{B}$, then

$$
dist_{\mathcal{B}}(f, \mathcal{H}^\infty) \approx \inf\left\{\epsilon > 0 : \sup_{w \in \mathbf{D}} \int_{\mathbf{D}} \frac{1_{\Omega_\epsilon(f)}(z)}{|1 - z\bar{w}|^2} dm(z) < \infty\right\}.
$$

Of course, the key issue is to prove that the left-hand-side distance is greater than or equal to a constant multiple of the right-hand-side infimum.

Note 3.6.2. Section 3.2 essentially consists of [WuXie1, Corollary 4] and the disc analogue of [WuXie2]. Note that the results have been extended to $p \in (0,2)$. So Theorem 3.2.3 (iii) reveals that in case of $p \in (1,2)$, $f^{(\frac{1-p}{2})}$ has radial limiting values on $\mathbf{T}$ even though $f \in \mathcal{Q}_p = \mathcal{B}$ may not have. Of course, Theorem 3.2.3 (ii) generalizes Theorem 1.1.3 (v) that has been also reproved in Wulan–Zhu [WulZh1] using the Schur Lemma, but also the corresponding result in [RocWu] for $p = 1$ and $\alpha > 0$.

Note 3.6.3. Section 3.3 comes basically from [WuXie1]. However, the argument presented in this section is motivated by [Zhu1, §4.3–4.4] and so offers a slightly different approach. In the cases of $\mathcal{BMOA}$ ($\mathcal{BMOA} = \mathcal{Q}_1$) and Bloch ($\mathcal{B} = \mathcal{Q}_p$, $p \in (1,2)$), see [Roc2, Theorem 2.10], [Zhu2, Theorems 5.29 and 3.23] and [Xi1, Theorem 3.2]. Up to this point, we should note that if $p \in (1,2)$, then

$$\sup_{j \in \mathbf{N}} |\lambda_j| < \infty \Leftrightarrow \left\| \sum_{j=1}^{\infty} |\lambda_j|^2 (1 - |z_j|^2)^p \delta_{z_j} \right\|_{\mathcal{CM}_p} < \infty.$$

Note 3.6.4. Section 3.4 is taken from an unpublished manuscript [Xi5], and has showed that the idea on interpolation by functions in Bergman spaces (cf. [Roc1] and Roc2) can be carried over to $\mathcal{Q}_p$, $p \in (0,2)$ including $\mathcal{BMOA}$ and $\mathcal{B}$; see also [Roc1], [Roc2], and [Sun], [BoNi], [Sei], [Sch] and [Xi1] for more information on the interpolations by values of functions in $\mathcal{BMOA}$ or $\mathcal{B}$.

Note 3.6.5. Section 3.5 is taken from [WiXi1]. Here it is perhaps appropriate to mention that the Cauchy–Schwarz inequality yields: for $p > 0$ and $f \in \mathcal{H}$,

$$\int_{\mathbf{D}} |f(z)f'(z)|(1 - |z|)^{p-1} dm(z) \le \int_{\mathbf{D}} |f'(z)|^2 (1 - |z|)^{p-1} dm(z),$$

where equality occurs at $f(z) \equiv z$. This inequality may be viewed as a holomorphic version of either the Opial inequality discussed in [AgPa] or the logarithmic Sobolev-type inequality presented in [Bec].

Chapter 4

Invariant Preduality through Hausdorff Capacity

As a characteristic of every holomorphic Q class, $\| \cdot \|_{\mathcal{Q}_p}$ is conformal invariant. So, it is natural and important to look into the structure of predual space of each $\mathcal{Q}_p$ under the invariant pairing $\int_{\mathbf{D}} f' \bar{g}' dm$. In this chapter, we do this job through the following aspects:

- Nonlinear Integrals and Maximal Operators;
- Adams Type Dualities;
- Quadratic Tent Spaces;
- Preduals under Invariant Pairing;
- Invariant Duals of Vanishing Classes.

4.1 Nonlinear Integrals and Maximal Operators

This section shows some applications of the notion of the nonlinear Choquet integral of a function with respect to the nonadditive Hausdorff capacity in maximal operators of the Hardy–Littlewood and nontangential types.

Given $p \in (0, 1]$, for a set $E \subseteq \mathbf{T}$ let

$$H_\infty^p(E) = \inf \left\{ \sum_{j=1}^{\infty} |I_j|^p : E \subseteq \bigcup_{j=1}^{\infty} I_j \right\},$$

be the p-dimensional Hausdorff capacity of E, where the infimum is taken over countably many open arcs $I_j \subseteq \mathbf{T}$ satisfying $E \subseteq \bigcup_j I_j$. If one had required the length of I_j to satisfy $|I_j| \leq \epsilon$ for some positive number ϵ, then one would have written $H_\epsilon^p(E)$. The classical Hausdorff measure of E of dimension p is $H_0^p(E) = \lim_{\epsilon \to 0} H_\epsilon^p(E)$. It is well known that $H_0^p(\cdot)$ is a outer measure, but not finite in general, whereas $H_\infty^p(\cdot)$ is. Moreover, $H_0^p(\cdot)$ and $H_\infty^p(\cdot)$ have the same null sets. On the other hand, if the infimum in the definition of $H_\infty^p(E)$ is taken over coverings of E by dyadic subarcs of $\mathbf{T}$, then the corresponding capacity, denoted $H_\infty^{p,d}(E)$, is called the dyadic p-dimensional Hausdorff capacity of E. Since each subarc I has two adjacent dyadic subarcs I' and I'' satisfying $I \subset I' \cup I''$ and $|I| = |I'| = |I''|$, we conclude

$$2^{-1} H_\infty^{p,d}(E) \le H_\infty^p(E) \le H_\infty^{p,d}(E).$$

In the above and the below, a dyadic subarc of $\mathbf{T}$ is an arc $I \subseteq \mathbf{T}$ of the form

$$\left\{ e^{i\theta} : j2^{1-k}\pi \le \theta < (j+1)2^{1-k}\pi \right\}, \quad k \in \{0\} \cup \mathbf{N}, \quad j = 0, 1, 2, \ldots, 2^k - 1.$$

Properties of the dyadic Hausdorff capacity needed include:

$$\lim_{j \to \infty} H_\infty^{p,d}(E_j) = H_\infty^{p,d}(E)$$

whenever E_j increases to E or E_j is compact and decreasing to E. And,

$$H_\infty^{p,d}(E_1 \cup E_2) + H_\infty^{p,d}(E_1 \cap E_2) \le H_\infty^{p,d}(E_1) + H_\infty^{p,d}(E_2).$$

At the same time, H_∞^p is a monotone and countably subadditive set function on the class of all subsets of $\mathbf{T}$ which vanishes on the empty set. Furthermore, H_∞^p is an outer capacity in the sense of

$$H_\infty^p(E) = \inf\{H_\infty^p(O) : \text{ open } O \supset E\}.$$

Associated with the Hausdorff capacity $H_\infty^p(\cdot)$ is the Choquet integral of a function f on $\mathbf{T}$:

$$\int_{\mathbf{T}} |f| dH_\infty^p = \int_0^\infty H_\infty^p(\{\zeta \in \mathbf{T} : |f(\zeta)| > t\}) dt.$$

Here the right-hand-side integral is the usual Lebesgue integral. As a matter of fact, $H_\infty^p(\{\zeta \in \mathbf{T} : f(\zeta) > t\})$ is nonincreasing with $t \ge 0$ and hence Lebesgue measurable. In particular,

$$\int_{\mathbf{T}} 1_E dH_\infty^p = H_\infty^p(E)$$

for $E \subseteq \mathbf{T}$, where 1_E is the characteristic function on E.

Below are four basic and useful properties of the Hausdorff capacity and its Choquet integral.

Lemma 4.1.1. *Given $p \in (0, 1]$, let f and $\{f_j\}_{j=1}^{\infty}$ be nonnegative functions on $\mathbf{T}$. Then*

(i) $\int_{\mathbf{T}} \kappa f dH_{\infty}^p = \kappa \int_{\mathbf{T}} f dH_{\infty}^p$ *for any constant $\kappa \geq 0$;*

(ii) $\int_{\mathbf{T}} f dH_{\infty}^p = 0 \iff f = 0$ H_{∞}^p-*a.e. on* $\mathbf{T}$; *that is,* $f = 0$ *on* $\mathbf{T}$ *except on* E *where* $H_{\infty}^p(E) = 0$;

(iii) $\int_{\mathbf{T}} (f(\zeta))^{\frac{1}{p}} |d\zeta| \leq \left(\int_{\mathbf{T}} f dH_{\infty}^p \right)^{\frac{1}{p}}$;

(iv) $\int_{\mathbf{T}} \sum_{j=1}^{\infty} f_j dH_{\infty}^p \leq 2 \sum_{j=1}^{\infty} \int_{\mathbf{T}} f_j dH_{\infty}^p$.

Proof. (i) and (ii) are straightforward. For (iii), we use the simple inequality

$$H_{\infty}^1(E) \leq \left(H_{\infty}^p(E) \right)^{\frac{1}{p}}, \quad E \subseteq \mathbf{T}$$

to derive

$$\int_{\mathbf{T}} (f(\zeta))^{\frac{1}{p}} |d\zeta| = p^{-1} \int_0^{\infty} H_{\infty}^1(\{\zeta \in \mathbf{T} : f(\zeta) > t\}) t^{\frac{1-p}{p}} dt$$

$$\leq \int_0^{\infty} \frac{d}{dt} \left(\int_0^t \left(H_{\infty}^1(\{\zeta \in \mathbf{T} : f(\zeta) > s\}) \right)^p ds \right)^{\frac{1}{p}} dt$$

$$= \left(\int_0^{\infty} \left(H_{\infty}^1(\{\zeta \in \mathbf{T} : f(\zeta) > s\}) \right)^p ds \right)^{\frac{1}{p}}$$

$$\leq \left(\int_0^{\infty} H_{\infty}^p(\{\zeta \in \mathbf{T} : f(\zeta) > s\}) ds \right)^{\frac{1}{p}}$$

$$= \left(\int_{\mathbf{T}} f dH_{\infty}^p \right)^{\frac{1}{p}}.$$

Finally, (iv) follows from the sublinearity of $H_{\infty}^{p,d}$. $\square$

Note that $H_{\infty}^p(\cdot)$ is not additive for $p \in (0, 1)$. This can be verified by dividing $\mathbf{T}$ into 2^m subarcs with equal length and letting $m \to \infty$. Of course, the inequality

$$\sum_{j=1}^{k} \int_{I_j} f dH_{\infty}^p \preceq \int_{\bigcup_j I_j} f dH_{\infty}^p, \quad f \geq 0$$

does not hold for any non-overlapping dyadic subarcs $\{I_j\}_{j=1}^{k}$. However, we can establish a weak form — the circular version of Melnikov's Covering Lemma.

Lemma 4.1.2. *Let $\{I_j\}$ be a collection of non-overlapping dyadic subarcs of $\mathbf{T}$. Then there is a subcollection $\{I_{j_k}\}$ such that*

(i) $\sum_{I_{j_k} \subseteq I} |I_{j_k}|^p \leq 2 |I|^p$ *for any dyadic subarc I of $\mathbf{T}$;*

(ii) $H_{\infty}^p \left(\bigcup_j I_j \right) \leq 2 \sum_k |I_{j_k}|^p$;

(iii) $\sum_k \int_{I_{j_k}} f dH_\infty^p \leq 2 \int_{\bigcup_k I_{j_k}} f dH_\infty^p$ *for any nonnegative function f on* **T**.

Proof. Let $j_1 = 1$. If $j_1, \ldots, j_k$ have been determined to ensure (i), then we define j_{k+1} as the first index such that $\{I_{j_1}, \ldots, I_{j_{k+1}}\}$ satisfies (i). Then $\{I_{j_k}\}$ forms a maximal subcollection of I_j obeying (i). We further show that this subcollection ensures (ii). To that end, choose j such that $j_k < j < j_{k+1}$ for some k. Then according to the choice of j_k, we choose a dyadic subarc $J_j \supseteq I_j$ such that

$$\sum_{I_{j_n} \subseteq J_j, n \leq k} |I_{j_n}|^p + |I_j|^p > 2|J_j|^p.$$

Consequently, we have

$$\sum_{I_{j_n} \subseteq J_j, n \leq k} |I_{j_n}|^p > |J_j|^p.$$

Clearly, we may assume that $\sum_k |I_{j_k}|^p$ is convergent — otherwise (ii) is trivial. Under this assumption, we see that the sequence $\{|J_j|^p\}_j$ is bounded and accordingly, we may consider the collection $\{K_n\}$ of the maximal arcs in the collection $\{J_n\}$, and then obtain

$$\bigcup_j I_j \subseteq \left(\bigcup_m I_{j_m}\right) \cup \left(\bigcup_m K_{j_m}\right).$$

This gives

$$
\begin{aligned}
H_\infty^p\left(\bigcup_j I_j\right) &\leq H_\infty^p\left(\bigcup_m I_{j_m}\right) + H_\infty^p\left(\bigcup_n K_n\right) \\
&\leq \sum_m |I_{j_m}|^p + \sum_m |K_{j_m}|^p \\
&\leq 2\sum_m |I_{j_m}|^p,
\end{aligned}
$$

as required in (ii).

To establish (iii), let $\{D_n\}$ be a sequence of dyadic subarcs of **T** such that

$$\{\zeta : \zeta \in \bigcup_k I_{j_k} \ \& \ f(\zeta) > t\} \subseteq \bigcup_n D_n, \quad t > 0.$$

Then using (i), we get

$$
\begin{aligned}
2\sum_n |D_n|^p &\geq \sum_n \sum_{I_{j_k} \subseteq D_n} |I_{j_k}|^p \\
&= \sum_n \sum_{I_{j_k} \subseteq D_n} H_\infty^p(I_{j_k}) \\
&\geq \sum_n \sum_{I_{j_k} \subseteq D_n} H_\infty^p(\{\zeta : \zeta \in I_{j_k} \ \& \ f(\zeta) > t\}) \\
&\geq \sum_k H_\infty^p(\{\zeta : \zeta \in I_{j_k} \ \& \ f(\zeta) > t\}).
\end{aligned}
$$

Consequently, we have by definition

$$2H_\infty^p(\{\zeta : \zeta \in \bigcup_k I_{j_k} \ \& \ f(\zeta) > t\}) \geq \sum_k H_\infty^p(\{\zeta : \zeta \in I_{j_k} \ \& \ f(\zeta) > t\}),$$

which, along with an integration over $(0, \infty)$ with respect to dt, gives the desired inequality. □

For a function $f \in \mathcal{L}^1(\mathbf{T})$, we write

$$Mf(\zeta) = \sup_{\zeta \in I} |I|^{-1} \int_I |f(\eta)||d\eta|$$

for the Hardy–Littlewood maximal function of f, where the supremum ranges over all arcs $I \subseteq \mathbf{T}$ containing ζ. As above, if the supremum is taken over all dyadic subarcs of $\mathbf{T}$, then we get a dyadic maximal function, denoted $M^d f$. It is clear that $M^d f(\zeta) \leq Mf(\zeta)$ but not the reverse. Even though, we have the inequality

$$H_\infty^p(\{\zeta \in \mathbf{T} : Mf(\zeta) > t\}) \leq 3^p H_\infty^p(\{\zeta \in \mathbf{T} : M^d f(\zeta) > 4^{-1}t\}),$$

and hence $M^d f$ controls Mf in the following sense:

$$\int_{\mathbf{T}} Mf(\zeta) dH_\infty^p(\zeta) \leq 2^2 3^p \int_{\mathbf{T}} M^d f(\zeta) dH_\infty^p(\zeta).$$

Theorem 4.1.3. *Let $q > 0$, $p \in (0, \min\{1, q\})$ and $\int_{\mathbf{T}} |f|^q dH_\infty^p < \infty$. Then:*

(i) $\int_{\mathbf{T}} (Mf)^q dH_\infty^p \preceq \int_{\mathbf{T}} |f|^q dH_\infty^p$.

(ii) $H_\infty^p(\{\zeta \in \mathbf{T} : Mf(\zeta) > t\}) \preceq t^{-p} \int_{\mathbf{T}} |f|^p dH_\infty^p$.

Proof. Without loss of generality, we may assume that f is nonnegative on $\mathbf{T}$.

(i) For each integer n, let $\{I_j^{(n)}\}_j$ be a family of non-overlapping dyadic subarcs of $\mathbf{T}$ such that

$$\{\zeta \in \mathbf{T} : 2^n < f(\zeta) \leq 2^{n+1}\} \subseteq \bigcup_j I_j^{(n)}$$

and

$$\sum_j |I_j^{(n)}|^p \leq 2H_\infty^p(\{\zeta \in \mathbf{T} : 2^n < f(\zeta) \leq 2^{n+1}\}).$$

If

$$J_n = \bigcup_j I_j^{(n)} \quad \text{and} \quad g = \sum_n 2^{(n+1)q} 1_{J_n},$$

then $f^q \leq g$. We consider two cases as follows.

Case 1: $q \in [1, \infty)$. For this, we have

$$(Mf)^q \leq M(f^q) \leq Mg \leq \sum_n 2^{(n+1)q} \sum_j M(1_{I_j^{(n)}}).$$

Note that if ζ_J is the center of a given subarc J of $\mathbf{T}$, then

$$\mathsf{M}(1_J)(\zeta) \preceq \min\{1, |\zeta - \zeta_J|^{-1}|J|\}, \quad \zeta \in \mathbf{T}.$$

So, this, along with Lemma 4.1.1 (iv), implies

$$
\begin{aligned}
\int_{\mathbf{T}} (\mathsf{M}f)^q dH_\infty^p
&\preceq \sum_n 2^{(n+1)q} \sum_j \int_{\mathbf{T}} \mathsf{M}(1_{I_j^{(n)}}) dH_\infty^p \\
&\preceq \sum_n 2^{(n+1)q} \sum_j |I_j^{(n)}|^p \\
&\preceq \sum_n 2^{(n+1)q} H_\infty^p(\{\zeta \in \mathbf{T} : 2^n < f(\zeta) \le 2^{n+1}\}) \\
&\preceq \sum_n \int_{2^{(n-1)q}}^{2^{nq}} H_\infty^p(\{\zeta \in \mathbf{T} : f(\zeta)^q > t\}) dt \\
&\preceq \int_{\mathbf{T}} f^q dH_\infty^p.
\end{aligned}
$$

Case 2: $p < q < 1$. As for this case, we employ $f \le \sum_n 2^{n+1} 1_{J_n}$ and again Lemma 4.1.1 (iv) to achieve

$$
\begin{aligned}
\int_{\mathbf{T}} (\mathsf{M}f)^q dH_\infty^p
&\le \int_{\mathbf{T}} \left(\sum_n 2^{(n+1)q} \sum_j (\mathsf{M}(1_{I_j^{(n)}}))^q \right) dH_\infty^p \\
&\preceq \sum_n 2^{(n+1)q} \sum_j \int_{\mathbf{T}} (\mathsf{M}(1_{I_j^{(n)}}))^q dH_\infty^p \\
&\preceq \sum_n 2^{(n+1)q} \sum_j |I_j^{(n)}|^p \\
&\preceq \int_{\mathbf{T}} f^q dH_\infty^p.
\end{aligned}
$$

(ii) For $t > 0$, let $\{I_j\}$ be the collection of maximal dyadic subarcs of $\mathbf{T}$ with $|I_j|^{-1} \int_{I_j} f > t$. Then $\{\zeta \in \mathbf{T} : \mathsf{M}^d f(\zeta) > t\} = \bigcup_j I_j$. Using Lemma 4.1.1 (iii) we get

$$|I_j|^p \le \left(t^{-1} \int_{I_j} f(\zeta)|d\zeta| \right)^p \preceq t^{-p} \int_{I_j} f^p dH_\infty^p.$$

Furthermore, applying Lemma 4.1.2 to $\{I_j\}$, we can find a subcollection $\{I_{j_k}\}$ of $\{I_j\}$ such that

$$H_\infty^p(\{\zeta \in \mathbf{T} : \mathsf{M}^d f(\zeta) > t\}) \preceq t^{-p} \sum_k \int_{I_{j_k}} f^p dH_\infty^p \preceq t^{-p} \int_{\bigcup_k I_{j_k}} f^p dH_\infty^p,$$

as desired. $\square$

For $\zeta \in \mathbf{T}$ let $\Lambda(\zeta)$ be the convex hull of $\{z \in \mathbf{D} : |z| < 2^{-1}\} \cup \{\zeta\}$ at $\zeta \in \mathbf{T}$. Associated with this region is the nontangential maximal function $\mathsf{N}(f)$ of a function f on $\mathbf{D}$. It is defined by

$$\mathsf{N}(f)(\zeta) = \sup_{z \in \Lambda(\zeta)} |f(z)|.$$

Theorem 4.1.4. *Let* $p \in (0, 1]$, $\bar{\mathbf{D}} = \mathbf{D} \cup \mathbf{T}$, *and* μ *be a nonnegative Radon measure on* $\mathbf{D}$*. Then the following statements are equivalent:*

(i) $\mu \in \mathcal{CM}_p$.

(ii) *There is a constant* $C > 0$ *such that*

$$\int_{\mathbf{D}} |f| d\mu \leq C \int_{\mathbf{T}} \mathsf{N}(f) dH_\infty^p$$

holds for every complex-valued function f on $\mathbf{D}$ with the right-hand-side integral being finite.

(iii) *For every* $q > p$*, there is a constant* $C > 0$ *such that*

$$\int_{\mathbf{D}} |f|^q d|\mu| \leq C \int_{\mathbf{T}} |f|^q dH_\infty^p$$

holds for every function f on $\bar{\mathbf{D}}$ with f being complex-valued harmonic on $\mathbf{D}$ and the right-hand-side integral being finite.

(iv) *For each* $t > 0$ *there is a constant* $C > 0$ *such that*

$$t^p \mu(\{z \in \mathbf{D} : |f(z)| > t\}) \leq C \int_{\mathbf{T}} |f|^p dH_\infty^p$$

holds for every function f on $\bar{\mathbf{D}}$ with f being complex-valued harmonic on $\mathbf{D}$ and the right-hand-side integral being finite.

Proof. We demonstrate the result by showing (i)$\Rightarrow$(ii)$\Rightarrow$(iii)$\Rightarrow$(i)$\Rightarrow$ and (iv)$\Leftrightarrow$(i).

(i)$\Rightarrow$(ii) Assume that μ is a p-Carleson measure. If $\int_{\mathbf{T}} \mathsf{N}(f) dH_\infty^p < \infty$, then for $t \geq 0$, let $\{I_j\}$ be a sequence of open subarcs on $\mathbf{T}$ such that

$$\{\zeta \in \mathbf{T} : \mathsf{N}(f)(\zeta) > t\} \subseteq \bigcup_j I_j.$$

Clearly, there is a sequence of disjoint open subarcs $\{J_k\}$ on $\mathbf{T}$ such that $\bigcup_j I_j = \bigcup_k J_k$. From this it follows that

$$J_k = \bigcup_{I_j \subseteq J_k} I_j \quad \text{and} \quad |J_k|^p \leq \sum_{I_j \subseteq J_k} |I_j|^p$$

hold for each k. Consequently,

$$\sum_k |J_k|^p \le \sum_k \sum_{I_j \subseteq J_k} |I_j|^p = \sum_j |I_j|^p.$$

If $|f(z)| > t$, there exists some J_k such that $|J_k| > 1 - |z|$ and $\mathsf{N}(f)(\zeta) > t$ when $\zeta \in J_k$, and thus $z \in S(J_k)$. Namely,

$$\{z \in \mathbf{D} : |f(z)| > t\} \subseteq \bigcup_k S(J_k).$$

Accordingly,

$$\mu(\{z \in \mathbf{D} : |f(z)| > t\}) \le \mu\left(\bigcup_k S(J_k)\right) \le \|\mu\|_{\mathcal{CM}_p} \sum_k |J_k|^p \le \|\mu\|_{\mathcal{CM}_p} \sum_j |I_j|^p.$$

Taking the infimum over all such coverings and integrating over $(0, \infty)$ with respect to dt, we obtain

$$
\begin{aligned}
\int_{\mathbf{D}} |f| d\mu &= \int_0^\infty \mu(\{z \in \mathbf{D} : |f(z)| > t\}) dt \\
&\le \|\mu\|_{\mathcal{CM}_p} \int_0^\infty H_\infty^p(\{\zeta \in \mathbf{T} : \mathsf{N}(f)(\zeta) > t\}) dt \\
&= \|\mu\|_{\mathcal{CM}_p} \int_{\mathbf{T}} \mathsf{N}(f) dH_\infty^p.
\end{aligned}
$$

(ii)$\Rightarrow$(iii) Suppose (ii) is valid. According to Theorem 4.1.4 (i), we read off

$$\int_{\mathbf{T}} (\mathsf{M}(f))^q dH_\infty^p \preceq \int_{\mathbf{T}} |f|^q dH_\infty^p, \quad q > p.$$

Since the function f in (ii) is complex-valued harmonic on $\mathbf{D}$ and its boundary value f exists on $\mathbf{T}$, we have the following well-known estimate:

$$\mathsf{N}(f)(\zeta) = \sup_{z \in \Lambda(\zeta)} |f(z)| \preceq \mathsf{M}(f)(\zeta), \quad \zeta \in \mathbf{T}.$$

By applying (ii) to $|f|^q$, $q > p$, we get a constant $C > 0$ such that

$$\int_{\mathbf{D}} |f|^q d\mu \le C \int_{\mathbf{T}} \mathsf{N}(|f|^q) dH_\infty^p \preceq C \int_{\mathbf{T}} \mathsf{M}(|f|^q) dH_\infty^p \preceq C \int_{\mathbf{T}} |f|^q dH_\infty^p.$$

So, (iii) follows.

(iii)$\Rightarrow$(i) Let $I \subseteq \mathbf{T}$ be an open arc. Consider the complex-valued harmonic function f_I generated by the characteristic function 1_I of I:

$$f_I(z) = \int_{\mathbf{T}} \frac{1_I(\zeta)(1 - |z|^2)}{|\zeta - z|^2} |d\zeta|, \quad z \in \mathbf{D}.$$

It is easy to see that if $z \in S(I)$, then $1 \preceq |f_I(z)|$. Because of $\int_{\mathbf{T}} 1_I dH_\infty^p \approx |I|^p$, (iii) implies

$$\mu\big(S(I)\big) \preceq \int_{\mathbf{D}} |f_I|^q d\mu \preceq C \int_{\mathbf{T}} 1_I dH_\infty^p \preceq C|I|^p$$

for a constant $C > 0$. Thus (i) follows.

(iv)$\Leftrightarrow$(i) If (iv) holds, then for the above-defined f_I we have

$$\mu\big(S(I)\big) \preceq \mu\big(\{z \in \mathbf{T} : 1 \preceq |f_I(z)|\}\big) \leq C \int_{\mathbf{T}} 1_I dH_\infty^p \preceq C|I|^p.$$

In other words, $\mu \in \mathcal{CM}_p$. Conversely, if (i) holds, then (ii) is true, and hence (iv) follows from

$$
\begin{aligned}
t^p \mu\big(\{z \in \mathbf{D} : |f(z)| > t\}\big) &\leq t^p \|\mu\|_{\mathcal{CM}_p} H_\infty^p\big(\{\zeta \in \mathbf{T} : N(f)(\zeta) > t\}\big) \\
&\preceq t^p \|\mu\|_{\mathcal{CM}_p} H_\infty^p\big(\{\zeta \in \mathbf{T} : M(f)(\zeta) > t\}\big) \\
&\preceq \|\mu\|_{\mathcal{CM}_p} \int_{\mathbf{T}} |f|^p dH_\infty^p.
\end{aligned}
$$

$\square$

4.2 Adams Type Dualities

For $\mathbf{X} = \mathbf{T}$ or $\bar{\mathbf{D}}$, we denote by $\mathcal{C}(\mathbf{X})$ and $\mathcal{M}(\mathbf{X})$ the space of continuous complex-valued functions and finite complex-valued Radon measures on $\mathbf{X}$, respectively. When $\mu \in \mathcal{M}(\mathbf{X})$, we use $|\mu|$ as the total variation of μ, which is a nonnegative measure determined by the property that if $d\mu = f d\nu$ where ν is a nonnegative measure, then $d|\mu| = |f| d\nu$. The Riesz Theorem states that the dual of $\mathcal{C}(\mathbf{X})$ is $\mathcal{M}(\mathbf{X})$.

Using this well-known dual result, we can establish the following Adams Duality Theorem.

Theorem 4.2.1. *Given $p \in (0,1]$. Let $\mathcal{L}^{1,p}(\mathbf{T})$ be the Morrey space of elements μ from $\mathcal{M}(\mathbf{T})$ for which*

$$\|\mu\|_{\mathcal{L}^{1,p}} = \sup_I |I|^{-p} |\mu|(I) < \infty,$$

where the supremum is taken over all subarcs I of $\mathbf{T}$. If $\mathcal{L}^1(H_\infty^p)$ stands for the completion of $\mathcal{C}(\mathbf{T})$ with respect to $\int_{\mathbf{T}} |f| dH_\infty^p$, then $\mathcal{L}^{1,p}(\mathbf{T})$ is isomorphic to the dual of $\mathcal{L}^1(H_\infty^p)$ under the pairing $\int_{\mathbf{T}} f d\mu$.

Proof. Because of $\mu \in \mathcal{L}^{1,p}(\mathbf{T})$, the linear functional $\mathsf{L}(f) = \int_{\mathbf{T}} f d\mu$ is well defined on $\mathcal{L}^1(H_\infty^p)$ since for any μ-measurable E we have

$$|\mu|(E) \leq \|\mu\|_{\mathcal{L}^{1,p}} H_\infty^p(E)$$

by using the definition of $H^p_\infty(\cdot)$. This in turn gives

$$|\mathsf{L}(f)| \leq \int_{\mathbf{T}} |f|d|\mu| \leq \|\mu\|_{\mathcal{L}^{1,p}} \int_{\mathbf{T}} |f|dH^p_\infty, \quad f \in \mathcal{L}^1(H^p_\infty).$$

As a result, we find that the norm $\|\mathsf{L}\|$ of L is not greater than $\|\mu\|_{\mathcal{L}^{1,p}}$.

Conversely, if L is a continuous linear functional on $\mathcal{L}^1(H^p_\infty)$, then there is a $\mu \in \mathcal{M}(\mathbf{T})$ such that L is expressed as $\mathsf{L}(f) = \int_{\mathbf{T}} fd\mu$ for $f \in \mathcal{C}(\mathbf{T})$, thanks to $\mathcal{C}(\mathbf{T}) \subseteq \mathcal{L}^1(H^p_\infty)$. But then for any $g \in \mathcal{C}(\mathbf{T})$,

$$\begin{aligned}
\left| \int_{\mathbf{T}} gd|\mu| \right| &\leq \int_{\mathbf{T}} |g|d|\mu| \\
&\leq \sup\left\{ fd\mu : f \in \mathcal{C}(\mathbf{T}) \& |f| \leq |g| \right\} \\
&\leq \|\mathsf{L}\| \sup\left\{ \int_{\mathbf{T}} |f|dH^p_\infty : f \in \mathcal{C}(\mathbf{T}) \& |f| \leq |g| \right\} \\
&\leq \|\mathsf{L}\| \int_{\mathbf{T}} |g|dH^p_\infty.
\end{aligned}$$

Now, given a subarc I of $\mathbf{T}$ and arbitrary small number $\epsilon > 0$, let I_ϵ be the subarc of $\mathbf{T}$ centered at the center of I and with length $|I| + \epsilon$, and choose a function $g \in \mathcal{C}(\mathbf{T})$ with $g = 1$ on I and $g = 0$ on $\mathbf{T} \setminus I_\epsilon$. Then

$$|\mu|(I) \leq \|\mathsf{L}\| H^p_\infty(I_\epsilon) \quad \text{which implies} \quad \|\mu\|_{\mathcal{L}^{1,p}} \leq \|\mathsf{L}\|. \qquad \square$$

Using Theorem 4.2.1, we develop a technique that is often easier to dominate the Choquet integrals against Hausdorff capacity.

Corollary 4.2.2. *For $p \in (0,1]$ let $\mathcal{L}^{1,p}_+(\mathbf{T})$ be the class of all nonnegative elements in $\mathcal{L}^{1,p}(\mathbf{T})$. If f is nonnegative and lower semicontinuous on $\mathbf{T}$, then*

$$\int_{\mathbf{T}} fdH^p_\infty \approx \sup\left\{ \int_{\mathbf{T}} fd\mu : \mu \in \mathcal{L}^{1,p}_+(\mathbf{T}) \& \|\mu\|_{\mathcal{L}^{1,p}} \leq 1 \right\}.$$

Proof. Let $\mathcal{L}^1(H^{p,d}_\infty)$ be the completion of $\mathcal{C}(\mathbf{T})$ under the norm $\int_{\mathbf{T}} |f|dH^{p,d}_\infty$. Then it is a Banach space, and hence the canonical mapping from $\mathcal{L}^1(H^{p,d}_\infty)$ into its second dual has norm

$$\int_{\mathbf{T}} |g|dH^{p,d}_\infty = \sup\left\{ \left| \int_{\mathbf{T}} gd\mu \right| : \mu \in \mathcal{L}^{1,p}_+(\mathbf{T}) \& \|\mu\|_{\mathcal{L}^{1,p}} \leq 1 \right\}, \quad g \in \mathcal{L}^1(H^{p,d}_\infty).$$

Now since f, nonnegative and lower semicontinuous on $\mathbf{T}$, can be approximated from below by a nonnegative and increasing sequence $\{f_j\}_{j=1}^\infty$ in $\mathcal{C}(\mathbf{T})$, we have

$$\begin{aligned}
\int_{\mathbf{T}} f_j dH^{p,d}_\infty &= \sup\left\{ \int_{\mathbf{T}} f_j d\mu : \mu \in \mathcal{L}^{1,p}_+(\mathbf{T}) \& \|\mu\|_{\mathcal{L}^{1,p}} \leq 1 \right\} \\
&\leq \sup\left\{ \int_{\mathbf{T}} fd\mu : \mu \in \mathcal{L}^{1,p}_+(\mathbf{T}) \& \|\mu\|_{\mathcal{L}^{1,p}} \leq 1 \right\}.
\end{aligned}$$

Using the limit

$$\lim_{j\to\infty} H_\infty^{p,d}(\{\zeta \in \mathbf{T} : f_j(\zeta) \geq t\}) = H_\infty^{p,d}(\{\zeta \in \mathbf{T} : f(\zeta) > t\}), \quad t > 0,$$

we obtain

$$\int_{\mathbf{T}} f dH_\infty^{p,d} \leq \sup\left\{ \int_{\mathbf{T}} f d\mu : \ \mu \in \mathcal{L}_+^{1,p}(\mathbf{T}) \ \& \ \|\mu\|_{\mathcal{L}^{1,p}} \leq 1 \right\} \leq 2 \int_{\mathbf{T}} f dH_\infty^{p,d},$$

as desired. □

Returning to $\mathbf{D}$, we have the following analogue of Theorem 4.2.1.

Theorem 4.2.3. *Given $p \in (0,1]$. Let CCM_p be the space of all $\mu \in \mathcal{M}(\bar{\mathbf{D}})$ whose total variation $|\mu|$ is in CM_p. If $\mathcal{LN}^1(H_\infty^p)$ is the completion of all functions $f \in \mathcal{C}(\bar{\mathbf{D}})$ with respect to*

$$\|f\|_{\mathcal{LN}^1(H_\infty^p)} = \int_{\mathbf{T}} \mathsf{N}(f) dH_\infty^p,$$

then CCM_p is isomorphic to the dual space of $\mathcal{LN}^1(H_\infty^p)$ under the pairing $\int_{\bar{\mathbf{D}}} f d\mu$.

Proof. Using the implication from (i) to (ii) of Theorem 4.1.4 we easily find that every $\mu \in CCM_p$ induces a bounded linear functional on $\mathcal{LN}^1(H_\infty^p)$. Conversely, suppose that L is a bounded linear functional on $\mathcal{LN}^1(H_\infty^p)$. Now since $\mathcal{C}(\bar{\mathbf{D}}) \subseteq \mathcal{LN}^1(H_\infty^p)$, we employ the above-mentioned Riesz Theorem to produce a $\mu \in \mathcal{M}(\bar{\mathbf{D}})$ such that

$$\mathsf{L}(f) = \int_{\bar{\mathbf{D}}} f d\mu \quad \text{and} \quad |\mathsf{L}(f)| \leq \|\mathsf{L}\| \|f\|_{\mathcal{LN}^1(H_\infty^p)} \quad \text{for} \ f \in \mathcal{C}(\bar{\mathbf{D}}).$$

But then for every $g \in \mathcal{C}(\bar{\mathbf{D}})$, we have

$$\left| \int_{\bar{\mathbf{D}}} g d|\mu| \right| \leq \sup\left\{ \int_{\mathbf{D}} f d\mu : \ f \in \mathcal{C}(\bar{\mathbf{D}}) \ \& \ |f| \leq |g| \right\} \leq \|\mathsf{L}\| \|g\|_{\mathcal{LN}^1(H_\infty^p)}.$$

To verify that $|\mu|$ is actually a p-Carleson measure on $\mathbf{D}$, we consider an open subarc I of $\mathbf{T}$ and assume that $I_\epsilon \subseteq \mathbf{T}$, for $\epsilon > 0$ small enough, is an open arc with the same center as I's and with length $|I| + \epsilon$, and take such a nonnegative $g \in \mathcal{C}(\bar{\mathbf{D}})$ that $g(z) = 1$ for $z \in S(I)$ and $g(z) = 0$ for $\bar{\mathbf{D}} \setminus S(I_\epsilon)$. This choice implies $\mathsf{N}(g) \preceq 1_{I_\epsilon}$ and

$$|\mu|(S(I)) \leq |\mu|(S(I_\epsilon)) = \left| \int_{\bar{\mathbf{D}}} g d|\mu| \right| \leq \|\mathsf{L}\| \|g\|_{\mathcal{LN}^1(H_\infty^p)} \preceq \|\mathsf{L}\| |I_\epsilon|^p.$$

Thus $|\mu| \in CM_p$. □

4.3   Quadratic Tent Spaces

We call $f \in \mathcal{T}_p^\infty$ provided f is Lebesgue measurable on $\mathbf{D}$ and satisfies

$$\|f\|_{\mathcal{T}_p^\infty} = \sup_{I \subseteq \mathbf{T}} \left(|I|^{-p} \int_{T(I)} |f(z)|^2 (1 - |z|^2)^p dm(z) \right)^{\frac{1}{2}} < \infty,$$

where the supremum runs over all open subarcs I of $\mathbf{T}$, and

$$T(E) = \{ re^{i\theta} \in \mathbf{D} : \ \text{dist}(e^{i\theta}, \mathbf{T} \setminus E) > 1 - r \}$$

is the tent over the set $E \subset \mathbf{T}$. Moreover, if

$$\|f\|_{\mathcal{T}_p^1} = \inf_\omega \left(\int_{\mathbf{D}} |f(z)|^2 (\omega(z))^{-1} (1 - |z|^2)^{-p} dm(z) \right)^{\frac{1}{2}} < \infty,$$

then we say $f \in \mathcal{T}_p^1$, where the infimum is taken over all nonnegative functions ω on $\mathbf{D}$ with $\|\omega\|_{\mathcal{LN}^1(H_\infty^p)} \leq 1$. Note that the last condition implies that if $z \in \mathbf{D} \setminus \{0\}$ and I_z is the subarc of $\mathbf{T}$ centered at $z/|z|$ with length $1 - |z|$, then

$$|\omega(z)| \leq \inf_{\zeta \in I_z} \mathsf{N}(\omega)(\zeta).$$

Furthermore, the last inequality is multiplied by 1_{I_z} and integrated in ζ with respect to dH_∞^p to produce

$$(1 - |z|)^p |\omega(z)| \precsim \|\omega\|_{\mathcal{LN}^1(H_\infty^p)} \precsim 1.$$

Thus, $\|f\|_{\mathcal{T}_p^1} = 0$ must yield $f = 0$ a.e. on $\mathbf{D}$.

Lemma 4.3.1. *Given $p \in (0, 1]$. If $\sum_{j=1}^\infty \|f_j\|_{\mathcal{T}_p^1} < \infty$, then $f = \sum_{j=1}^\infty f_j \in \mathcal{T}_p^1$ with $\|f\|_{\mathcal{T}_p^1} \leq 2 \sum_{j=1}^\infty \|f_j\|_{\mathcal{T}_p^1}$.*

Proof. Without loss of generality, we may assume $\|f_j\|_{\mathcal{T}_p^1} > 0$ for all $j = 1, 2, \ldots$. For any $\epsilon \in (0, 1)$ we take $\omega_j \geq 0$ such that $\|\omega_j\|_{\mathcal{LN}^1(H_\infty^p)} \leq 1$ and

$$\int_{\mathbf{D}} |f_j(z)|^2 (\omega_j(z))^{-1} (1 - |z|^2)^{-p} dm(z) \leq (1 + \epsilon) \|f_j\|_{\mathcal{T}_p^1}^2.$$

Using the Cauchy–Schwarz inequality we derive

$$|f|^2 \leq \left(\sum_{j=1}^\infty \|f_j\|_{\mathcal{T}_p^1} \omega_j \right) \left(\sum_{j=1}^\infty (\omega_j \|f_j\|_{\mathcal{T}_p^1})^{-1} |f_j|^2 \right).$$

Let

$$\omega = \left(2 \sum_{j=1}^\infty \|f_j\|_{\mathcal{T}_p^1} \right)^{-1} \left(\sum_{j=1}^\infty \omega_j \|f_j\|_{\mathcal{T}_p^1} \right).$$

Then

$$\|\omega\|_{\mathcal{LN}^1(H^p_\infty)} \leq \int_{\mathbf{T}} \mathsf{N}(\omega) dH^{p,d}_\infty$$

$$= \left(2\sum_{j=1}^{\infty} \|f_j\|_{\mathcal{T}^1_p}\right)^{-1} \int_{\mathbf{T}} \left(\sum_{j=1}^{\infty} \|f_j\|_{\mathcal{T}^1_p} \mathsf{N}(\omega_j)\right) dH^{p,d}_\infty$$

$$\leq 1.$$

At the same time, by Fatou's Lemma we have

$$\int_{\mathbf{D}} |f(z)|^2 \big(\omega(z)\big)^{-1} (1-|z|^2)^{-p} dm(z)$$

$$\leq 2\left(\sum_{j=1}^{\infty} \|f_j\|_{\mathcal{T}^1_p}\right) \sum_{j=1}^{\infty} \left(\|f_j\|^{-1}_{\mathcal{T}^1_p} \int_{\mathbf{D}} |f_j(z)|^2 \big(\omega_j(z)\big)^{-1} (1-|z|^2)^{-p} dm(z)\right)$$

$$= 2(1+\epsilon)\left(\sum_{j=1}^{\infty} \|f_j\|_{\mathcal{T}^1_p}\right)^2,$$

thereupon deriving the desired estimate. $\qquad\square$

It is an immediate consequence of Lemma 4.3.1 that $\mathcal{T}^1_p$ is complete under the quasi-norm $\|\cdot\|_{\mathcal{T}^1_p}$. To better understand the structure of $\mathcal{T}^1_p$, we introduce the concept of a $\mathcal{T}^1_p$-atom. Given $p \in (0,1]$. A function a on $\mathbf{D}$ is called a $\mathcal{T}^1_p$-atom provided there exists a subarc I of $\mathbf{T}$ such that a is supported in the tent $T(I)$ and satisfies

$$\int_{T(I)} |a(z)|^2 (1-|z|)^{-p} dm(z) \leq |I|^{-p}.$$

Theorem 4.3.2. *Let $p \in (0,1]$. Then:*

(i) *$f \in \mathcal{T}^1_p$ if and only if there is a sequence of $\mathcal{T}^1_p$-atoms $\{a_j\}$ and an ℓ^1-sequence $\{\lambda_j\}$ such that $f = \sum_{j=1}^{\infty} \lambda_j a_j$. Moreover,*

$$\|f\|_{\mathcal{T}^1_p} \approx \|f\|_{\mathcal{T}^1_p} = \inf\left\{\sum_{j=1}^{\infty} |\lambda_j| : f = \sum_{j=1}^{\infty} \lambda_j a_j\right\},$$

where the infimum is taken over all possible atomic decompositions of $f \in \mathcal{T}^1_p$. Consequently, $\mathcal{T}^1_p$ is a Banach space under the norm $\|\cdot\|_{\mathcal{T}^1_p}$.

(ii) *$\mathcal{T}^\infty_p$ is isomorphic to the dual of $\mathcal{T}^1_p$ under the pairing $\langle f, g \rangle = \int_{\mathbf{D}} f\bar{g} dm$.*

Proof. (i) Given a $\mathcal{T}^1_p$-atom a. By definition there is an arc $I \subset \mathbf{T}$ centered at ζ_I such that the support of a is contained in $T(I)$ and

$$\int_{T(I)} |a(z)|^2 (1-|z|)^{-p} dm(z) \leq |I|^{-p}.$$

Now for $\epsilon > 0$, put

$$\omega(z) = |I|^{-p} \min\left\{1, \left(\frac{|I|}{2|z - \zeta_I|}\right)^{p+\epsilon}\right\}.$$

Note that $\omega(z) = |I|^{-p}$ as $|z - \zeta_I| \leq |I|/2$ and $\omega(z)$ falls off as $z \in \mathbf{D} \setminus S(I)$. Thus, for each $\zeta \in \mathbf{T}$ we derive

$$\mathsf{N}(\omega)(\zeta) \preceq |I|^{-p} \min\left\{1, \left(\frac{|I|}{2|\zeta - \zeta_I|}\right)^{p+\epsilon}\right\}.$$

As a result, we have

$$
\begin{aligned}
\int_{\mathbf{T}} \mathsf{N}(\omega) dH_\infty^p &\leq \int_{|\zeta - \zeta_I| \leq |I|/2} \mathsf{N}(\omega) dH_\infty^p + \int_{|\zeta - \zeta_I| > |I|/2} \mathsf{N}(\omega) dH_\infty^p \\
&\preceq 1 + |I|^\epsilon \int_0^{|I|^{-(p+\epsilon)}} t^{-p/(p+\epsilon)} dt \\
&\preceq 1.
\end{aligned}
$$

Since $\omega \approx |I|^{-p}$ on $T(I)$, we conclude

$$\int_{T(I)} |a(z)|^2 \big(\omega(z)\big)^{-1} (1 - |z|)^{-p} dm(z) \approx |I|^p \int_{T(I)} |a(z)|^2 (1 - |z|)^{-p} dm(z) \preceq 1.$$

Accordingly, $a \in \mathcal{T}_p^1$ with $\|a\|_{\mathcal{T}_p^1} \preceq 1$. This, plus Lemma 4.3.1, further yields that if $\{a_j\}$ is a sequence of $\mathcal{T}_p^1$-atoms and $\sum_{j=1}^\infty |\lambda_j| < \infty$, then $\sum_{j=1}^\infty \lambda_j a_j$ is convergent in $\|\cdot\|_{\mathcal{T}_p^1}$ to a function $f \in \mathcal{T}_p^1$ with $\|f\|_{\mathcal{T}_p^1} \preceq \sum_{j=1}^\infty |\lambda_j|$.

Conversely, suppose $f \in \mathcal{T}_p^1$. If $\|f\|_{\mathcal{T}_p^1} = 0$, then $f = 0$ a.e. on $\mathbf{D}$ and hence there is nothing more to argue. If $\|f\|_{\mathcal{T}_p^1} \neq 0$, then there exists an $\omega \geq 0$ on $\mathbf{D}$ such that

$$\int_{\mathbf{D}} |f(z)|^2 \big(\omega(z)\big)^{-1} (1 - |z|)^{-p} dm(z) \leq 2\|f\|_{\mathcal{T}_p^1}^2.$$

For each integer k let $E_k = \{\zeta \in \mathbf{T} : N\omega(\zeta) > 2^k\}$ and choose a sequence of open arcs $\{I_{j,k}\}$ on $\mathbf{T}$ such that

$$E_k \subseteq \bigcup_j I_{j,k} \quad \text{and} \quad \sum_j |I_{j,k}|^p \leq 2H_\infty^p(E_k).$$

Consequently, there exists a sequence of disjoint open subarcs $\{J_{j,k}\}$ of $\mathbf{T}$ such that $\bigcup_j J_{j,k} = \bigcup_j I_{j,k}$. This implies $J_{j,k} = \bigcup_{I_{l,k} \subseteq J_{j,k}} I_{j,k}$ and so by $p \in (0,1]$,

$$\sum_j |J_{j,k}|^p \leq \sum_j \sum_{I_{l,k} \subseteq J_{j,k}} |I_{l,k}|^p = \sum_j |I_{j,k}|^p.$$

Note that $E_k \subseteq \bigcup_j J_{j,k}$ and $\{J_{j,k}\}$ are disjoint for each k. So $T(E_k) \subseteq \bigcup_j S(J_{j,k})$. Now, putting

$$T_{j,k} = S(J_{j,k}) \setminus \bigcup_{m>k} \bigcup_n S(J_{n,m}),$$

we derive that $\{T_{j,k}\}$ are disjoint for all different choices of j, k but also that for any $l \in \mathbf{N}$ the following inclusion holds:

$$\bigcup_{k=-l}^{l} \bigcup_j T_{j,k} \supseteq T(E_l) \setminus \bigcup_{m>l} \bigcup_n S(J_{n,m}).$$

This inclusion implies

$$\bigcup_{j,k} T_{j,k} \supseteq \bigcup_k T(E_k) \setminus \bigcap_l \bigcup_{m>l} \bigcup_n S(J_{n,m}).$$

From the definition of E_k and $\int_{\mathbf{T}} \mathsf{N}(\omega) dH_\infty^p \le 1$ it follows that

$$\bigcup_k T(E_k) = \{z \in \mathbf{D} : \omega(z) > 0\}$$

and

$$\int_{\bigcup_{m>l} \bigcup_n S(J_{n,m})} dm \preceq \sum_{m>l} \sum_n |J_{n,m}|^p \preceq \sum_{m>l} H_\infty^p(E_m) \to 0 \text{ as } l \to \infty.$$

Noticing also that ω is allowed to be 0 only when $f = 0$, we get $f = \sum_{j,k} f 1_{T_{j,k}}$ a.e. on $\mathbf{D}$, where $1_{T_{j,k}}$ is the characteristic function of $T_{j,k}$. Now, if

$$a_{j,k} = f 1_{T_{j,k}} \left(|J_{j,k}|^p \int_{T_{j,k}} |f(z)|^2 (1 - |z|)^{-p} dm(z) \right)^{-\frac{1}{2}}$$

and

$$\lambda_{j,k} = \left(|J_{j,k}|^p \int_{T_{j,k}} |f(z)|^2 (1 - |z|)^{-p} dm(z) \right)^{\frac{1}{2}},$$

then

$$f = \sum_{j,k} \lambda_{j,k} f a_{j,k}.$$

It is easy to see that $a_{j,k}$ is a T_p^1-atom, but also $\{\lambda_{j,k}\}$ is ℓ^1-summable due to the following estimate established by

$$z \in T_{j,k} \Rightarrow \omega(z) \le 2^{k+1}$$

and the Cauchy–Schwarz inequality:

$$\sum_{j,k} |\lambda_{j,k}| \leq \sum_{j,k} 2^{\frac{k+1}{2}} |J_{j,k}|^{\frac{p}{2}} \left(\int_{T_{j,k}} |f(z)|^2 (\omega(z))^{-1} (1-|z|)^{-p} dm(z) \right)^{\frac{1}{2}}$$

$$\leq \left(\sum_{j,k} 2^{k+1} |J_{j,k}|^p \right)^{\frac{1}{2}} \left(\sum_{j,k} \int_{T_{j,k}} |f(z)|^2 (\omega(z))^{-1} (1-|z|)^{-p} dm(z) \right)^{\frac{1}{2}}$$

$$\preceq \|f\|_{T_p^1} \left(\sum_k 2^k \sum_j |J_{j,k}|^p \right)^{\frac{1}{2}}$$

$$\preceq \|f\|_{T_p^1} \left(\sum_k 2^k H_\infty^p(E_k) \right)^{\frac{1}{2}}$$

$$\preceq \|f\|_{T_p^1} \left(\int_{\mathbf{T}} \mathsf{N}(\omega) dH_\infty^p \right)^{\frac{1}{2}}.$$

By Lemma 4.3.1 it follows that $\sum_{j,k} \lambda_{j,k} f a_{j,k}$ converges to f in $\|\cdot\|_{T_p^1}$. Since $\|\cdot\|_{T_p^1}$ is a norm, Lemma 4.3.1 is applied to yield that T_p^1 is a Banach space under this norm.

(ii) Let $f \in T_p^1$, $g \in T_p^\infty$ and $d\mu_{g,p}(z) = |g(z)|^2 (1-|z|^2)^p dm(z)$. Then $\mu_{g,p} \in \mathcal{CM}_p$ and hence it follows from Theorem 4.1.4 and its proof that

$$\int_{\mathbf{D}} |g(z)|^2 \omega(z)(1-|z|^2)^p dm(z) \leq \|g\|_{T_p^\infty}^2 \|\omega\|_{\mathcal{LN}^1(H_\infty^p)} \leq \|g\|_{T_p^\infty}^2$$

holds for all ω in the definition of $f \in T_p^1$. An application of the Cauchy–Schwarz inequality gives

$$\int_{\mathbf{D}} |f(z)g(z)| dm(z) \leq \left(\int_{\mathbf{D}} |f(z)|^2 (\omega(z))^{-1} (1-|z|^2)^{-p} dm(z) \right)^{\frac{1}{2}} \|g\|_{T_p^\infty}.$$

Thus every $g \in T_p^\infty$ induces a bounded linear functional on T_p^1 via the above pairing $\langle f, g \rangle = \int_{\mathbf{D}} f\bar{g} dm$.

Conversely, let L be a continuous linear functional on T_p^1 and fix an open arc $I \subseteq \mathbf{T}$ centered at ζ_I. If f is supported in $T(I)$ with $f \in \mathcal{L}^2(T(I))$, then

$$\int_{T(I)} |f(z)|^2 (1-|z|^2)^{-p} dm(z) \preceq |I|^{-p} \int_{T(I)} |f|^2 dm.$$

Now for $\epsilon > 0$, set

$$\omega(z) = |I|^{-p} \min\left\{ 1, \left(\frac{|I|}{2|z - \zeta_I|} \right)^{p+\epsilon} \right\}.$$

As proved in the first part of (i) we once again obtain

$$\mathsf{N}(\omega)(\zeta) \preceq |I|^{-p} \min\left\{ 1, \left(\frac{|I|}{2|\zeta - \zeta_I|} \right)^{p+\epsilon} \right\}, \quad \zeta \in \mathbf{T}.$$

This yields

$$\int_{\mathbf{T}} \mathsf{N}(\omega) dH_\infty^p \ \preceq \ \int_{|\varsigma-\varsigma_I|\leq|I|/2} \mathsf{N}(\omega) dH_\infty^p + \int_{|\varsigma-\varsigma_I|>|I|/2} \mathsf{N}(\omega) dH_\infty^p$$

$$\preceq \ 1 + |I|^\epsilon \int_0^{|I|^{-(p+\epsilon)}} t^{-p/(p+\epsilon)} dt$$

$$\preceq \ 1.$$

From the fact that the above-selected ω satisfies $\omega^{-1} \approx |I|^p$ on $T(I)$ it follows that

$$\|f\|_{\mathcal{T}_p^1}^2 \leq \int_{T(I)} |f(z)|^2(\omega(z))^{-1}(1-|z|^2)^{-p} dm(z) \preceq \int_{T(I)} |f|^2 dm,$$

and so that $f \in \mathcal{T}_p^1$. Hence L induces a continuous linear functional on $\mathcal{L}^2(T(I))$, and acts with some function $g_I \in \mathcal{L}^2(T(I))$ via the $\mathcal{L}^2$-inner product. Taking an increasing family of closed subarcs $\{I_j\}$ of $\mathbf{T}$ such that $\mathbf{T} = \bigcup_j I_j$, we thus get a function g such that

$$\mathsf{L}(f) = \int_{\mathbf{D}} f\bar{g} dm \qquad \text{for all} \quad f \in \mathcal{T}_p^1 \quad \text{supported in} \quad T(I).$$

To establish $g \in \mathcal{T}_p^\infty$, let $f(z) = (1-|z|^2)^p \overline{g(z)} 1_{T(I)}(z)$. Then

$$\mathsf{L}(f) = \int_{T(I)} |g(z)|^2 (1-|z|^2)^p dm(z)$$

$$\preceq \ \|\mathsf{L}\| \left(|I|^p \int_{T(I)} |g(z)|^2 (1-|z|^2)^p dm(z) \right)^{\frac{1}{2}},$$

and hence

$$\left(|I|^{-p} \int_{T(I)} |g(z)|^2 (1-|z|^2)^p dm(z) \right)^{\frac{1}{2}} \preceq \|\mathsf{L}\|.$$

In other words, $g \in \mathcal{T}_p^\infty$ with $\|g\|_{\mathcal{T}_p^\infty} \preceq \|\mathsf{L}\|$. $\qquad\qquad\square$

4.4 Preduals under Invariant Pairing

Recall that a holomorphic function f on $\mathbf{D}$ is of the Sobolev class $\mathcal{H}'$ (denoted $\mathcal{W}_1^1$ sometimes, see [ArFi]) if f' belongs to the Hardy space $\mathcal{H}^1$, that is,

$$\|f\|_{\mathcal{H}'} = \|f'\|_{\mathcal{H}^1} = \sup_{r\in(0,1)} \int_{\mathbf{T}} |f'(r\varsigma)||d\varsigma| < \infty.$$

A famous duality theorem of Fefferman-type is that $\mathcal{BMOA}$ is isomorphic to the dual of $\mathcal{H}'$ with respect to the invariant pairing

$$\langle f,g\rangle_{inv} = \int_{\mathbf{D}} f'\overline{g'}dm = \lim_{r\to 1}\int_0^r\int_0^{2\pi} f'(re^{i\theta})\overline{g'(re^{i\theta})}rd\theta dr.$$

Meanwhile, the dual of $\mathcal{D}$ under $\langle\cdot,\cdot\rangle_{inv}$ is isomorphic to $\mathcal{D}$. Here and henceforth, for simplicity we prefer to work modulo constants, and by invariance we mean

$$\langle f\circ\phi, g\circ\phi\rangle_{inv} = \langle f,g\rangle_{inv}, \quad \phi\in Aut(\mathbf{D}).$$

Thanks to

$$\mathcal{H}'\subset\mathcal{D}\subset\mathcal{Q}_p\subset\mathcal{BMOA}, \quad p\in(0,1),$$

the predual of $\mathcal{Q}_p$ under $\langle\cdot,\cdot\rangle_{inv}$ should contain $\mathcal{H}'$ but be contained in $\mathcal{D}$. This observation leads to the discovery of a new kind of holomorphic space.

Given $p\in(0,1]$. Let $\mathcal{P}_p$ be the class of all holomorphic functions f on $\mathbf{D}$ with

$$\|f\|_{\mathcal{P}_p} = \inf_\omega\left(\int_{\mathbf{D}}|(zf(z))''|^2(\omega(z))^{-1}(1-|z|^2)^{2-p}dm(z)\right)^{\frac{1}{2}} < \infty,$$

where the infimum ranges over all nonnegative functions ω on $\mathbf{D}$ with

$$\|\omega\|_{\mathcal{LN}^1(H_\infty^p)}\leq 1.$$

Using the implication

$$\|\omega\|_{\mathcal{LN}^1(H_\infty^p)}\leq 1 \Rightarrow \omega(z)\preceq(1-|z|^2)^{-p},$$

we derive that if $f\in\mathcal{P}_p$, then

$$\int_{\mathbf{D}}\left(|(zf(z))''|(1-|z|^2)\right)^2 dm(z)\preceq\|f\|^2_{\mathcal{P}_p}.$$

This, along with Lemma 4.3.1, implies readily that $\mathcal{P}_p$ is a quasi-Banach space.

Theorem 4.4.1. *Let $p\in(0,1]$. Then $\mathcal{Q}_p$ is isomorphic to the dual of $\mathcal{P}_p$ under $\langle\cdot,\cdot\rangle_{inv}$.*

Proof. On the one hand, let $f\in\mathcal{P}_p$ and $g\in\mathcal{Q}_p$. Note that $\langle\cdot,\cdot\rangle_{inv}$ has another expression

$$\langle f,g\rangle_{inv} = \int_{\mathbf{D}}(zf(z))''\overline{g'(z)}(1-|z|^2)dm(z).$$

So, using the Cauchy–Schwarz inequality, we get

$$|\langle f,g\rangle_{inv}|$$

$$\preceq \left(\int_{\mathbf{D}}|(zf(z))''|^2(\omega(z))^{-1}(1-|z|^2)^{2-p}dm(z)\right)^{\frac{1}{2}}\|\omega\|^{\frac{1}{2}}_{\mathcal{LN}^1(H_\infty^p)}\|g\|_{\mathcal{Q}_p,1}$$

$$\preceq \|g\|_{\mathcal{Q}_p,1}\left(\int_{\mathbf{D}}|(zf(z))''|^2(\omega(z))^{-1}(1-|z|^2)^{2-p}dm(z)\right)^{\frac{1}{2}}$$

for any ω required in $\mathcal{P}_p$. Thus, each member of $\mathcal{Q}_p$ induces a bounded linear functional on $\mathcal{P}_p$.

Conversely, suppose L is a bounded linear functional on $\mathcal{P}_p$. Let

$$D(f)(z) = (zf(z))''(1 - |z|^2), \quad z \in \mathbf{D}.$$

Then D is an isometric map from $\mathcal{P}_p$ into T_p^1. Since T_p^1 is a Banach space under $\|\cdot\|_{T_p^1}$, it follows from the Hahn–Banach Extension Theorem that we can select a function $h \in T_p^\infty$ such that

$$\mathsf{L}(f) = \int_{\mathbf{D}} (zf(z))''(1 - |z|^2)\overline{h(z)}dm(z), \quad f \in \mathcal{P}_p.$$

Note again that

$$\int_{\mathbf{D}} |f'(z)|^2 dm(z) \;\preceq\; \int_{\mathbf{D}} |(zf(z))''|^2(1 - |z|^2)^2 dm(z)$$

$$\preceq\; \int_{\mathbf{D}} |(zf(z))''|^2(\omega(z))^{-1}(1 - |z|^2)^{2-p}dm(z)$$

for any ω used in $\mathcal{P}_p$ and consequently $\mathcal{P}_p$ is a subspace of $\mathcal{D}$. Thus we get by an application of the reproducing formula for $\mathcal{D}$,

$$(zf(z))'' = 2\pi^{-1}\int_{\mathbf{D}} \frac{(wf(w))''(1 - |w|^2)}{(1 - \bar{w}z)^3}dm(w), \quad f \in \mathcal{P}_p.$$

Accordingly,

$$
\begin{aligned}
\mathsf{L}(f) &= 2\pi^{-1}\int_{\mathbf{D}} (wf(w))'' \left(\int_{\mathbf{D}} \frac{\overline{h(z)}(1 - |z|^2)}{(1 - \bar{w}z)^3}dm(z) \right)(1 - |w|^2)dm(w) \\
&= \langle f, g\rangle_{inv},
\end{aligned}
$$

where

$$g(w) = 2\pi^{-1}\int_0^w \left(\int_{\mathbf{D}} \frac{h(z)(1 - |z|^2)}{(1 - u\bar{z})^3}dm(z) \right) du.$$

The argument will be done if we can prove $g \in \mathcal{Q}_p$. To do so, it suffices to show that $(1 - |w|)^p|g'(w)|^2 dm(w)$ is a p-Carleson measure. But, this follows from an application of Lemma 3.1.2 to the inequality

$$|g'(w)| \leq \int_{\mathbf{D}} \frac{|h(z)|(1 - |z|^2)}{|1 - w\bar{z}|^3}dm(z),$$

and the fact that $|h(z)|^2(1 - |z|^2)^p dm(z)$ is a p-Carleson measure. $\square$

To see that for $p \in (0,1)$, $\mathcal{P}_p$ lies properly between $\mathcal{H}'$ and $\mathcal{D}$, we need an estimate for the radial maximal operator of the generalized Poisson integral of an $\mathcal{H}^1$-function on $\mathbf{T}$ with respect to the Hausdorff capacity.

Lemma 4.4.2. *Given $p \in (0,1]$ let $f \in \mathcal{H}^1(\mathbf{T})$. If*

$$I_p f(z) = \int_{\mathbf{T}} \frac{f(\eta)(1 - |z|^2)^{2-p}}{|z - \eta|^2} |d\eta|, \quad z \in \bar{\mathbf{D}},$$

then

$$\int_{\mathbf{T}} \mathsf{N}(I_p f) dH_\infty^p \preceq \int_{\mathbf{T}} \sup_{r \in (0,1)} |I_p f(r\zeta)| dH_\infty^p(\zeta) \preceq \|f\|_{\mathcal{H}^1}.$$

Proof. We begin with proving the first estimate. Note that $I_p f(z)(1 - |z|^2)^{p-1}$ is harmonic on $\mathbf{D}$. So if $z = re^{i\theta}$ and

$$D_z = \{se^{it} \in \mathbf{D} : |s - r| < 1 - r \ \& \ |t - \theta| < 1 - r\},$$

then we have the following sub-mean-value inequality for $q \in (0,1)$:

$$|I_p f(z)|^q \preceq (1 - |z|)^{-2} \int_{|w-z|<1-r} |I_p f(w)|^q dm(w).$$

Consequently, if $z \in \Lambda(\lambda)$ where $|z| \geq 1/2$ and $\lambda = e^{iu} \in \mathbf{T}$, and $\zeta \in \mathbf{T}$, then

$$
\begin{aligned}
|I_p f(z)|^q &\preceq (1 - |z|)^{-2} \int_{D_z} |I_p f(se^{it})|^q dm(se^{it}) \\
&\preceq (1 - |z|)^{-2} \int_{\theta-(1-|z|)}^{\theta+(1-|z|)} \int_{2|z|-1}^1 |I_p f(se^{it})|^q s\, ds\, dt \\
&\preceq (1 - |z|)^{-1} \int_{\theta-(1-|z|)}^{\theta+(1-|z|)} \sup_{s \in (0,1)} |I_p f(se^{it})|^q dt.
\end{aligned}
$$

Now that $z \in \Lambda(\lambda)$ implies $|\theta - u| \leq c(1 - |z|)$ for some constant $c > 1$, we derive

$$\big(\theta - (1 - |z|), \theta + (1 - |z|)\big) \subseteq \big(u - (1+c)(1 - |z|), u + (1+c)(1 - |z|)\big)$$

and then

$$\mathsf{N}(I_p(f))(\zeta) \preceq \left(\mathsf{M}\Big(\sup_{s \in (0,1)} |I_p f(s\cdot)|^q \Big)(\zeta) \right)^{\frac{1}{q}}, \quad \zeta \in \mathbf{T}.$$

Again, M is the Hardy–Littlewood maximal operator. Applying Theorem 4.1.3 (i) and $q^{-1} > 1$ to the last inequality, we get

$$\int_{\mathbf{T}} \mathsf{N}(I_p(f))(\zeta) dH_\infty^p(\zeta) \preceq \int_{\mathbf{T}} \sup_{s \in (0,1)} |I_p f(s\zeta)| dH_\infty^p(\zeta).$$

For the second estimate, we should recognize that f can be written as $\sum_j \lambda_j a_j$ on $\mathbf{T}$ where $\sum_j |\lambda_j| \preceq \|f\|_{\mathcal{H}^1}$ and a_j is an $\mathcal{H}^1$-atom on $\mathbf{T}$. Recall that an $\mathcal{H}^1$-atom is a function $a : \mathbf{T} \to \mathbf{C}$ which is supported on an open subarc I of $\mathbf{T}$, and satisfies

$$\|a\|_\infty = \|a\|_{\mathcal{L}^\infty(\mathbf{T})} < |I|^{-1} \quad \text{and} \quad \int_{\mathbf{T}} a(\zeta) |d\zeta| = 0.$$

So, we have by Lemma 4.1.1 (iv),

$$\int_{\mathbf{T}} \sup_{s\in(0,1)} \|l_p f(s\zeta)| dH_\infty^p(\zeta) \preceq \sum_j |\lambda_j| \int_{\mathbf{T}} \sup_{s\in(0,1)} \|l_p a_j(s\zeta)| dH_\infty^p(\zeta)$$

$$\preceq \sum_j |\lambda_j|$$

$$\preceq \|f\|_{\mathcal{H}^1},$$

provided that the following is verified:

$$\int_{\mathbf{T}} \sup_{s\in(0,1)} \|l_p a(s\zeta)| dH_\infty^p(\zeta) \preceq 1 \quad \text{for any } \mathcal{H}^1\text{-atom } a \text{ on } \mathbf{T}.$$

In so doing, we may assume that the $\mathcal{H}^1$-atom a has the support I centered at ζ_I. Using $\|a\|_\infty < |I|^{-1}$, we are now able to derive that for $s \in (0,1)$,

$$\|l_p a(s\zeta)| \leq \int_I \frac{|a(\eta)|(1-s^2)^{2-p}}{|\zeta - s\eta|^2} |d\eta| \preceq |I|^{-1} \int_I \frac{|d\eta|}{|\zeta - \eta|^p} \preceq |I|^{-p}, \quad \zeta \in \mathbf{T},$$

which implies

$$\sup_{s\in(0,1)} \|l_p a(s\zeta)| \preceq |I|^{-p}, \quad \zeta \in \mathbf{T}.$$

Meanwhile, using $\int_{\mathbf{T}} a(\eta)|d\eta| = 0$, $\|a\|_\infty < |I|^{-1}$ and the mean value theorem for derivatives, we further obtain that for $s \in (0,1)$,

$$\|l_p a(s\zeta)| = (1-s^2)^{2-p} \left| \int_{\mathbf{T}} \left(\frac{a(\eta)}{|\zeta - s\eta|^2} - \frac{a(\eta)}{|\zeta - s\zeta_I|^2} \right) |d\eta| \right|$$

$$\leq \|a\|_\infty \int_I \left| \frac{(1-s^2)^{2-p}}{|\zeta - s\eta|^2} - \frac{(1-s^2)^{2-p}}{|\zeta - s\zeta_I|^2} \right| |d\eta|$$

$$\preceq |I|^{2-p} |\zeta - \zeta_I|^{-2} \quad \text{as} \quad |\zeta - \zeta_I| > 2|I|.$$

Accordingly,

$$\sup_{s\in(0,1)} \|l_p a(s\zeta)| \preceq |I|^{2-p} |\zeta - \zeta_I|^{-2} \quad \text{as} \quad |\zeta - \zeta_I| > 2|I|.$$

Therefore, if $\mu \in \mathcal{L}_+^{1,p}$, then

$$\int_{\mathbf{T}} \sup_{s\in(0,1)} \|l_p a(s\zeta)| d\mu(\zeta)$$

$$= \left(\int_{|\zeta-\zeta_I|\leq 2|I|} + \int_{|\zeta-\zeta_I|>2|I|} \right) \sup_{s\in(0,1)} \|l_p a(s\zeta)| d\mu(\zeta)$$

$$\preceq 1 + |I|^{2-p} \int_0^{(2|I|)^{-2}} \mu(\{\zeta \in \mathbf{T} : |\zeta - \zeta_I|^{-2} > t\}) dt$$

$$\preceq 1.$$

Therefore, the desired estimate follows from Corollary 4.2.2. $\qquad\square$

Theorem 4.4.3. *Let $p \in (0,1]$ and μ be a nonnegative Radon measure on $\mathbf{D}$. Then:*

(i) *There is a constant $C > 0$ such that $\int_{\mathbf{D}} |\mathsf{I}_{\mathsf{p}}(f)| d\mu \leq C \|f\|_{\mathcal{H}^1}$ holds for any $f \in \mathcal{H}^1$ if and only if $\mu \in \mathcal{CM}_p$.*

(ii) *$\mathcal{H}' \subseteq \mathcal{P}_p \subseteq \mathcal{D}$ holds with*

$$\|f\|_{\mathcal{D}} \preceq \|f\|_{\mathcal{P}_p} \preceq \|f\|_{\mathcal{H}'}, \quad f \in \mathcal{H}'.$$

In particular, $\mathcal{H}' = \mathcal{P}_1$ and $\mathcal{P}_p \neq \mathcal{D}$.

Proof. (i) If $\mu \in \mathcal{CM}_p$, then Lemma 4.4.2 and the argument for the implication (i)$\Rightarrow$(ii) in Theorem 4.1.4 leads to

$$\int_{\mathbf{D}} |\mathsf{I}_{\mathsf{p}}(f)| d\mu \preceq \|\mu\|_{\mathcal{CM}_p} \int_{\mathbf{T}} \mathsf{N}(\mathsf{I}_p(f)) dH^p_{\infty} \preceq \|\mu\|_{\mathcal{CM}_p} \|f\|_{\mathcal{H}^1} \quad f \in \mathcal{H}^1.$$

Conversely, suppose that μ satisfies the above-assigned inequality. Given a subarc $I \subset \mathbf{T}$ with center ζ_I, let $g(z) = (1 - (1 - |I|)\overline{\zeta_I} z)^{-2}$. Then $\|g\|_{\mathcal{H}^1} \approx |I|^{-1}$. Using the formula

$$\mathsf{I}_p(g)(z) = 2\pi (1 - |z|^2)^{1-p} (1 - (1 - |I|)\overline{\zeta_I} z)^{-2},$$

we derive

$$\mu(S(I)) |I|^{-1-p} \preceq \int_{\mathbf{D}} |\mathsf{I}_p(g)| d\mu \leq C \|g\|_{\mathcal{H}^1} \preceq C |I|^{-1},$$

thus implying $\mu \in \mathcal{CM}_p$.

(ii) Because $\mathcal{P}_p \subseteq \mathcal{D}$ has been established above, we here show $\mathcal{H}' \subseteq \mathcal{P}_p$. Let $f \in \mathcal{H}'$ with $\|f\|_{\mathcal{H}'} > 0$ and

$$\omega(z) = |(zf(z))'|(1 - |z|^2)^{1-p} \|f\|_{\mathcal{H}'}^{-1}.$$

Then $(zf(z))' \in \mathcal{H}^1$ and hence it is applied to produce

$$\int_{\mathbf{D}} |(zf(z))''|^2 (\omega(z))^{-1} (1 - |z|^2)^{2-p} dm(z)$$

$$\leq \|f\|_{\mathcal{H}'} \int_{\mathbf{D}} \frac{|(zf(z))''|^2}{|(zf(z))'|} (1 - |z|^2) dm(z)$$

$$\preceq \|f\|_{\mathcal{H}'} \|zf(z)\|_{\mathcal{H}'}$$

$$\preceq \|f\|_{\mathcal{H}'}^2.$$

In the meantime, since $f \in \mathcal{H}'$, $g(z) = (zf(z))'$ can be represented by the Poisson integral of its boundary value function $g(\zeta) = f(\zeta) + \zeta f'(\zeta)$, $\zeta \in \mathbf{T}$:

$$(zf(z))' = \frac{1}{2\pi} \int_{\mathbf{T}} g(\zeta) \Big(\frac{1 - |z|^2}{|\zeta - z|^2}\Big) |d\zeta|.$$

This formula implies $|l_p g(z)| = 2\pi(1-|z|^2)^{1-p}|(zf(z))'|$. So, by Lemma 4.4.2 we have

$$
\begin{aligned}
\|\omega\|_{\mathcal{LN}^1(H_\infty^p)} &\preceq \|f\|_{\mathcal{H}'}^{-1} \int_{\mathbf{T}} \mathsf{N}(l_p g) dH_\infty^p \\
&\preceq \|f\|_{\mathcal{H}'}^{-1} \int_{\mathbf{T}} \sup_{s\in(0,1)} |l_p g(s\zeta)| dH_\infty^p(\zeta) \\
&\preceq \|f\|_{\mathcal{H}'}^{-1} \|g\|_{\mathcal{H}^1} \\
&\preceq 1,
\end{aligned}
$$

thereupon producing $f \in \mathcal{P}_p$.

Of course, $\mathcal{P}_p \neq \mathcal{D}$, for if it were not, then it would follow that $\mathcal{Q}_p = \mathcal{D}$, a contradiction, since the duals of $\mathcal{P}_p$ and $\mathcal{D}$ are $\mathcal{Q}_p$ and $\mathcal{D}$ respectively under the invariant paring $\langle \cdot, \cdot \rangle_{inv}$.

In order to see $\mathcal{P}_p \neq \mathcal{H}'$ as $p \in (0,1)$, it suffices to prove $\mathcal{H}' = \mathcal{P}_1$ due to $\mathcal{BMOA} \neq \mathcal{Q}_p$ and Theorem 4.4.1. To this end, we handle two situations. On the one hand, if $f \in \mathcal{H}'$, then by taking

$$
\omega(z) = |(zf(z))'| \|f\|_{\mathcal{H}'}^{-1}
$$

(naturally, $\|f\|_{\mathcal{H}'} > 0$ is assumed, otherwise, there is nothing to do) we apply the proof above to establish

$$
\int_{\mathbf{D}} |(zf(z))''|^2 (\omega(z))^{-1}(1-|z|^2) dm(z) \preceq \|f\|_{\mathcal{H}'}^2.
$$

Note that

$$
\|\omega\|_{\mathcal{LN}^1(H_\infty^1)} \approx \|\mathsf{N}(\omega)\|_{\mathcal{L}^1(\mathbf{T})} \preceq \|(zf(z))'\|_{\mathcal{H}^1} \|f\|_{\mathcal{H}'}^{-1} \preceq 1.
$$

So, $f \in \mathcal{P}_1$. Namely, $\mathcal{H}' \subseteq \mathcal{P}_1$.

On the other hand, if $f \in \mathcal{P}_1$, then for any nonnegative ω with

$$
\|\mathsf{N}(\omega)\|_{\mathcal{L}^1(\mathbf{T})} \approx \|\omega\|_{\mathcal{LN}^1(H_\infty^1)} \preceq 1,
$$

we can take an account of the well-known square function characterization of $\mathcal{H}^1$, the Cauchy–Schwarz inequality and Fubini's Theorem to derive

$$
\begin{aligned}
\|zf(z)\|_{\mathcal{H}'} &= \|(zf(z))'\|_{\mathcal{H}^1} \\
&\preceq \int_{\mathbf{T}} \left(\int_{\Lambda(\zeta)} |(zf(z))''|^2 dm(z) \right)^{\frac{1}{2}} |d\zeta| \\
&\preceq \int_{\mathbf{T}} \left(\int_{\Lambda(\zeta)} |(zf(z))''|^2 (\omega(z))^{-1} dm(z) \right)^{\frac{1}{2}} (\mathsf{N}(\omega)(\zeta))^{\frac{1}{2}} |d\zeta| \\
&\preceq \left(\int_{\mathbf{D}} |(zf(z))''|^2 (\omega(z))^{-1}(1-|z|^2) dm(z) \right)^{\frac{1}{2}} \|\mathsf{N}(\omega)\|_{\mathcal{L}^1(\mathbf{T})}^{\frac{1}{2}}.
\end{aligned}
$$

Hence we get $\|zf(z)\|_{\mathcal{H}'} \preceq \|f\|_{\mathcal{P}_1} < \infty$. Because of $\mathcal{P}_1 \subseteq \mathcal{D}$, a further calculation shows $f \in \mathcal{H}'$ and then $\mathcal{P}_1 \subseteq \mathcal{H}'$. Therefore $\mathcal{P}_1 = \mathcal{H}'$. $\qquad\square$

4.5 Invariant Duals of Vanishing Classes

Note that the inclusion $\mathcal{D} \subset \mathcal{Q}_p,\, p > 0$ can be improved by imbedding the vanishing Q class — $\mathcal{Q}_{p,0}$ — all functions $f \in \mathcal{Q}_p$ with $\lim_{|w|\to 1} E_p(f,w) = 0$, between $\mathcal{D}$ and $\mathcal{Q}_p$. So it seems reasonable and possible to use $\mathcal{Q}_{p,0}$ to give another look at the predual structure of $\mathcal{Q}_p$, but also to explore more properties on the dual of $\mathcal{Q}_{p.0}$.

To begin with, we consider the dyadic partition of $\mathbf{D}$ by letting $\mathcal{E} = \{E_n\}_{n\in\mathbf{N}}$ be the family of the sets

$$\left\{ z \in \mathbf{D}:\; 2^{-(k+1)} < 1 - |z| \le 2^{-k} \;\&\; 2^{-(k+1)}\pi j \le \arg z < 2^{-(k+1)}\pi(j+1) \right\},$$

where $k \in \mathbf{N} \cup \{0\}$ and $0 \le j < 2^{k+2}$. And, for each $n \in \mathbf{N}$ we choose a_n to be the center of E_n.

Lemma 4.5.1. *Given $p \in (0,1)$. For $n \in \mathbf{N}$ let $\mathcal{X}_n$ respectively $\mathcal{Y}_n$ consist of those functions g, respectively f, holomorphic on $\mathbf{D}$ for which*

$$\|g\|_{\mathcal{X}_n} = \left(\int_{\mathbf{D}} |g'(z)|^2 (1 - |\sigma_{a_n}(z)|^2)^p dm(z) \right)^{\frac{1}{2}} < \infty,$$

respectively,

$$\|f\|_{\mathcal{Y}_n} = \left(\int_{\mathbf{D}} |(zf(z))''|^2 (1 - |\sigma_{a_n}(z)|^2)^{-p} (1 - |z|^2)^2 dm(z) \right)^{\frac{1}{2}} < \infty.$$

Then the dual of $\mathcal{X}_n$ is isomorphic to $\mathcal{Y}_n$ under $\langle \cdot, \cdot \rangle_{inv}$. Moreover,

$$\|f\|_{\mathcal{Y}_n} \approx \sup\{|\langle f, g \rangle_{inv}| : \|g\|_{\mathcal{X}_n} \le 1\}, \quad f \in \mathcal{Y}_n,$$

where the constants involved in the foregoing and following estimates are independent of f and a_n.

Proof. For $g \in \mathcal{X}_n$ and $f \in \mathcal{Y}_n$ we use the Cauchy–Schwarz inequality to produce

$$|\langle g, f \rangle_{inv}| = \left| \int_{\mathbf{D}} g'(z)\overline{(zf(z))''}(1 - |z|^2)dm(z) \right| \le \|g\|_{\mathcal{X}_n}\|f\|_{\mathcal{Y}_n};$$

that is, f induces a continuous linear functional on $\mathcal{X}_n$.

On the other hand, suppose that L is a continuous linear functional on $\mathcal{X}_n$. Then L can be extended, with preserving the norm, to a continuous linear functional on $\mathcal{L}^2\big(\mathbf{D}, (1 - |\sigma_{a_n}(z)|^2)^p dm(z)\big)$ and consequently, there exists a function $h \in \mathcal{L}^2\big(\mathbf{D}, (1 - |\sigma_{a_n}(z)|^2)^{-p}(1 - |z|^2)^2 dm(z)\big)$ such that

$$\mathsf{L}(g) = \langle g, h \rangle = \pi^{-1} \int_{\mathbf{D}} g'(z)\overline{h(z)}(1 - |z|^2)dm(z), \quad g \in \mathcal{X}_n,$$

and

$$\|L\| = \left(\int_{\mathbf{D}} |h(z)|^2 (1 - |\sigma_{a_n}(z)|^2)^{-p}(1 - |z|^2)^2 dm(z) \right)^{\frac{1}{2}}.$$

Note that if $g \in \mathcal{X}_n$, then $g'(z)$ has the reproducing formula

$$g'(z) = 2\pi^{-1} \int_{\mathbf{D}} \frac{g'(w)(1 - |w|^2)}{(1 - \bar{w}z)^3} dm(w), \quad z \in \mathbf{D}.$$

Accordingly, if $g \in \mathcal{X}_n$, then by Fubini's Theorem,

$$L(g) = \pi^{-1} \int_{\mathbf{D}} g'(w) \overline{(wf(w))''} (1 - |w|^2) dm(w),$$

where f is an element in $\mathcal{H}$ determined by

$$(wf(w))'' = 2\pi^{-1} \int_{\mathbf{D}} h(z)(1 - \bar{z}w)^{-3}(1 - |z|^2) dm(z), \quad w \in \mathbf{D}.$$

Observe that the function f is holomorphic in $\mathbf{D}$ and that

$$\int_{\mathbf{D}} |h(z)|^2 (1 - |\sigma_{a_n}(z)|^2)^{-p}(1 - |z|^2)^2 dm(z) < \infty.$$

So it remains to prove

$$\int_{\mathbf{D}} |(wf(w))''|^2 \frac{|1 - \bar{a}w|^{2p}}{(1 - |w|^2)^{p-2}} dm(w) \preceq \int_{\mathbf{D}} |h(z)|^2 \frac{|1 - \bar{a}z|^{2p}}{(1 - |z|^2)^{p-2}} dm(z), \quad a \in \mathbf{D},$$

where the constant depends only on p. To do so, we introduce the functions F and H by

$$h(w) = H(w)(1 - |w|^2)^{-\beta} \quad \text{and} \quad (zf(z))'' = F(z)(1 - |z|^2)^{-\beta},$$

where $\beta \in (1 + p, 3 - p)$. Then the required demonstration is equivalent to showing

$$\int_{\mathbf{D}} \frac{|F(w)|^2 |1 - \bar{a}w|^{2p}}{(1 - |w|^2)^{p+2\beta-2}} dm(w) \preceq \int_{\mathbf{D}} \frac{|H(z)|^2 |1 - \bar{a}z|^{2p}}{(1 - |z|^2)^{p+2\beta-2}} dm(z), \quad a \in \mathbf{D},$$

where

$$F(w) = 2\pi^{-1}(1 - |w|^2)^\beta \int_{\mathbf{D}} H(z)(1 - |z|^2)^{1-\beta}(1 - \bar{z}w)^{-3} dm(z).$$

For $a \in \mathbf{D}$, put

$$d\mu_a(z) = (1 - |z|^2)^{2-p-2\beta} |1 - \bar{a}z|^{2p} dm(z)$$

and consider the linear operator $\mathsf{T} : H \to F$. If we can prove the operator norm estimates:

$$\|\mathsf{T}\|_{\mathcal{L}^q(\mathbf{D}, d\mu_a) \to \mathcal{L}^q(\mathbf{D}, d\mu_a)} \le C_q \quad \text{for} \quad q = 1, \infty,$$

where C_1 and C_∞ are constants independent of $a \in \mathbf{D}$, then the required estimate follows from the Riesz–Thorin Theorem.

The case $q = \infty$ follows from

$$|F(w)| \le 2\pi^{-1}(1-|w|^2)^\beta \|H\|_{\mathcal{L}^\infty(\mathbf{D},d\mu_a)} \int_{\mathbf{D}}(1-|z|^2)^{1-\beta}|1-\bar{z}w|^{-3}dm(z)$$

$$\preceq \|H\|_{\mathcal{L}^\infty(\mathbf{D},d\mu_a)}.$$

However, in the case of $q = 1$, we note that the formula of F implies

$$\int_{\mathbf{D}}|F|\,d\mu_a$$
$$\le 2\pi^{-1}\int_{\mathbf{D}}(1-|z|^2)^{1-\beta}|H(z)|\left(\int_{\mathbf{D}}\frac{|1-\bar{a}w|^{2p}}{(1-|w|^2)^{p+\beta-2}|1-\bar{z}w|^3}dm(w)\right)dm(z).$$

Hence it is enough to establish

$$K(a) = \int_{\mathbf{D}}\frac{|1-\bar{a}w|^{2p}}{(1-|w|^2)^{p+\beta-2}|1-\bar{z}w|^3}dm(w) \le C(1-|z|^2)^{1-p-\beta}|1-\bar{a}z|^{2p},$$

where C is independent of a and z. To that end, we make a change of variable $w = \sigma_z(\lambda)$ for a given $z \in \mathbf{D}$, obtaining

$$1-|\sigma_z(\lambda)|^2 = \frac{(1-|\lambda|^2)(1-|z|^2)}{|1-\bar{z}\lambda|^2}, \quad 1-\bar{a}\sigma_z(\lambda) = \frac{1-\bar{a}z+\lambda(\bar{a}-\bar{z})}{1-\bar{z}\lambda},$$

and

$$\begin{aligned}
K(a) &= (1-|z|^2)^{1-p-\beta}\int_{\mathbf{D}}\frac{|1-\bar{a}z+\lambda(\bar{a}-\bar{z})|^{2p}}{|1-\bar{z}\lambda|^{5-2\beta}(1-|\lambda|^2)^{\beta+p-2}}dm(\lambda) \\
&\le \frac{2^{2p}|1-\bar{a}z|^{2p}}{(1-|z|^2)^{p+\beta-1}}\int_{\mathbf{D}}|1-\bar{z}\lambda|^{2\beta-5}(1-|\lambda|^2)^{2-\beta-p}dm(\lambda) \\
&\preceq \frac{|1-\bar{a}z|^{2p}}{(1-|z|^2)^{p+\beta-1}},
\end{aligned}$$

as desired. In the above estimates we have used the inequalities:

$$1+p < \beta < 3-p \quad \text{and} \quad |1-\bar{a}z+\lambda(\bar{a}-\bar{z})| \le |1-\bar{a}z|+|\bar{a}-\bar{z}| \le 2|1-\bar{a}z|. \qquad \square$$

On the basis of Lemma 4.5.1, we find a dual of the vanishing Q space.

Theorem 4.5.2. *For $p \in (0,1)$ let $\mathcal{R}_{p,1}$ be the class of those holomorphic functions f on $\mathbf{D}$ for which $f = \sum_{n=1}^\infty f_n$, $f_n \in \mathcal{H}$ with*

$$S_{a_n}(f_n) = \left(\int_{\mathbf{D}}|(zf_n(z))''|^2(1-|\sigma_{a_n}(z)|^2)^{-p}(1-|z|^2)^2dm(z)\right)^{\frac{1}{2}} < \infty$$

and

$$\|f\|_{\mathcal{R}_{p,1}} = \inf \sum_{n=1}^{\infty} S_{a_n}(f_n),$$

where the infimum is taken over all the foregoing representations for which the convergence is uniform on compact subsets of $\mathbf{D}$. *Then the dual of* $\mathcal{Q}_{p,0}$ *is isomorphic to* $\mathcal{R}_{p,1}$ *under* $\langle \cdot, \cdot \rangle_{inv}$. *More precisely, every function* $f \in \mathcal{R}_{p,1}$ *induces a continuous linear functional on* $\mathcal{Q}_{p,0}$ *by that* $\langle g, f \rangle_{inv}$ *for any* $g \in \mathcal{Q}_{p,0}$. *Conversely, if* L *is a continuous linear functional on* $\mathcal{Q}_{p,0}$, *then there exists* $f \in \mathcal{R}_{p,1}$ *such that* $\mathsf{L}(g) = \langle g, f \rangle_{inv}$ *for any* $g \in \mathcal{Q}_{p,0}$. *Moreover,*

$$\|f\|_{\mathcal{R}_{p,1}} \approx \sup \left\{ |\langle g, f \rangle_{inv}| : g \in \mathcal{Q}_{p,0}, \|g\|_{\mathcal{Q}_p} \leq 1 \right\}, \quad f \in \mathcal{R}_{p,1}.$$

Proof. If $g \in \mathcal{Q}_{p,0}$, then $g \in \mathcal{Q}_p$. Thus, $f \in \mathcal{R}_{p,1}$ and Cauchy–Schwarz's inequality imply

$$|\langle g, f \rangle_{inv}| \leq \pi^{-1} \int_{\mathbf{D}} |g'(z)(zf(z))''|(1 - |z|^2)dm(z) \preceq \sum_{n=1}^{\infty} S_{a_n}(f_n)\|g\|_{\mathcal{Q}_{p,1}},$$

and consequently, f induces a continuous linear functional $\mathsf{L}(g) = \langle g, f \rangle_{inv}$ on $\mathcal{Q}_{p,0}$.

To prove that for every continuous linear functional L on $\mathcal{Q}_{p,0}$ there exists an $f \in \mathcal{R}_{p,1}$ such that $\mathsf{L}(g) = \langle g, f \rangle_{inv}$ for any $g \in \mathcal{Q}_{p,0}$, we use

$$1 - |\sigma_a(z)|^2 \approx 1 - |\sigma_{a_n}(z)|^2, \quad a \in E_n, \ z \in \mathbf{D},$$

to get that the supremum in the definition of the $\| \cdot \|_{\mathcal{Q}_{p,1}}$ can be taken over $\{a_n\}_{n=1}^{\infty}$. Suppose now that each $\mathcal{X}_n$ consists of those functions $g \in \mathcal{H}$ for which

$$\|g\|_{\mathcal{X}_n} = \left(\int_{\mathbf{D}} |g'(z)|^2(1 - |\sigma_{a_n}(z)|^2)^p dm(z) \right)^{\frac{1}{2}} < \infty.$$

Denote by $\mathcal{X}$ the direct c_0-sum of $\{\mathcal{X}_n\}$; that is, the space of holomorphic function sequences $\{g_n\}_1^{\infty}$ on $\mathbf{D}$ such that $g_n \in \mathcal{X}_n$ for every $n \in \mathbf{N}$ and $\lim_{n \to \infty} \|g_n\|_{\mathcal{X}_n} = 0$. Equipping $\mathcal{X}$ with the norm $\|\{g_n\}_1^{\infty}\|_{\mathcal{X}} = \sup_{n \in \mathbf{N}} \|g_n\|_{\mathcal{X}_n}$, we see that $\mathcal{Q}_{p,0}$ is a normed subspace of $\mathcal{X}$ under this new norm. Analogously we define $\mathcal{Y}$ to be the ℓ^1-sum of the spaces $\mathcal{Y}_n$, where each $\mathcal{Y}_n$ is the class of all holomorphic functions f on $\mathbf{D}$ with

$$\|f\|_{\mathcal{Y}_n} = \left(\int_{\mathbf{D}} |(zf(z))''|^2(1 - |\sigma_{a_n}(z)|^2)^{-p}(1 - |z|^2)^2 \, dm(z) \right)^{\frac{1}{2}} < \infty.$$

Accordingly, the dual of $\mathcal{X}$ is isometrically isomorphic to the ℓ^1-sum of the dual spaces $\mathcal{X}_n^*$ of $\mathcal{X}_n$. Using Lemma 4.5.1, we see that $\mathcal{X}_n^*$ can be replaced by $\mathcal{Y}_n$ and so the dual of $\mathcal{X}$ is isomorphic to $\mathcal{Y}$ under the pairing

$$\sum_{n=1}^{\infty} \langle g_n, f_n \rangle_{inv}, \quad f_n \in \mathcal{Y}_n, \ g_n \in \mathcal{X}_n.$$

Suppose L is a continuous linear functional on $\mathcal{Q}_{p,0}$. Using Lemma 4.5.1 with the Hahn–Banach extension of L to $\mathcal{X}$ we obtain such an $f_n \in \mathcal{Y}_n$ that

$$\mathsf{L}(g) = \sum_{n=1}^{\infty} \langle g, f_n \rangle_{inv}, \quad g \in \mathcal{Q}_{p,0},$$

and the norm $\|\mathsf{L}\|$ of L satisfies $\sum_{n=1}^{\infty} \|f_n\|_{\mathcal{Y}_n} \preceq \|\mathsf{L}\|$. Finally, if $f = \sum_{n=1}^{\infty} f_n$, then this series converges uniformly on compact sets of $\mathbf{D}$. Hence

$$f \in \mathcal{H}, \quad \|f\|_{\mathcal{Y}} \leq \sum_{n=1}^{\infty} \|f_n\|_{\mathcal{Y}_n} \quad \text{and} \quad \mathsf{L}(g) = \langle g, f \rangle_{inv} \quad \text{for} \quad g \in \mathcal{Q}_{p,0}.$$

This proves $f \in \mathcal{R}_{p,1}$ with $\|f\|_{\mathcal{R}_{p,1}} \preceq \|\mathsf{L}\|$, as desired. $\qquad\square$

Quite naturally, we are led to connect $\mathcal{R}_{p,1}$ with another space defined in terms of Hausdorff capacity.

Theorem 4.5.3. *For $p \in (0,1)$ let $\mathcal{R}_{p,2}$ be the class of all $f \in \mathcal{H}$ with*

$$\|f\|_{\mathcal{R}_{p,2}} = \inf_{\omega \geq 0, \; \|\omega\|_{\mathcal{L}\mathcal{N}^1(H^p_\infty)} \leq 1} \left(\int_{\mathbf{D}} |f'(z)|^2 \big(\omega(z)\big)^{-1} (1 - |z|^2)^{-p} dm(z) \right)^{\frac{1}{2}} < \infty.$$

Then the dual of $\mathcal{R}_{p,2}$ is isomorphic to $\mathcal{Q}_p$ under $\langle \cdot, \cdot \rangle_{inv}$. Consequently,

$$\|f\|_{\mathcal{R}_{p,2}} \preceq \sup \left\{ |\langle f, g \rangle_{inv}| : \; g \in \mathcal{Q}_p, \; \|g\|_{\mathcal{Q}_p} \leq 1 \right\}, \quad f \in \mathcal{R}_{p,2}.$$

Proof. As proved in Theorem 4.4.1, we have that if $f \in \mathcal{R}_{p,2}$ and $g \in \mathcal{Q}_p$, then

$$|\langle f, g \rangle_{inv}| \preceq \left(\int_{\mathbf{D}} |f'(z)|^2 \big(\omega(z)\big)^{-1} (1 - |z|^2)^{-p} dm(z) \right)^{\frac{1}{2}} \|g\|_{\mathcal{Q}_{p,1}}.$$

As a result, $|\langle f, g \rangle_{inv}| \preceq \|f\|_{\mathcal{R}_{p,2}} \|g\|_{\mathcal{Q}_{p,1}}$; that is, $g \in \mathcal{Q}_p$ induces a continuous linear functional on $\mathcal{R}_{p,2}$.

Conversely, suppose L is a continuous linear functional on $\mathcal{R}_{p,2}$. Then it follows from Theorem 4.3.2 (i) and the Hahn–Banach Extension Theorem that there exists a function g on $\mathbf{D}$ such that $d\mu_{g,p} = |g(z)|^2 (1 - |z|^2)^p dm(z)$ is a p-Carleson measure on $\mathbf{D}$ and

$$\mathsf{L}(f) = \int_{\mathbf{D}} f' \bar{g} \, dm, \quad f \in \mathcal{R}_{p,2}.$$

Also, it is easy to check that each $f \in \mathcal{R}_{p,2}$ has the reproducing formula

$$f'(z) = \pi^{-1} \int_{\mathbf{D}} f'(w)(1 - \bar{w}z)^{-2} dm(w), \quad z \in \mathbf{D}.$$

Thus,

$$L(f) = \int_{\mathbf{D}} f'(w) \overline{\left(\pi^{-1} \int_{\mathbf{D}} g(z)(1 - \bar{z}w)^{-2} dm(z) \right)} dm(w) = \langle f, h \rangle_{inv},$$

where

$$h(w) = \int_0^w \left(\pi^{-1} \int_{\mathbf{D}} g(z)(1 - \bar{z}\zeta)^{-2} dm(z) \right) d\zeta, \quad w \in \mathbf{D}.$$

Thanks to $\mu_{g,p} \in \mathcal{CM}_p$ and

$$|h'(w)| \leq \pi^{-1} \int_{\mathbf{D}} |g(z)||1 - \bar{z}w|^{-2} dm(z), \quad w \in \mathbf{D},$$

we conclude from Lemma 3.1.2 that $|h'(w)|^2 (1 - |w|^2)^p dm(w)$ is a p-Carleson measure on $\mathbf{D}$ and then $h \in \mathcal{Q}_p$ which defines L.

Applying a corollary of the Hahn–Banach Extension Theorem (cf. [DeV, p. 48, Corollary 2]) to $\mathcal{R}_{p,2}$, we see that if $f \in \mathcal{R}_{p,2}$ is nonconstant, then there exists L, a continuous linear functional on $\mathcal{R}_{p,2}$, such that

$$\|L\| = 1 \quad \text{and} \quad \|f\|_{\mathcal{R}_{p,2}} = L(f).$$

With the help of the foregoing argument, we can find a function $g \in \mathcal{Q}_p$ such that

$$\|g\|_{\mathcal{Q}_p} \leq 1 \quad \text{and} \quad L(f) = \langle f, g \rangle_{inv}, \quad f \in \mathcal{R}_{p,2}.$$

This clearly implies the inequality required in Theorem 4.5.3.   □

In order to explore the relationship among $\mathcal{P}_p$, $\mathcal{R}_{p,1}$ and $\mathcal{R}_{p,2}$, we need a density result as follows.

Lemma 4.5.4. *Given $p \in (0,1)$.*

(i) *If $[\mathcal{Q}_{p,0}]^*$ stands for the space of those $f \in \mathcal{H}$ with*

$$\sup\{|\langle f, g \rangle_{inv}| : \ g \in \mathcal{Q}_{p,0}, \ \|g\|_{\mathcal{Q}_{p,1}} \leq 1\} < \infty,$$

then $\mathcal{R}_{p,2}$ is a dense subset of $[\mathcal{Q}_{p,0}]^*$.

(ii) *If $g \in \mathcal{Q}_p$ and $g_r(z) = g(rz)$ for $r \in (0,1)$, then there exists a sequence $\{r_n\}$ in $(0,1)$ such that*

$$\lim_{n \to \infty} r_n = 1 \quad \text{and} \quad \lim_{n \to \infty} \langle f, g_{r_n} \rangle_{inv} = \langle f, g \rangle_{inv}, \quad f \in \mathcal{R}_{p,2}.$$

Proof. (i) Since $\mathcal{Q}_{p,0} \subset \mathcal{Q}_p$, we conclude by Theorem 4.5.3 that

$$\sup \{|\langle g, f \rangle_{inv}| : \ g \in \mathcal{Q}_{p,0}, \ \|g\|_{\mathcal{Q}_{p,1}} \leq 1\} \preceq \|f\|_{\mathcal{R}_{p,2}}.$$

This means $\mathcal{R}_{p,2} \subseteq \mathcal{R}_{p,1}$.

Secondly, we prove that $\mathcal{R}_{p,2}$ contains all the polynomials. To do so, taking $\omega = 1$ on $\mathbf{D}$, we have $\int_{\mathbf{T}} \mathsf{N}(\omega)dH^p_\infty \preceq 1$. If f is a polynomial, then $\|f\|_{\mathcal{H}^\infty} < \infty$. Accordingly,

$$\|f\|_{\mathcal{R}_{p,2}} \lesssim \|f'\|_{\mathcal{H}^\infty} \left(\int_{\mathbf{D}} (1 - |z|^2)^{-p} dm(z) \right)^{\frac{1}{2}} \preceq \|f'\|_{\mathcal{H}^\infty}, \quad p \in (0,1).$$

Thirdly, we prove that the polynomials are dense in $\mathcal{R}_{p,1}$. For this, let $f \in \mathcal{R}_{p,1}$. Then

$$f = \sum_{n=1}^\infty f_n \quad \text{and} \quad \sum_{n=1}^\infty S_{a_n}(f_n) < \infty, \quad \text{where} \quad f_n \in \mathcal{Y}_n.$$

Since the dual of $\mathcal{X}_n$ is isomorphic to $\mathcal{Y}_n$, and the hypothesis that $\langle g, f \rangle_{inv} = 0$ for every polynomial g obviously implies $f' = 0$, we derive that the polynomials are dense in each $\mathcal{Y}_n$. Consequently, for any $\epsilon > 0$, there is a sequence of polynomials P_n such that $\sum_{n=1}^\infty S_{a_n}(f_n - P_n) < \epsilon$. Choose $j \geq 1$ so that $\sum_{n=j+1}^\infty S_{a_n}(f_n) < \epsilon$, and define the polynomial P by $P = \sum_{n=1}^j P_j$. Then

$$\|f - P\|_{\mathcal{R}_{p,1}} \leq \sum_{n=1}^j S_{a_n}(f_n - P_n) + \sum_{n=j+1}^\infty S_{a_n}(f_n) < 2\epsilon.$$

Combining the above-proved results and Theorem 4.5.3, we get that $\mathcal{R}_{p,2}$ is dense in $\mathcal{R}_{p,1}$, hence proving the desired assertion.

(ii) Note that $\|g_r\|_{\mathcal{Q}_{p,1}} \leq \|g\|_{\mathcal{Q}_{p,1}}$ (cf. [WiXi2]). So g_r can be treated as a bounded family in the dual space of $\mathcal{R}_{p,2}$ under $\langle \cdot, \cdot \rangle_{inv}$. Furthermore, via the Banach–Alaoglu Theorem, there exists an increasing sequence $\{r_n\}$ in $(0,1)$ and a function $h \in \mathcal{Q}_p$ such that

$$\lim_{n\to\infty} r_n = 1 \quad \text{and} \quad \lim_{n\to\infty} \langle f, g_{r_n} \rangle_{inv} = \langle f, h \rangle_{inv}, \quad f \in \mathcal{R}_{p,2}.$$

Taking $f(z) = z^j$, $j \geq 0$, we get $h = g$. $\square$

Now we are in a position to show that $\mathcal{R}_{p,2}$ or $\mathcal{P}_p$ is another description of either $\mathcal{R}_{p,1}$, and conversely, $\mathcal{R}_{p,1}$ suggests a sort of decomposition theorem for $\mathcal{P}_p$ or $\mathcal{R}_{p,2}$.

Theorem 4.5.5. *Given $p \in (0,1)$. Then:*

(i) *$\mathcal{R}_{p,2}$ is isomorphic to the dual of $\mathcal{Q}_{p,0}$ under $\langle \cdot, \cdot \rangle_{inv}$.*

(ii) *$\mathcal{R}_{p,2} = \mathcal{R}_{p,1} = \mathcal{P}_p$.*

(iii) *The second dual of $\mathcal{Q}_{p,0}$ is isomorphic to $\mathcal{Q}_p$ under $\langle \cdot, \cdot \rangle_{inv}$.*

Proof. Since (ii) and (iii) follow from (i), Theorems 4.5.3, 4.5.2 and 4.4.1, we just check (i). As in Lemma 4.5.4 (ii), we define $g_w(z) = g(wz)$ for $g \in \mathcal{Q}_p$ and $w \in \bar{\mathbf{D}}$ and then get $g_w \in \mathcal{Q}_p$ with $\|g_w\|_{\mathcal{Q}_p,1} \leq \|g\|_{\mathcal{Q}_p,1}$, but also $g_w \in \mathcal{Q}_{p,0}$ for every $w \in \mathbf{D}$. For the sake of convenience, denote by $\mathcal{B}_{\mathcal{Q}_p}$ and $\mathcal{B}_{\mathcal{Q}_{p,0}}$ the closed unit balls in $\mathcal{Q}_p$ and $\mathcal{Q}_{p,0}$ respectively. Then

$$\sup_{g \in \mathcal{B}_{\mathcal{Q}_p}} |\langle f, g \rangle_{inv}| = \sup_{g \in \mathcal{B}_{\mathcal{Q}_{p,0}}} |\langle f, g \rangle_{inv}|, \quad f \in \mathcal{R}_{p,2}.$$

Indeed, suppose $f \in \mathcal{R}_{p,2}$. Then by Lemma 4.5.4 (ii) we have that if $g \in \mathcal{Q}_p$, then

$$\sup_{w \in \mathbf{D}} |\langle f, g_w \rangle_{inv}| = \sup_{\zeta \in \bar{\mathbf{D}}} |\langle f, g_\zeta \rangle_{inv}|.$$

With this, we further get

$$
\begin{aligned}
\sup_{g \in \mathcal{B}_{\mathcal{Q}_p}} |\langle f, g \rangle_{inv}| &= \sup_{g \in \mathcal{B}_{\mathcal{Q}_p}} \sup_{w \in \bar{\mathbf{D}}} |\langle f, g_w \rangle_{inv}| \\
&\leq \sup_{g \in \mathcal{B}_{\mathcal{Q}_p}} \sup_{w \in \mathbf{D}} |\langle f, g_w \rangle_{inv}| \\
&\leq \sup_{g \in \mathcal{B}_{\mathcal{Q}_{p,0}}} |\langle f, g \rangle_{inv}| \\
&\leq \sup_{g \in \mathcal{B}_{\mathcal{Q}_p}} |\langle f, g \rangle_{inv}|,
\end{aligned}
$$

thereupon verifying the above-desired equation. This equation, along with the estimate in Theorem 4.5.3, yields

$$\|f\|_{\mathcal{R}_{p,2}} \preceq \sup_{g \in \mathcal{B}_{\mathcal{Q}_{p,0}}} |\langle f, g \rangle_{inv}| \preceq \|f\|_{\mathcal{R}_{p,2}}, \quad f \in \mathcal{R}_{p,2},$$

which in turn tells us that $\mathcal{R}_{p,2}$ is isomorphic to a subspace of the invariant dual space of $\mathcal{Q}_{p,0}$. Using Lemma 4.5.4 (i) we derive that $\mathcal{R}_{p,2}$ is isomorphic to the dual space of $\mathcal{Q}_{p,0}$ under $\langle \cdot, , \cdot \rangle_{inv}$. $\square$

4.6 Notes

Note 4.6.1. Section 4.1 may be regarded as an expanded version of [Xi4, Subsection 2.1]. Even though, we would like to make a few remarks on this section. The properties (i)–(ii)–(iii) of Lemma 4.1.1 are elementary, and Lemma 4.1.1 (iii) is the unit circle version of [ChaDXi, Theorem (i)] that sharpens [OrVe, Lemma 3]. Lemma 4.1.2 is the circular variation of [Mel, Lemmas 1–2], and its proof is a modification of the argument for [OrVe, Lemma 2]. Theorem 4.1.3 is the counterpart of [OrVe, Theorem] which extends [Ad2, Theorem A]. Theorem 4.1.4 (iv) is a new contribution to [Xi4, Section 2.1]. In the proof of Theorem 4.1.4 (i)$\Rightarrow$(ii) we have consulted [AhJe].

Note 4.6.2. Section 4.2 is an outcome of understanding the circular version of [Ad2, Section 3] and [Xi4, Theorem 2.2]. Motivated by Theorem 4.2.1 we may pose a natural predual problem as follows. For $p \in (0,1)$, let $\mathcal{L}_{fun}^{1,p}(\mathbf{T})$ be the Morrey space of complex-valued functions f on $\mathbf{T}$ such that

$$\sup_{I \subseteq \mathbf{T}} |I|^{-p} \int_I |f(\zeta) - f_I| |d\zeta| < \infty.$$

What is the predual of $\mathcal{L}_{fun}^{1,p}(\mathbf{T})$ under $\int_{\mathbf{T}} g(\zeta)\overline{f(\zeta)}|d\zeta|$?

Note 4.6.3. Section 4.3 is a by-product of modifying [Xi4, Theorem 2.3] and transferring [DaXi, Section 5] to the unit circle. See also [AlvJMi] and [CoiMeSt] for some related topics.

Note 4.6.4. Section 4.4 is basically taken from [Xi4, Section 2.3]. From the formula

$$\langle f, g \rangle_{inv} = \int_{\mathbf{D}} \left(zf(z) \right)'' \overline{g'(z)} (1 - |z|^2) dm(z)$$

it follows readily that the extremal holomorphic Besov space $\mathcal{B}_1 = \mathcal{B}_1^{1,1}$ comprising all $f \in \mathcal{H}$ with

$$\int_{\mathbf{D}} |(zf(z))''| dm(z) < \infty$$

has the Bloch space $\mathcal{B}$ as the invariant dual; see also [ArFi] and [ArFiPe1]. Of course, $\mathcal{B}_1 \subset \mathcal{H}'$ — because $f \in \mathcal{B}_1$ has a representation $f = \sum_{j=1}^\infty \lambda_j \sigma_{z_j}(z)$ and

$$\int_{\mathbf{T}} |\sigma'_{z_j}(\zeta)| |d\zeta| = \int_{\mathbf{T}} |d\sigma_{z_j}(\zeta)| = 2\pi,$$

where $\sum_{j=1}^\infty |\lambda_j| < \infty$ and $\{z_j\}$ is a sequence in $\mathbf{D}$.

In the proof of Lemma 4.4.2 we have borrowed ideas used in [Ad3, Theorem 2.2], [Ad4, p. 33], [Pa1, Theorem 7.1.8], [Or, Lemma 1.1] and [Kr, Theorems 2.1–2.2]. Similarly, we can prove that if

$$J_p f(\zeta) = \int_{\mathbf{T}} \frac{f(\eta)}{|\zeta - \eta|^p} |d\eta|, \quad p \in (0,1),$$

then

$$\int_{\mathbf{T}} \mathsf{M}(J_p f) dH_\infty^p \preceq \|f\|_{\mathcal{H}^1}, \quad f \in \mathcal{H}^1.$$

In the very first part of the proof of Theorem 4.4.3 we have used a trick from [Bae2, p. 27]. Additionally, it is worth mentioning the geometric nature of Theorem 4.4.3. To be more specific, let f be a conformal map of $\mathbf{D}$ onto a domain $\mathbf{\Omega} \subset \mathbf{C}$ with the boundary $\partial\mathbf{\Omega}$. Then it follows from Theorem 4.4.3 that

$$(\text{Area}(\mathbf{\Omega}))^{\frac{1}{2}} = \left(\int_{\mathbf{D}} |f'|^2 dm \right)^{\frac{1}{2}} \preceq \|f\|_{\mathcal{P}_p} \preceq \int_{\mathbf{T}} |f'(\zeta)| |d\zeta| = \text{Length}(\partial\mathbf{\Omega}),$$

improving the well-known isoperimetric inequality without sharp constant.

Note 4.6.5. Section 4.5 is established via a modification of the main techniques used in [PaXi] and [WiXi2]. Lemma 4.5.1 is crucial and can be also proved via Schur's Lemma. In fact, it is enough to show the following boundedness:

$$\int_{\mathbf{D}} \left(\int_{\mathbf{D}} |1 - \bar{a}z|^p (1 - |w|^2)^{\frac{2-p}{2}} |h(w)| K(z,w) dm(w) \right)^2 dm(z)$$

$$\preceq \int_{\mathbf{D}} \left(|1 - \bar{a}z|^p (1 - |z|^2)^{\frac{2-p}{2}} |h(z)| \right)^2 dm(z),$$

where the constant does not depend on a, and

$$K(z,w) = \left| \frac{1 - \bar{a}z}{1 - \bar{a}w} \right|^p \left(\frac{(1 - |z|^2)^{\frac{2-p}{2}} (1 - |w|^2)^{\frac{p}{2}}}{|1 - \bar{w}z|^3} \right).$$

Selecting numbers γ, β such that $p < \beta < 1$ and $-1 - \frac{p}{2} < \gamma < -\beta - \frac{p}{2}$, we do some basic integral estimates with the change of variables: $w = \sigma_z(\lambda)$ and $z = \sigma_w(\lambda)$ to deduce

$$\int_{\mathbf{D}} K(z,w)(1 - |w|^2)^\gamma |1 - \bar{a}w|^\beta dm(w) \preceq (1 - |z|^2)^\gamma |1 - \bar{a}z|^\beta, \quad z \in \mathbf{D}$$

and

$$\int_{\mathbf{D}} K(z,w)(1 - |z|^2)^\gamma |1 - \bar{a}z|^\beta dm(w) \preceq (1 - |w|^2)^\gamma |1 - \bar{a}w|^\beta \quad w \in \mathbf{D}.$$

These inequalities justify the required conditions of Schur's Lemma and then give the desired boundedness.

The definition of $\mathcal{R}_{p,1}$ can extend to $p \in [1,2)$. Using the Cauchy–Schwarz inequality, we derive that if $f = \sum_n f_n \in \mathcal{R}_{p,1}$, $p > 1$, then $f \in \mathcal{B}_1$ with

$$\int_{\mathbf{D}} |(zf(z))''| dm(z) \leq \sum_{n=1}^{\infty} S_{a_n}(f_n) \left(\int_{\mathbf{D}} (1 - |\sigma_{a_n}(z)|^2)^p (1 - |z|^2)^{-2} dm(z) \right)^{\frac{1}{2}}$$

$$\preceq \sum_{n=1}^{\infty} S_{a_n}(f_n).$$

Note that $\mathcal{B}_1$ is the minimal Möbius-invariant space. So it is our conjecture that $\mathcal{R}_{p,1} = \mathcal{B}_1$ for $p \in (1,2)$ and $\mathcal{R}_{1,1} = \mathcal{H}'$ with equivalent seminorms. The key issue to settle this conjecture seems to see whether or not Lemma 4.5.1 allows an extension from $p \in (0,1)$ to $p \in [1,2)$. Nevertheless, we can employ the characterization of $\mathcal{Q}_{p,0}$ in terms of the second order derivative (cf. [AuNoZh]) and the methods of proving Lemma 4.5.1 and Theorem 4.5.2 to show that $\mathcal{R}_{p,1}$, $p \in (0,1)$ consists of those $f \in \mathcal{H}$ for which $f = \sum_{n=1}^{\infty} f_n$, $f_n \in \mathcal{H}$ is convergent uniformly on compacta of $\mathbf{D}$, and is equipped with the seminorm

$$\inf \sum_{n=1}^{\infty} \left(\int_{\mathbf{D}} |f_n'(z)|^2 (1 - |\sigma_{a_n}(z)|^2)^{-p} dm(z) \right)^{\frac{1}{2}} < \infty,$$

where the infimum ranges over all above-given representations f.

If $\mathcal{VMOA}$ and $\mathcal{B}_0$ denote the spaces of all $f \in \mathcal{H}$ with $\lim_{|w| \to 1} E_1(f, w) = 0$ and $\lim_{|w| \to 1}(1 - |w|^2)|f'(w)| = 0$ respectively, then it is known that the invariant duals of these two spaces are isomorphic to $\mathcal{H}'$ and $\mathcal{B}_1$ respectively; see also [ArFiPe1].

Additionally, [Wu4] has claimed that the first and second duals of a $\mathcal{Q}_{p,0}$-type space are found under an appropriate pairing.

Chapter 5

Cauchy Pairing with Expressions and Extremities

We have seen from the previous chapter that each holomorphic Q class exists as a dual space with respect to the invariant pairing $\langle \cdot, \cdot \rangle_{inv}$, however the dual space may have different preduals. In this chapter, we introduce another dual pairing, which is variant under $Aut(\mathbf{D})$, but allows us to handle all the cases of the preduals and duals for the holomorphic Q classes. This pairing is the so-called Cauchy pairing:

$$\langle f, g \rangle = \lim_{r \to 1} \int_{\mathbf{T}} f(r\zeta) \overline{g(r\zeta)} |d\zeta|.$$

One more advantage of the Cauchy pairing is that it enables us to settle some natural questions on the expressions and extremes of the related holomorphic function spaces. The further details distribute the following sections:

- Background on Cauchy Pairing;
- Cauchy Duality by Dot Product;
- Atom-like Representations;
- Extreme Points of Unit Balls.

5.1 Background on Cauchy Pairing

Although we intend later to work almost exclusively in the Cauchy pairing attached to $\mathcal{Q}_p$, it is most convenient to start abstractly.

Suppose $\mathcal{X} \subseteq \mathcal{H}$ is a Banach space equipped with the norm $\| \cdot \|_{\mathcal{X}}$ and contains $\mathcal{H}(\bar{\mathbf{D}})$ (comprising all holomorphic functions on a neighborhood of $\bar{\mathbf{D}}$) as a dense subspace. If $\mathcal{X}^*$ stands for the dual space of $\mathcal{X}$ — that is — the space of all complex-valued continuous/bounded linear functionals on $\mathcal{X}$, then the following action with respect to the running variable w:

$$\Theta(\mathsf{L})(z) = \overline{\mathsf{L}\big((1 - \bar{z}w)^{-1}\big)}, \quad \mathsf{L} \in \mathcal{X}^*, \quad z \in \mathbf{D},$$

defines a one-to-one mapping from $\mathcal{X}^*$ into $\mathcal{H}$, and hence this induces an isometric copy: $\Theta(\mathcal{X}^*)$ of $\mathcal{X}^*$, called the Cauchy dual of $\mathcal{X}$, via the norm

$$\|\Theta(\mathsf{L})\|_{\Theta(\mathcal{X}^*)} = \|\mathsf{L}\|_{\mathcal{X}^*} = \sup\{|\mathsf{L}(f)| : \ \|f\|_{\mathcal{X}} \leq 1\}.$$

This concept comes from the fact that any $\mathsf{L} \in \mathcal{X}^*$ can be written as a Cauchy pairing between $f \in \mathcal{X}$ and $g = \Theta(\mathsf{L})$:

$$\mathsf{L}(f) = (2\pi)^{-1}\langle f, g\rangle = f(0)\overline{g(0)} + \pi^{-1}\lim_{r \to 1}\int_{\mathbf{D}} f'(rz)\overline{g'(rz)}(-\log|z|^2)dm(z).$$

This is actually a derivative representation of the Cauchy pairing, and reveals that the Cauchy pairing is not invariant under $Aut(\mathbf{D})$; that is,

$$\langle f \circ \sigma_w, g \circ \sigma_w\rangle \neq \langle f, g\rangle \quad \text{unless} \ \ w = 0.$$

According to the Cauchy pairing, we naturally find out

$$\Theta(\mathcal{D}^*) \cong \mathcal{A}^{2,0} \quad \text{and} \quad \Theta\big((\mathcal{A}^{2,0})^*\big) \cong \mathcal{D}.$$

At the same time,

$$\Theta(\mathcal{VMOA}^*) \cong \mathcal{H}^1 \quad \text{and} \quad \Theta\big((\mathcal{H}^1)^*\big) \cong \mathcal{BMOA}$$

are the well-known Fefferman–Sarason's dualities. In addition, for $q \in [1, \infty)$ and $\beta \in (-1, \infty)$ let

$$\mathcal{D}^{q,\beta} = \{f \in \mathcal{H} : \ \|f\|_{\mathcal{D}^{q,\beta}} = |f(0)| + \|f'\|_{\mathcal{A}^{q,\beta}} < \infty\}$$

be the Dirichlet type space, then we reach Anderson–Clunie–Pommerenke's dualities (cf. [AnClPo]):

$$\Theta(\mathcal{B}_0^*) \cong \mathcal{D}^{1,0} \quad \text{and} \quad \Theta\big((\mathcal{D}^{1,0})^*\big) \cong \mathcal{B}.$$

In the above and the below, the symbol "$\cong$" means "isomorphic". The preceding examples lead to the following general result.

Theorem 5.1.1. *Let $(\mathcal{X}, \|\cdot\|_{\mathcal{X}})$ be a Banach space with $\mathcal{H}(\bar{\mathbf{D}}) \subseteq \mathcal{X} \subseteq \mathcal{H}$. Then $\Theta(\mathcal{X}^*)$ lies between $\mathcal{H}(\bar{\mathbf{D}})$ and $\mathcal{H}$. Moreover, if $\mathcal{H}(\bar{\mathbf{D}})$ is dense in $\Theta(\mathcal{X}^*)$, then $\Theta\big((\Theta(\mathcal{X}^*))^*\big)$ comprises $f \in \mathcal{H}$ for which there is a sequence $\{f_j\}$ in $\mathcal{X}$ with $\sup_{j \in \mathbf{N}} \|f_j\|_{\mathcal{X}} < \infty$ and $f_j(z) \to f(z)$ for each $z \in \mathbf{D}$. Consequently,*

$$\|f\|_{\Theta\big((\Theta(\mathcal{X}^*))^*\big)} = \|f\|_{\mathcal{X}^{**}} = \inf\{\limsup\|f_j\|_{\mathcal{X}}\},$$

where the infimum ranges over all sequences $\{f_j\}$ described above.

Proof. For the first assertion, we consider $\mathsf{L}_j \in \mathcal{X}^*$ which is determined by

$$\mathsf{L}_j(f) = f^{(j)}(0)/j!, \quad j \in \mathbf{N} \cup \{0\}, \quad f \in \mathcal{X};$$

we then obtain

$$|\mathsf{L}_j(f)| \leq \sup_{|z|=r} r^{-j}|f(z)| \preceq r^{-j}\|f\|_{\mathcal{X}}, \quad r \in (0,1),$$

hence implying $\limsup_j \|\mathsf{L}_j\|_{\mathcal{X}^*}^{j^{-1}} \leq 1$. Assuming $g \in \mathcal{H}(\bar{D})$ with the Taylor expansion $g(z) = \sum_{j=0}^{\infty} a_j z^j$, we derive that $\sum_{j=0}^{\infty} a_j \mathsf{L}_j$ converges in $\mathcal{X}^*$. Note that $\Theta(\mathsf{L}_j) = z^j$. So

$$g = \sum_{j=0}^{\infty} a_j \Theta(\mathsf{L}_j) = \Theta\left(\sum_{j=0}^{\infty} a_j \mathsf{L}_j\right) \in \Theta(\mathcal{X}^*).$$

On the other hand, since $\sup_{z \in K} \|(1-\bar{z}w)^{-1}\|_{\mathcal{X}} < \infty$ holds for any compact subset K of $\mathbf{D}$, we conclude $\Theta(\mathcal{X}^*) \subseteq \mathcal{H}$.

Regarding the second assertion, suppose $\mathcal{H}(\bar{\mathbf{D}})$ is dense in $\Theta(\mathcal{X}^*)$. Since Θ is injective, we conclude that $\Theta\big((\Theta(\mathcal{X}^*))^*\big)$ is isometrically isomorphic to $\mathcal{X}^{**}$, the second dual of $\mathcal{X}$, but also the canonical imbedding $\mathcal{X} \mapsto \mathcal{X}^{**}$ gives $\mathcal{X} \subseteq \Theta\big((\Theta(\mathcal{X}^*))^*\big)$. Now for $f \in \Theta\big((\Theta(\mathcal{X}^*))^*\big)$, we can use the hypothesis to find a sequence $\{f_j\}$ such that it is convergent to f in the topology of weak-star and

$$\|f_j\|_{\mathcal{X}} \leq \|f\|_{\Theta\big((\Theta(\mathcal{X}^*))^*\big)}.$$

Consequently,

$$\begin{aligned} f(z) &= (2\pi)^{-1}\langle f, (1-\bar{z}w)^{-1}\rangle \\ &= (2\pi)^{-1} \lim_{j \to \infty} \langle f_j, (1-\bar{z}w)^{-1}\rangle \\ &= \lim_{j \to \infty} f_j(z), \quad z \in \mathbf{D}. \end{aligned}$$

Meanwhile, if $f \in \mathcal{H}$ and if $\{f_j\}$ is a bounded sequence in $\mathcal{X}$ which converges pointwisely to f on $\mathbf{D}$, then each functional $\mathsf{L}_{f_j}(\Phi) = \Phi(f_j)$ belongs to $(\Theta(\mathcal{X}^*))^*$ and satisfies

$$\overline{\mathsf{L}_{f_j}\big((1-\bar{z}w)^{-1}\big)} = f_j(z) \to f(z), \quad z \in \mathbf{D}.$$

This, together with the density of $\mathcal{H}(\bar{\mathbf{D}})$ in $\Theta(\mathcal{X}^*)$, yields that $\{\mathsf{L}_{f_j}\}$ is convergent weak-star to a functional $\mathsf{L} \in (\Theta(\mathcal{X}^*))^*$ with

$$\overline{\mathsf{L}\big((1-\bar{z}w)^{-1}\big)} = f(z), \quad z \in \mathbf{D} \quad \text{and} \quad \|\mathsf{L}\|_{(\Theta(\mathcal{X}^*))^*} \leq \limsup_{j \to \infty} \|f_j\|_{\mathcal{X}},$$

as desired. $\qquad\square$

In many situations, we identify a Banach space $\mathcal{X}$ isomorphically with an accessible sequence space, say ℓ^q, $q \in [1, \infty)$ and c_0, in order to figure out the Cauchy dual $\Theta(\mathcal{X}^*)$ of $\mathcal{X}$. More precisely, let $\mathcal{Y}$ be one of these sequence spaces and $\Phi : \mathcal{X} \mapsto \mathcal{Y}$ be bounded from above and below. Then there is a sequence $\{L_j\}$ in $\mathcal{X}^*$ such that $\Phi(f) = \{L_j(f)\}$ for any $f \in \mathcal{X}$, and hence the adjoint operator $\Phi^* : \mathcal{Y}^* \mapsto \mathcal{X}^*$, given by $\Phi^*(s_j) = \sum_j s_j L_j$, is bounded and onto. Of course, $L \in \mathcal{X}^*$ can be expressed as

$$L = \sum_j s_j L_j \quad \text{with} \quad \{s_j\} \in \mathcal{Y}^* \quad \text{and} \quad \|L\|_{\mathcal{X}^*} \approx \inf \|\{s_j\}\|_{\mathcal{Y}^*}$$

where the infimum ranges over all sequences $\{s_j\} \in \mathcal{Y}^*$ with the foregoing representation. This amounts to saying that $f \in \Theta(\mathcal{X}^*)$ can be written as

$$f = \sum_j s_j f_j \quad \text{with} \quad \{s_j\} \in \mathcal{Y}^*, \quad f_j = \Theta(L_j) \quad \text{and} \quad \|f\|_{\Theta(\mathcal{X}^*)} \approx \inf \|\{s_j\}\|_{\mathcal{Y}^*},$$

where the infimum ranges over all sequences $\{s_j\} \in \mathcal{Y}^*$ with the last representation.

To gain a solid understanding of the previous principle, let us give one more example which will be used later on.

Example 5.1.2. Given $\alpha \in (0,1)$, let $\mathcal{A}^{-\alpha}$ and $\mathcal{A}_0^{-\alpha}$ be the spaces of $f \in \mathcal{H}$ with

$$\|f\|_{\mathcal{A}^{-\alpha}} = \sup_{z \in \mathbf{D}} (1 - |\lambda|^2)^\alpha |f(z)| < \infty \quad \text{and} \quad \lim_{|z| \to 1} (1 - |\lambda|^2)^\alpha |f(z)| = 0,$$

respectively. Then

$$\Theta\big((\mathcal{A}_0^{-\alpha})^*\big) \cong \mathcal{D}^{1,-\alpha} \quad \text{and} \quad \Theta\big((\mathcal{D}^{1,-\alpha})^*\big) \cong \mathcal{A}^{-\alpha}.$$

If $W = \{w_j\}$ is a discrete subset of $\mathbf{D}$ such that

$$\inf_{w_j \in W} |z - w_j| < \kappa(1 - |z|), \quad z \in \mathbf{D}$$

holds for a sufficiently small constant $\kappa \in (0,1)$, then:

(i) $\|f\|_{\mathcal{A}^{-\alpha}} \approx \sup_{w_j \in W} (1 - |w_j|)^\alpha |f(w_j)|, \quad f \in \mathcal{A}^{-\alpha}$.

(ii) The mapping $\Phi : f \mapsto \{(1 - |w_j|)^\alpha f(w_j)\}_{w_j \in W}$ is an isomorphism between $\mathcal{A}_0^{-\alpha}$ and a closed subspace of c_0.

(iii) Any element $f \in \mathcal{D}^{1,-\alpha}$ can be written as

$$f(z) = \sum_{w_j \in W} c_\lambda (1 - |w_j|)^\alpha (1 - \overline{w_j} z)^{-1} \quad \text{with} \quad \|f\|_{\mathcal{D}^{1,-\alpha}} \approx \inf \sum_{w_j \in W} |c_{w_j}|,$$

where the infimum ranges over all sequences for the representation of f above.

Proof. For the Cauchy duality relations, we first prove $\Theta\big((\mathcal{D}^{1,-\alpha})^*\big) \cong \mathcal{A}^{-\alpha}$. On the one hand, if $g \in \mathcal{A}^{-\alpha}$ then

$$\sup_{z \in \mathbf{D}}(1 - |z|)^{1+\alpha}|g'(z)| \precsim \|g\|_{\mathcal{A}^{-\alpha}}$$

and hence g produces an element of $\Theta\big((\mathcal{D}^{1,-\alpha})^*\big)$ thanks to the above-mentioned formula of $\langle f, g \rangle$ involving the derivatives of f and g, plus the following estimate

$$\left| \int_{\mathbf{D}} f'(rz)\overline{g'(rz)}(-\log|z|^2)dm(z) \right|$$

$$\leq \left(\sup_{z \in \mathbf{D}}(-\log|z|^2)^{1+\alpha}|g'(z)| \right) \int_{\mathbf{D}} |f'(rz)|(-\log|z|^2)^{-\alpha}dm(z)$$

$$\precsim \|g\|_{\mathcal{A}^{-\alpha}}\|f\|_{\mathcal{D}^{1,-\alpha}}.$$

Now, suppose $\mathsf{L} \in \Theta\big((\mathcal{D}^{1,-\alpha})^*\big)$. Since $f \in \mathcal{D}^{1,-\alpha}$ can be approximated by f_r in norm, i.e., $\lim_{r \to 1} \|f_r - f\|_{\mathcal{D}^{1,-\alpha}} = 0$, we can consider

$$f_r(z) = f(rz) = (2\pi)^{-1} \int_{\mathbf{T}} f_r(\zeta)(1 - z\bar{\zeta})^{-1}|d\zeta|.$$

Clearly, we have

$$\lim_{r \to 1} \mathsf{L}(f_r) = \lim_{r \to 1}(2\pi)^{-1} \int_{\mathbf{T}} f_r(\zeta)\mathsf{L}\big((1 - zr\bar{\zeta})^{-1}\big)|d\zeta|,$$

where L is treated as acting with respect to the running variable z. Since L is bounded, we conclude that for $w \in \mathbf{D}$,

$$\left| \mathsf{L}\big((1 - \bar{w}z)^{-1}\big) \right| \leq \|\mathsf{L}\|_{\Theta\big((\mathcal{D}^{1,-\alpha})^*\big)} \|(1 - \bar{w}z)^{-1}\|_{\mathcal{D}^{1,-\alpha}} \precsim \frac{\|\mathsf{L}\|_{\Theta\big((\mathcal{D}^{1,-\alpha})^*\big)}}{(1 - |w|)^\alpha}.$$

This yields

$$g(w) = (2\pi)^{-1}\overline{\mathsf{L}\big((1 - \bar{w}z)^{-1}\big)} \in \mathcal{A}^{-\alpha} \quad \text{with} \quad \mathsf{L}(f) = \lim_{r \to 1} \mathsf{L}(f_r) = (2\pi)^{-1}\langle f, g \rangle.$$

Next, we give a proof for $\Theta\big((\mathcal{A}_0^{-\alpha})^*\big) \cong \mathcal{D}^{1,-\alpha}$. If $g \in \mathcal{D}^{1,-\alpha}$ then it produces an element of $\Theta\big((\mathcal{A}_0^{-\alpha})^*\big)$ as proved above. Conversely, suppose $\mathsf{L} \in \Theta\big((\mathcal{A}_0^{-\alpha})^*\big)$. For each $j \in \mathbf{N} \cup \{0\}$ let $f_j(z) = z^j$, $z \in \mathbf{D}$. Then $f_j \in \mathcal{A}_0^{-\alpha}$ with $\|f_j\|_{\mathcal{A}^{-\alpha}} = 1$. Assuming $b_j = \overline{\mathsf{L}(f_j)}$, we derive

$$|b_j| \leq \|\mathsf{L}\|_{\Theta\big((\mathcal{A}_0^{-\alpha})^*\big)} \|f_j\|_{\mathcal{A}^{-\alpha}} = \|\mathsf{L}\|_{\Theta\big((\mathcal{A}_0^{-\alpha})^*\big)},$$

yielding that $g(z) = \sum_{j=0}^{\infty} b_j z^j$ has a radius of convergence greater than or equal to 1 and hence is holomorphic on $\mathbf{D}$. Furthermore, we show

$$g \in \mathcal{D}^{1,-\alpha} \quad \text{with} \quad \|g\|_{\mathcal{D}^{1,-\alpha}} \precsim \|\mathsf{L}\|_{\Theta\big((\mathcal{A}_0^{-\alpha})^*\big)}.$$

To this end, we observe that $g_r(z) = g(rz)$ belongs to $\mathcal{D}^{1,-\alpha}$ for each $r \in (0,1)$. So, using the previous duality argument we have

$$\|g_r\|_{\mathcal{D}^{1,-\alpha}} \preceq \sup \left\{ \frac{(2\pi)^{-1}|\langle g_r, f \rangle|}{\|f\|_{\mathcal{A}^{-\alpha}}} : f \in \mathcal{A}^{-\alpha} \setminus \{0\} \right\}.$$

Since $\mathsf{L} \in \Theta\big((\mathcal{A}_0^{-\alpha})^*\big)$, we conclude that if $f \in \mathcal{A}^{-\alpha}$ with $f(z) = \sum_{j=0}^{\infty} a_j z^j$, then

$$\|f_r\|_{\mathcal{A}^{-\alpha}} \leq \|f\|_{\mathcal{A}^{-\alpha}} \quad \text{and} \quad \mathsf{L}(f_r) = \sum_{j=0}^{\infty} a_j \overline{b_j} r^j = (2\pi)^{-1}\langle f, g_r \rangle.$$

Accordingly,

$$(2\pi)^{-1}|\langle g_r, f \rangle| = (2\pi)^{-1}|\langle f, g_r \rangle| = |\mathsf{L}(f_r)| \leq \|\mathsf{L}\|_{\Theta\big((\mathcal{A}_0^{-\alpha})^*\big)} \|f\|_{\mathcal{A}^{-\alpha}}.$$

This, along with the above norm estimate on g_r, gives

$$\|g_r\|_{\mathcal{D}^{1,-\alpha}} \preceq \|\mathsf{L}\|_{\Theta\big((\mathcal{A}_0^{-\alpha})^*\big)}.$$

An application of Fatou's Lemma to the last inequality yields

$$g \in \mathcal{D}^{1,-\alpha} \quad \text{with} \quad \|g\|_{\mathcal{D}^{1,-\alpha}} \preceq \|\mathsf{L}\|_{\Theta\big((\mathcal{A}_0^{-\alpha})^*\big)}.$$

In the meantime, $f \in \mathcal{A}_0^{-\alpha}$ is equivalent to $\lim_{r \to 1} \|f - f_r\|_{\mathcal{A}^{-\alpha}} = 0$. Therefore,

$$\mathsf{L}(f) = \lim_{r \to 1} \mathsf{L}(f_{r^2}) = \lim_{r \to 1} \sum_{j=0}^{\infty} a_j \overline{b_j} r^{2j} = (2\pi)^{-1}\langle f, g \rangle, \quad f \in \mathcal{A}_0^{-\alpha},$$

as desired.

The assertions (i) and (ii) can be verified by the following two implications for a sufficiently small $\kappa \in (0, 1/2)$:

$$|z - w_j| < \kappa(1 - |z|) \Rightarrow 1 - \kappa < (1 - |w_j|)(1 - |z|)^{-1} < 1 + \kappa$$

and

$$|\sigma_{w_j}(z)| < \kappa \Rightarrow |(1 - |z|)^\alpha f(z) - (1 - |w_j|)^\alpha f(w_j)| \preceq \|f\|_{\mathcal{A}^{-\alpha}} |\sigma_{w_j}(z)|.$$

Of course, (iii) is a straightforward consequence of (i), (ii), and an application of the foregoing general principle to the Cauchy dualities for $\mathcal{D}^{1,-\alpha}$, $\mathcal{A}^{-\alpha}$, and $\mathcal{A}_0^{-\alpha}$. $\qquad \square$

5.2 Cauchy Duality by Dot Product

However, a careful look at the invariant dual space of $\mathcal{Q}_{p,0}$, $p \in (0,1)$ (see e.g., Theorem 4.5.2) reveals that it seems more natural to consider certain vector-valued sequence spaces than the Carleson measure type sequence spaces in Theorem 3.4.2. In so doing, recall that given a Banach space $\mathcal{Y}$ over $\mathbf{C}$, $\ell^q(\mathcal{Y})$ (where $q \in [1, \infty)$) and $c_0(\mathcal{Y})$ are the spaces of sequences $\{\|y_j\|_{\mathcal{Y}}\}$ that belong to ℓ^q and c_0 respectively. Of course, their norms are defined respectively by

$$\|\{y_j\}\|_{\ell^q(\mathcal{Y})} = \left(\sum_j \|y_j\|_{\mathcal{Y}}^q \right)^{q^{-1}} \quad \text{and} \quad \|\{y_j\}\|_{c_0(\mathcal{Y})} = \sup_j \|y_j\|_{\mathcal{Y}}.$$

Classically, the dual of $c_0(\mathcal{Y})$ is $\ell^1(\mathcal{Y}^*)$, namely, to every continuous linear functional Λ on $c_0(\mathcal{Y})$ there corresponds a unique sequence $\{s_j\} \in \ell^1(\mathcal{Y}^*)$ such that

$$\Lambda(\{y_j\}) = \sum_j s_j(y_j), \quad \{y_j\} \in c_0(\mathcal{Y}) \quad \text{with} \quad \|\Lambda\|_{(c_0(\mathcal{Y}))^*} = \sum_j \|s_j\|_{\mathcal{Y}^*}.$$

If again $\mathcal{X}$ is the given space in Theorem 5.1.1, if Φ is an isomorphism from $\mathcal{X}$ onto some closed subspace of $c_0(\mathcal{Y})$ and Φ_j expresses the projection of Φ on the j-th coordinate, then $\mathsf{L} \in \mathcal{X}^*$ can be written as

$$\mathsf{L}(f) = \sum_j s_j(\Phi_j(f)), \quad \{s_j\} \in \ell^1(\mathcal{Y}^*) \quad \text{with} \quad \|\mathsf{L}\|_{\mathcal{X}^*} \approx \inf \sum_j \|s_j\|_{\mathcal{Y}^*},$$

where the infimum ranges over all sequences $\{s_j\}$ obeying the form of $\mathsf{L}(f)$ represented above.

Theorem 5.2.1. *Given $p \in (0, \infty)$, let $W = \{w_j\}$ be a discrete subset of $\mathbf{D}$ such that there exists a constant $\kappa \in (0,1)$ ensuring $\inf_j |z - w_j| < \kappa(1 - |z|)$, $z \in \mathbf{D}$. For $q > 0$, $j \in \mathbf{N} \cup \{0\}$, and $f \in \mathcal{Q}_{p,0}$ set*

$$\Phi_{q,0}f(z) = f(0), \quad \Phi_{q,j}f(z) = (1 - |w_j|^2)^{\frac{q}{2}} \int_0^z \frac{f'(w)}{(1 - \overline{w_j}w)^{\frac{p+q}{2}}} dw, \quad j \in \mathbf{N}.$$

Then:

(i) $\Phi_q(f) = \{\Phi_{q,j}(f)\}$ *is an isomorphism from $\mathcal{Q}_{p,0}$ onto a closed subspace of $c_0(\mathcal{D}^{2,p})$.*

(ii) *Each element $g \in \Theta(\mathcal{Q}_{p,0}^*)$ can be represented as $g = \overline{a_0} + \sum_{j=1}^{\infty} \overline{a_j} g_j$ where $\{a_j\} \in \ell^1$ and*

$$g_j(z) = \frac{z(1 - |w_j|^2)^{\frac{q}{2}}}{\pi} \int_{\mathbf{D}} \frac{h_j'(w)(-\log|w|^2)}{(1 - w_j\overline{w})^{\frac{p+q}{2}}(1 - \overline{w}z)^2} dm(w) \text{ with } \|h_j\|_{\mathcal{D}^{2,2-p}} \approx 1.$$

In this case

$$\|g\|_{\Theta(\mathcal{Q}_{p,0}^*)} \approx \inf \|\{a_j\}\|_{\ell^1},$$

where the infimum is taken over all complex-valued sequences $\{a_j\}$ satisfying the previously-mentioned representation of g.

Proof. (i) This follows from the vanishing version of Lemma 3.1.1, the definition of $\mathcal{Q}_{p,0}$ and the following implication for $z \in \mathbf{D}$:

$$\frac{|w_j - w|}{1 - |w|} < \kappa \Rightarrow \frac{1 - |w|}{1 - |w_j|} < (1 - \kappa)^{-1} \quad \text{and} \quad \left|\frac{1 - \overline{w_j}z}{1 - \bar{w}z}\right| \le 1 + \frac{|w_j - w|}{|1 - \bar{w}z|} < 1 + \kappa.$$

(ii) By the above-stated general duality principle we see that any $\mathsf{L} \in \mathcal{Q}_{p,0}^*$ can be written as

$$\mathsf{L}(f) = a_0 f(0) + \sum_{j=1}^{\infty} a_j s_j\big(\Phi_{q,j}(f)\big),$$

where

$$s_j \in (\mathcal{D}^{2,p})^* \quad \text{with} \quad \|s_j\|_{(\mathcal{D}^{2,p})^*} = 1 \quad \text{and} \quad \|\{a_j\}\|_{\ell^1} < \infty.$$

Note that to each $s_j \in (\mathcal{D}^{2,p})^*$ with $\|s_j\|_{(\mathcal{D}^{2,p})^*} = 1$ there corresponds a function $h_j \in \mathcal{D}^{2,2-p}$ such that $\|h_j\|_{\mathcal{D}^{2,2-p}} \approx 1$ and

$$s_j(g) = g(0)\overline{h_j(0)} + \pi^{-1} \int_{\mathbf{D}} g'(z)\overline{h_j'(z)}(-\log|z|^2)dm(z), \quad g \in \mathcal{D}^{2,p}.$$

This fact, along with a routine calculation with $\overline{\mathsf{L}\big((1 - \bar{z}w)\big)}$, yields the desired result. $\qquad\qquad\qquad\qquad\qquad\qquad\qquad\qquad\qquad\qquad\qquad\qquad\qquad\qquad\square$

Obviously, Theorem 5.2.1 is not good enough from a function-theoretic viewpoint. In what follows we characterize the Cauchy dual of $\mathcal{Q}_{p,0}$ in terms of the so-called weak factorization. Given two Banach spaces $\mathcal{U}, \mathcal{V} \subseteq \mathcal{H}$, let the dot product $\mathcal{U} \odot \mathcal{V}$ be the class of all functions $f \in \mathcal{H}$ of the form $f(z) = \sum_{j=1}^{\infty} u_j(z)v_j(z)$, $z \in \mathbf{D}$, where $u_j \in \mathcal{U}, v_j \in \mathcal{V}$, and the sum converges on compacta of $\mathbf{D}$. $\mathcal{U} \odot \mathcal{V}$ becomes a Banach space equipped with the norm

$$\|f\|_{\mathcal{U} \odot \mathcal{V}} = \inf\Big\{\sum_{j=1}^{\infty} \|u_j\|_{\mathcal{U}}\|v_j\|_{\mathcal{V}} : \quad f = \sum_{j=1}^{\infty} u_j v_j\Big\}.$$

It is easy to see that if $\mathcal{H}(\bar{\mathbf{D}})$ is dense in both $\mathcal{U}$ and $\mathcal{V}$ so is it in $\mathcal{U} \odot \mathcal{V}$.

To identify the Cauchy dual of a vanishing Q class with a dot product, we need a more dedicate treatment for the integrals used in Theorem 5.2.1.

Lemma 5.2.2. *Given $\alpha \in [0, \infty)$, $\beta \in (-1, \infty)$ and $-1 < \gamma < 2(1 + \beta)$, let $K \in \mathcal{H}$ be such that $K(z) = \sum_{j=0}^{\infty} k_j z^j$ with $|k_j - k_{j-1}| \preceq j^{\alpha}$ for $j \in \mathbf{N}$. Define*

$$\mathsf{H}_{\lambda,K}f(z) = \int_{\mathbf{D}} \left(\frac{1 - \bar{\lambda}z}{1 - \lambda\bar{w}}\right) K(z\bar{w})(-\log|w|^2)^{\beta} f(w)dm(w), \quad \lambda \in \mathbf{D}, \ f \in \mathcal{D}^{2,\gamma}.$$

If $\delta = \gamma - 2(\beta - \alpha) > -1$, then $\mathsf{H}_{\lambda,K}$ is a bounded operator from $\mathcal{D}^{2,\gamma}$ to $\mathcal{D}^{2,\delta}$ with $\sup_{\lambda \in \mathbf{D}} \|\mathsf{H}_{\lambda,K}\|_{\mathcal{D}^{2,\gamma} \mapsto \mathcal{D}^{2,\delta}} < \infty.$

Proof. Case 1: $\delta > 1$. With this condition, we have that if $f(z) = \sum_{j=0}^{\infty} a_j z^j \in \mathcal{D}^{2,\delta}$, then

$$\|f\|_{\mathcal{D}^{2,\delta}}^2 \approx \int_{\mathbf{D}} |f(z)|^2 (1 - |z|^2)^{\delta-2} dm(z) \approx \sum_{j=0}^{\infty} (1+j)^{1-\delta} |a_j|^2.$$

Since all polynomials are dense in $\mathcal{D}^{2,\gamma}$, it suffices to demonstrate that there is a constant $C > 0$ independent of $\lambda \in \mathbf{D}$ such that $\|\mathsf{H}_{\lambda,K} f\|_{\mathcal{D}^{2,\delta}} \leq C \|f\|_{\mathcal{D}^{2,\gamma}}$ holds for any polynomial f. To do so, note that $\|\mathsf{H}_{\lambda,K} f\|_{\mathcal{D}^{2,\delta}}$ is a subharmonic function of $\lambda \in \mathbf{D}$ and extendable continuously to $\bar{\mathbf{D}}$. Thus it is enough to consider $\|\mathsf{H}_{1,K} f\|_{\mathcal{D}^{2,\delta}}$ thanks to

$$\mathsf{H}_{\lambda,K} f(z) = \mathsf{H}_{1,K} f_\lambda(\bar{\lambda} z) \quad \text{where} \quad \lambda \in \mathbf{T} \quad \text{and} \quad f_\lambda(z) = f(\lambda z).$$

To that end, noticing

$$c_{j,\beta} = 2\pi \int_0^1 r^{2j} (-\log r^2)^\beta dr \approx (j+1)^{-(1+\beta)}, \quad j \in \mathbf{N} \cup \{0\},$$

we get that if $f(z) = \sum_{j=0}^{\infty} a_j z^j$ is a polynomial, then

$$\mathsf{H}_{1,K} f(z) = (1-z) \sum_{n=0}^{\infty} c_{n,\beta} a_n \sum_{j=0}^{n} k_j z^j = F_1(z) + F_2(z) + F_3(z),$$

where

$$F_1(z) = \sum_{j=1}^{\infty} (k_j - k_{j-1}) z^j \sum_{n=j}^{\infty} c_{n,\beta} a_n, \quad F_2(z) = k_0 \sum_{j=0}^{\infty} c_{j,\beta} a_j,$$

and

$$F_3(z) = -\sum_{j=0}^{\infty} k_j z^{j+1} c_{j,\beta} a_j.$$

Taking $\eta > \gamma - 1$, $\zeta > 1$ and $\eta + \zeta = 2(\beta - 1)$, we employ the Cauchy–Schwarz inequality, the estimate of $c_{j,\beta}$ and the assumption on K to produce

$$\|F_1\|_{\mathcal{D}^{2,\delta}}^2 \preceq \sum_{j=1}^{\infty} \frac{|k_j - k_{j-1}|^2}{(j+1)^{\delta-1}} \left| \sum_{n=j}^{\infty} c_{n,\beta} a_n \right|^2$$

$$\preceq \sum_{j=1}^{\infty} \frac{|k_j - k_{j-1}|^2}{(j+1)^{\delta-1}} \left(\Big(\sum_{n=j}^{\infty} \frac{|a_n|^2}{(n+1)^\eta} \Big) \Big(\sum_{n=j}^{\infty} \frac{1}{(n+1)^\zeta} \Big) \right)$$

$$\preceq \sum_{j=1}^{\infty} \frac{|k_j - k_{j-1}|^2}{(j+1)^{\delta-1}} j^{1-\zeta} \Big(\sum_{n=j}^{\infty} \frac{|a_n|^2}{(n+1)^\eta} \Big)$$

$$\preceq \sum_{n=1}^{\infty} \frac{|a_n|^2}{(n+1)^\eta} \sum_{j=1}^{n} (j+1)^{2+2\alpha-\zeta-\delta}$$

$$\preceq \|f\|_{\mathcal{D}^{2,\gamma}}^2.$$

Owing to $-1 < \gamma < 2(1+\beta)$, the Cauchy–Schwarz inequality yields

$$|F_2(z)|^2 \leq |k_0|^2 \Big(\sum_{j=0}^{\infty} c_{j,\beta}^2 (1+j)^{\delta-1} \Big) \Big(\sum_{j=0}^{\infty} (1+j)^{1-\delta} |a_j|^2 \Big) \preceq |k_0|^2 \|f\|_{\mathcal{D}^{2,\delta}}^2.$$

Furthermore, the condition on K gives $|k_j| \preceq (1+j)^{1+\alpha}$ for $j \in \{0\} \cup \mathbf{N}$ and thus

$$\|F_3\|_{\mathcal{D}^{2,\delta}}^2 \preceq \sum_{j=0}^{\infty} c_{j,\beta}^2 (1+j)^{1-\delta} |k_j|^2 |a_j|^2 \preceq \|f\|_{\mathcal{D}^{2,\gamma}}^2.$$

The above three estimates imply $\|\mathsf{H}_{1,K} f\|_{\mathcal{D}^{2,\delta}} \preceq \|f\|_{\mathcal{D}^{2,\gamma}}^2$.

Case 2: $\delta \in (-1, 1]$. Under this condition, we define

$$K_1(z) = zK'(z) = \sum_{j=0}^{\infty} k_{j,1} z^j \quad \text{and} \quad K_2(z) = z^2 K''(z) = \sum_{j=0}^{\infty} k_{j,2} z^j.$$

Using the assumption on K once again, we derive

$$|k_{j,1} - k_{j-1,1}| \preceq j^{1+\alpha} \quad \text{and} \quad |k_{j,2} - k_{j-1,2}| \preceq j^{2+\alpha} \quad \text{for} \quad j \in \mathbf{N},$$

and so that if $f \in \mathcal{D}^{2,\gamma}$, then

$$\begin{aligned}
\|\mathsf{H}_{\lambda,K} f\|_{\mathcal{D}^{2,\delta}}^2 &\approx |\mathsf{H}_{\lambda,K} f(0)|^2 + \|(\mathsf{H}_{\lambda,K} f)'\|_{\mathcal{D}^{2,2+\delta}}^2 \\
&\preceq \|f\|_{\mathcal{D}^{2,\gamma}}^2 + \int_{\mathbf{D}} |(\mathsf{H}_{\lambda,K} f)''(z)|^2 (1-|z|^2)^{2+\delta} dm(z) \\
&\preceq \|f\|_{\mathcal{D}^{2,\gamma}}^2 + \int_{\mathbf{D}} |\mathsf{H}_{\lambda,K_1} f(z)|^2 (1-|z|^2)^{\delta} dm(z) \\
&\quad + \int_{\mathbf{D}} |\mathsf{H}_{\lambda,K_2} f(z)|^2 (1-|z|^2)^{2+\delta} dm(z) \\
&\preceq \|f\|_{\mathcal{D}^{2,\gamma}}^2 + \|\mathsf{H}_{\lambda,K_1} f\|_{\mathcal{D}^{2,2+\delta}}^2 + \|\mathsf{H}_{\lambda,K_2} f\|_{\mathcal{D}^{2,4+\delta}}^2 \\
&\preceq \|f\|_{\mathcal{D}^{2,\gamma}}^2,
\end{aligned}$$

as desired.   □

Now we are ready to establish the following factorization theorem for the Cauchy dual of a vanishing Q class.

Theorem 5.2.3. *Given $p \in (0, 2)$, let $W = \{w_j\}$ be a discrete subset of $\mathbf{D}$ such that there exists a constant $\kappa \in (0, 1)$ ensuring $\inf_j |z - w_j| < \kappa(1 - |z|)$ for any $z \in \mathbf{D}$. If $\kappa > 0$ is small enough, then:*

(i) $\Theta(\mathcal{Q}_{p,0}^*) \cong \mathcal{D}^{1, -\frac{2-p}{2}} \odot \mathcal{D}^{2, 2-p}$. *Equivalently, $f \in \mathcal{D}^{1, \frac{p-2}{2}} \odot \mathcal{D}^{2, 2-p}$ if and only if*

$$f = \sum_j c_j \frac{(1-|w_j|^2)^{1-\frac{p}{2}}}{1 - \overline{w_j} z} v_j, \quad \|f\|_{\Theta(\mathcal{Q}_{p,0}^*)} \approx \|\{c_j\}\|_{\ell^1} < \infty,$$

where $v_j \in \mathcal{D}^{2, 2-p}$ with $\|v_j\|_{\mathcal{D}^{2, 2-p}} = 1$.

(ii) *The polynomials are dense in $\Theta(\mathcal{Q}_{p,0}^*)$ and so $\Theta\big((\Theta(\mathcal{Q}_{p,0}^*))^*\big) \cong \mathcal{Q}_p$.*

Proof. (i) Suppose $f \in \mathcal{Q}_{p,0}$ and $g \in \mathcal{D}^{1,-\frac{2-p}{2}} \odot \mathcal{D}^{2,2-p}$. So to prove that $\langle \cdot, g \rangle$ generates a continuous linear functional on $\mathcal{Q}_{p,0}$, we just need to check that

$$\int_{\mathbf{D}} |f'(z)(uv)'(z)|(-\log|z|^2)dm(z) \precsim \|f\|_{\mathcal{Q}_p,1}\|u\|_{\mathcal{D}^{1,-\frac{2-p}{2}}}\|v\|_{\mathcal{D}^{2,2-p}}$$

for any $u \in \mathcal{D}^{1,-\frac{2-p}{2}}$ and $v \in \mathcal{D}^{2,2-p}$. Nevertheless, Example 5.1.2 tells us that it is enough to demonstrate that the last inequality holds for

$$u(z) = (1 - |w_j|)^{1-\frac{p}{2}}(1 - \overline{w_j}z)^{-1},$$

as $\kappa > 0$ is sufficiently small. This can be done by handling the following two integrals:

$$I_1(f) = \int_{\mathbf{D}} |f'(z)|(1 - |w_j|)^{1-\frac{p}{2}}|1 - \overline{w_j}z|^{-2}|v(z)|(-\log|z|^2)dm(z)$$

and

$$I_2(f) = \int_{\mathbf{D}} |f'(z)|(1 - |w_j|)^{1-\frac{p}{2}}|1 - \overline{w_j}z|^{-1}|v'(z)|(-\log|z|^2)dm(z).$$

In order to control $I_1(z)$, we employ Lemma 3.1.1 with

$$d\mu_{f,p}(z) = |f'(z)|^2(1 - |z|^2)^p dm(z) \in \mathcal{CM}_p,$$

the Cauchy–Schwarz inequality and the fact that $(1 - |z|^2)$ and $(-\log|z|^2)$ give equivalent Bergman spaces to get

$$\begin{aligned}
I_1(f) &\precsim \left(\int_{\mathbf{D}} \frac{(1 - |w_j|^2)^{2-p}}{|1 - \overline{w_j}z|^2} d\mu_{f,p}(z)\right)^{\frac{1}{2}} \left(\int_{\mathbf{D}} |v(w)|^2(1 - |z|^2)^{-p}dm(z)\right)^{\frac{1}{2}} \\
&\precsim \|f\|_{\mathcal{Q}_p,1}\|v\|_{\mathcal{D}^{2,2-p}}.
\end{aligned}$$

Similarly, for $I_2(f)$ we have

$$\begin{aligned}
I_2(f) &\precsim \left(\int_{\mathbf{D}} \frac{(1 - |w_j|^2)^{2-p}}{|1 - \overline{w_j}z|^2} d\mu_{f,p}(z)\right)^{\frac{1}{2}} \left(\int_{\mathbf{D}} |v'(w)|^2(1 - |z|^2)^{2-p}dm(z)\right)^{\frac{1}{2}} \\
&\precsim \|f\|_{\mathcal{Q}_p,1}\|v\|_{\mathcal{D}^{2,2-p}}.
\end{aligned}$$

Conversely, assuming $g \in \Theta(\mathcal{Q}_{p,0}^*)$, we can use Theorem 5.2.1 (ii) to obtain $g = \overline{a_0} + \sum_{j=1}^{\infty} \overline{a_j}g_j$ where $\|\{a_j\}\|_{\ell^1} \precsim \|g\|_{\Theta(\mathcal{Q}_{p,0}^*)}$ and

$$g_j(z) = \frac{z(1 - |w_j|^2)^{1-\frac{p}{2}}}{\pi} \int_{\mathbf{D}} \frac{h_j'(w)(-\log|w|^2)}{(1 - w_j\overline{w})^{\frac{p+q}{2}}(1 - \overline{w}z)^2}dm(w)$$

with $\|h_j\|_{\mathcal{D}^{2,2-p}} \approx 1$. Clearly, if $K(z) = (1-z)^{-2}$, $\alpha = 0, \beta = 1, \gamma = 4 - p$ in Lemma 5.2.2, then

$$g_j(z) = \frac{z(1 - |w_j|^2)^{1-\frac{p}{2}}}{\pi(1 - \overline{w_j}z)} H_{w_j, K} h_j'(z)$$

and hence $g \in \mathcal{D}^{1,-\frac{2-p}{2}} \odot \mathcal{D}^{2,2-p}$ because of Lemma 5.2.2 with

$$\int_{\mathbf{D}} \frac{(1 - |z|^2)^{-\frac{2-p}{2}}}{|1 - \overline{w_j}z|^2} dm(z) \preceq (1 - |w_j|)^{\frac{p-2}{2}} \quad \text{and} \quad \|H_{w_j, K} h_j'\|_{\mathcal{D}^{2,2-p}} \preceq \|h_j\|_{\mathcal{D}^{2,2-p}}.$$

(ii) Since the polynomials are dense in $\mathcal{D}^{1,-\frac{2-p}{2}} \odot \mathcal{D}^{2,2-p}$ and $\mathcal{Q}_{p,0}$, we conclude from (i) and Theorem 5.1.1 that the desired assertion is true. $\qquad \square$

Corollary 5.2.4. *Given $p \in (0,2)$. Then:*

(i) $\Theta(\mathcal{Q}_{p_1,0}^*) \subseteq \Theta(\mathcal{Q}_{p_2,0}^*)$ *for $0 < p_2 \le p_1 < 2$. In particular, $\Theta(\mathcal{Q}_{1,0}^*) \cong \mathcal{H}^1$ and $\Theta(\mathcal{Q}_{p,0}^*) \cong \mathcal{D}^{1,0}$ for all $p \in (1,2)$.*

(ii) $\|f \circ \sigma_w\|_{\Theta(\mathcal{Q}_{p,0}^*)} \approx \|f\|_{\Theta(\mathcal{Q}_{p,0}^*)}$ *for $f \in \Theta(\mathcal{Q}_{p,0}^*)$ and $w \in \mathbf{D}$.*

Proof. (i) and (ii) can be demonstrated by using Theorem 5.2.3 with some simple calculations with $\mathcal{D}^{1,-\frac{2-p}{2}} \odot \mathcal{D}^{2,2-p}$. $\qquad \square$

5.3   Atom-like Representations

For the dyadic closed intervals $[2\pi k 2^{-n}, 2\pi(k+1)2^{-n}]$ we write $S_{n,k}$ for the corresponding Carleson boxes. As before, we have that for $p \in (0, \infty)$ and $f \in \mathcal{H}$,

$$f \in \mathcal{Q}_p \Leftrightarrow \sup_{0 \le k \le 2^n, n \in \mathbf{N}} 2^{-np} \int_{S_{n,k}} |f'(z)|^2 (1 - |z|^2)^p dm(z) < \infty$$

and

$$f \in \mathcal{Q}_{p,0} \Leftrightarrow \lim_{n \to \infty} \sup_{0 \le k \le 2^n} 2^{-np} \int_{S_{n,k}} |f'(z)|^2 (1 - |z|^2)^p dm(z) = 0.$$

However, if $p \in (0,1)$, then for $f \in \mathcal{H}$,

$$f \in \mathcal{Q}_p \Leftrightarrow \|f\|_{\mathcal{Q}_p,3}^2 = \sup_{I \subseteq [0,2\pi]} |I|^{-p} \int_I \int_I \frac{|f(e^{it}) - f(e^{is})|^2}{|e^{it} - e^{is}|^{2-p}} dt ds < \infty$$

and

$$f \in \mathcal{Q}_{p,0} \Leftrightarrow \limsup_{|I| \to 0} |I|^{-p} \int_I \int_I \frac{|f(e^{it}) - f(e^{is})|^2}{|e^{it} - e^{is}|^{2-p}} dt ds = 0.$$

Here $|I| = b - a$ if I is an interval with endpoints a and b.

Lemma 5.3.1. *Given $p \in (0,1)$ and $f \in \mathcal{H}$. Then there is a sequence $\{I_j\}$ of intervals contained in $[0, 2\pi]$ such that $\lim_{j \to \infty} |I_j| = 0$ and*

$$f \in \mathcal{Q}_p \Leftrightarrow \sup_{j \in \mathbf{N}} |I_j|^{-p} \int_{I_j} \int_{I_j} \frac{|f(e^{it}) - f(e^{is})|^2}{|e^{it} - e^{is}|^{2-p}} dt ds < \infty$$

and

$$f \in \mathcal{Q}_{p,0} \Leftrightarrow \limsup_{j \to \infty} |I_j|^{-p} \int_{I_j} \int_{I_j} \frac{|f(e^{it}) - f(e^{is})|^2}{|e^{it} - e^{is}|^{2-p}} dt ds = 0.$$

Proof. Without loss of generality, we may assume $f(0) = 0$. It suffices to prove the first equivalence. To do so, for $f \in \mathcal{Q}_p$ and $0 \leq a < b \leq 2\pi$ let

$$F_f(a, b) = (b - a)^{-p} \int_a^b \int_a^b \frac{|f(e^{it}) - f(e^{is})|^2}{|e^{it} - e^{is}|^{2-p}} dt ds.$$

We are about to show that the family of functions

$$\mathcal{F} = \{F_f : f \in \mathcal{Q}_p, \; \|f\|_{\mathcal{Q}_p,3} = 1\}$$

is equicontinuous on any δ-triangle

$$\Delta_\delta = \{(a, b) : \; 0 \leq a \leq b - \delta < b \leq 2\pi\}.$$

By so doing, we observe that for $(a_1, b), (a_2, b) \in \Delta_\delta, a_1 < a_2$ the following estimate holds:

$$\begin{aligned}
|F_f(a_1, b) - F_f(a_2, b)| &\leq \left(1 - \left(\frac{b - a_1}{b - a_2}\right)^p\right) F_f(a_1, b) \\
&\quad + \frac{2}{\delta^p} \int_{a_1}^{a_2} \int_{a_1}^b \frac{|f(e^{it}) - f(e^{is})|^2}{|e^{it} - e^{is}|^{2-p}} dt ds \\
&= \mathrm{T}_1 + \mathrm{T}_2.
\end{aligned}$$

Evidently, if a_2 is so close to a_1 that $(a_1, b), (a_2, b) \in \Delta_\delta$, then T_1 tends to 0 uniformly for $f \in \mathcal{Q}_p$ with $\|f\|_{\mathcal{Q}_p,3} = 1$. To handle T_2, we choose $b_2 \in (a_2, b)$ and assume

$$J_1 = \int_{a_1}^{b_2} \int_{a_1}^{b_2} \frac{|f(e^{it}) - f(e^{is})|^2}{|e^{it} - e^{is}|^{2-p}} dt ds \quad \text{and} \quad J_2 = \int_{a_1}^{a_2} \int_{b_2}^b \frac{|f(e^{it}) - f(e^{is})|^2}{|e^{it} - e^{is}|^{2-p}} dt ds.$$

Clearly, $J_1 \leq (b_2 - a_1)^p$. To estimate J_2, we note that there is a constant $\kappa > 0$ such that

$$|e^{it} - e^{is}| \geq \kappa(b_2 - a_2), \quad (t, s) \in [b_2, b] \times [a_1, a_2].$$

Using $q > 1$, the Hölder inequality and the inclusion $\mathcal{Q}_p \subset \mathcal{BMOA} \subset \mathcal{L}^q(\mathbf{T})$, we derive

$$J_2 \;\leq\; \frac{\left((a_2 - a_1)(b - b_2)\right)^{\frac{q-1}{q}}}{\left(\kappa(b_2 - a_2)\right)^{2-p}} \left(\int_0^{2\pi} \int_0^{2\pi} |f(e^{it}) - f(e^{is})|^{2q} dt ds \right)^{\frac{1}{q}}$$

$$\preceq\; \frac{\left((a_2 - a_1)(b - b_2)\right)^{\frac{q-1}{q}}}{\left(\kappa(b_2 - a_2)\right)^{2-p}} \|f\|^2_{\mathcal{Q}_p,3}.$$

Accordingly, if a_2 tends to a_1 with $(a_2 - a_1)^{\frac{q-1}{q}}(b_2 - a_2)^{2-p} \to 0$ for a suitable b_2, then J_1 and J_2 approach 0 uniformly on the unit sphere of $(\mathcal{Q}_p, \|\cdot\|_{\mathcal{Q}_p,3})$. The equicontinuity in the other variable can be verified similarly. To close the proof, we employ a compactness argument to find that for each $k \in \mathbf{N}$ there are a δ_k and an $S_k \subset \Delta_{\delta_{k+1}} \setminus \Delta_{\delta_k}$ such that

$$\inf_{(x,y)\in S_k} |F_f(a, b) - F_f(x, y)| < 3^{-1}, \quad (a, b) \in \Delta_{\delta_k}, \quad \|f\|_{\mathcal{Q}_p,3} = 1.$$

Now, enumerating the collection of intervals with endpoints in $\cup S_k$, we obtain a sequence of intervals $I_j = (a_j, b_j)$ such that $|I_j| = b_j - a_j \to 0$ and

$$\inf_{j\in\mathbf{N}} |F_f(a, b) - F_f(a_j, b_j)| < 2^{-1}\|f\|^2_{\mathcal{Q}_p,3}, \quad (a, b) \in \Delta_{\delta_k}, \quad f \in \mathcal{Q}_p.$$

This proves the desired assertion. $\qquad\qquad\qquad\qquad\qquad\qquad\qquad\qquad\quad\square$

In what follows we give two descriptions of $\Theta(\mathcal{Q}_{p,0}^*)$ using atom-like functions of different types. The atom-like function of the first type is from $\mathbf{D}$. More precisely, given $p \in (0, \infty)$ we consider a function $u \in \mathcal{H}$ on a Carleson box $S(I)$ based on an arc $I \subseteq \mathbf{T}$ satisfying

$$|I|^p \int_{S(I)} |u(z)|^2(1 - |z|^2)^p dm(z) \leq 1.$$

The desired atom-like function is of the form

$$a(z) = z \int_{S(I)} \frac{u(w)}{(1 - z\bar{w})^2}(1 - |w|^2)^p dm(w).$$

Note that if $\mathsf{L} \in \mathcal{Q}_{p,0}^*$ is given by

$$\mathsf{L}(f) = \int_{S(I)} f'(w)\overline{u(w)}(1 - |w|^2)^p dm(w), \quad f \in \mathcal{Q}_{p,0},$$

then

$$a(z) = \overline{\mathsf{L}\left((1 - \bar{z}w)^{-1}\right)}$$

and hence
$$a \in \Theta(\mathcal{Q}_{p,0}^*) \quad \text{with} \quad \|a\|_{\Theta(\mathcal{Q}_{p,0}^*)} = \|\mathsf{L}\|_{\mathcal{Q}_{p,0}^*} \preceq 1.$$

The atom-like function of the second type is based on **T**. Given $p \in (0,1)$ and I a subinterval of $[0, 2\pi]$, let v be a Lebesgue measurable function on I such that

$$|I|^p \int_I \int_I \frac{|v(t) - v(s)|^2}{|e^{it} - e^{is}|^{2-p}} dt ds = |I|^p \int_I \int_I |v(t) - v(s)|^2 |e^{it} - e^{is}|^{p-2} dt ds \leq 1$$

and

$$b(z) = z \int_I \int_I \frac{(v(t) - v(s))(e^{-it} - e^{-is})}{(1 - ze^{-it})(1 - ze^{-is})} |e^{it} - e^{is}|^{p-2} dt ds.$$

This function $b(z)$ is called the atom-like function of the second type. It is indeed the element of $\Theta(\mathcal{Q}_{p,0}^*)$ generated by the following L, i.e.,

$$b(z) = \overline{\mathsf{L}\big((1 - \bar{z}e^{it})^{-1}\big)},$$

where L is an element of $\mathcal{Q}_{p,0}^*$ determined by

$$\mathsf{L}(f) = \int_I \int_I \overline{(v(t) - v(s))} \big(f(e^{it}) - f(e^{is})\big) |e^{it} - e^{is}|^{p-2} dt ds, \quad f \in \mathcal{Q}_{p,0}.$$

Theorem 5.3.2. (i) *If $p \in (0, 2)$, then every function $f \in \Theta(\mathcal{Q}_{p,0}^*)$ can be written as*

$$f = f(0) + \sum_{j=1}^{\infty} \overline{c_j} a_j, \quad \{c_j\} \in \ell^1,$$

where $\{a_j\}$ are atom-like functions of the first type, and

$$\|f\|_{\Theta(\mathcal{Q}_{p,0}^*)} \approx |f(0)| + \inf\Big\{\sum_{j=1}^{\infty} |c_j|\Big\}, \quad f \in \Theta(\mathcal{Q}_{p,0}^*),$$

for which the infimum is taken over all above-mentioned representations of f.

(ii) *If $p \in (0, 1)$, then every function $f \in \Theta(\mathcal{Q}_{p,0}^*)$ can be written as*

$$f = f(0) + \sum_{j=1}^{\infty} \overline{c_j} b_j, \quad \{c_j\} \in \ell^1,$$

where $\{b_j\}$ are atom-like functions of the second type, and

$$\|f\|_{\Theta(\mathcal{Q}_{p,0}^*)} \approx |f(0)| + \inf\Big\{\sum_{j=1}^{\infty} |c_j|\Big\}, \quad f \in \Theta(\mathcal{Q}_{p,0}^*),$$

for which the infimum is taken over all above-mentioned representations of f.

Proof. (i) We introduce a mapping Ψ from $\mathcal{Q}_{p,0}$ to $c_0\big(\mathcal{L}^2(\mathbf{D},(1-|z|^2)^p dm(z))\big)$ as follows. Fix an enumeration $\{J_j\}_{j=1}^\infty$ of the dyadic subarcs of $\mathbf{T}$ and let $\{S_j\}_{j=1}^\infty$ be the sequence of Carleson boxes based on $\{J_j\}_{j=1}^\infty$. For $f \in \mathcal{Q}_{p,0}$ set

$$\Psi_{1,0}f = f(0), \quad \Psi_{1,j}f = |J_j|^{-\frac{p}{2}} 1_{S_j} f', \quad j \in \mathbf{N}.$$

Then

$$\Psi_1 = \{\Psi_{1,j}\}_{j=1}^\infty : \quad \mathcal{Q}_{p,0} \mapsto c_0\big(\mathcal{L}^2(\mathbf{D},(1-|z|^2)^p dm(z))\big)$$

is bounded from above and below. Consequently, $\Psi(\mathcal{Q}_{p,0})$ is a closed subspace of $c_0\big(\mathcal{L}^2(\mathbf{D},(1-|z|^2)^p dm(z))\big)$. This fact and that general duality principle right before Theorem 5.2.1 give that for $\mathsf{L} \in \mathcal{Q}_{p,0}^*$ and $f \in \mathcal{Q}_{p,0}$, $\mathsf{L}(f)$ can be represented as

$$\mathsf{L}(f) \;=\; c_0 f(0) + \sum_{j=1}^\infty c_j \int_{\mathbf{D}} \overline{g_j(z)} \Psi_{1,j}f(z)(1-|z|^2)^p dm(z)$$

$$\;=\; c_0 f(0) + \sum_{j=1}^\infty c_j |J_j|^{-\frac{p}{2}} \int_{S_j} \overline{g_j(z)} f'(z)(1-|z|^2)^p dm(z),$$

where $\{c_j\} \in \ell^1$ and

$$g_j \in \mathcal{L}^2\big(\mathbf{D},(1-|z|^2)^p dm(z)\big) \text{ with } \|g_j\|_{\mathcal{L}^2\big(\mathbf{D},(1-|z|^2)^p dm(z)\big)} = 1.$$

Moreover, we can select $\{c_j\}$ and $\{g_j\}$ such that $\|\{c_j\}\|_{\ell^1} \preceq \|\mathsf{L}\|_{\mathcal{Q}_{p,0}^*}$. Note that $\mathcal{H} \cap \mathcal{L}^2(S_j,(1-|z|^2)^p dm(z))$ is a closed subspace of $\mathcal{L}^2(S_j,(1-|z|^2)^p dm(z))$. So an application of the Riesz Representation Theorem produces a function $u_j \in \mathcal{H}$ such that for $f \in \mathcal{Q}_{p,0}$,

$$|J_j|^{-\frac{p}{2}} \int_{S_j} \overline{g_j(z)} f'(z)(1-|z|^2)^p dm(z) = \int_{S_j} \overline{u_j(z)} f'(z)(1-|z|^2)^p dm(z).$$

Letting

$$a_j(z) = z \int_{S_j} \frac{u_j(w)}{(1-z\bar{w})^2}(1-|w|^2)^p dm(w)$$

and evaluating $\overline{\mathsf{L}\big((1-\bar{z}w)^{-1}\big)}$, we derive the desired representation result.

(ii) The argument is similar. The key issue is to introduce a map from $\mathcal{Q}_{p,0}$ to $c_0\big(\mathcal{L}^2([0,2\pi] \times [0,2\pi], |e^{it} - e^{is}|^{p-2} dt ds)\big)$. In so doing, for $f \in \mathcal{Q}_{p,0}$, $p \in (0,1)$ and the sequence of intervals $\{I_j\}$ determined by Lemma 5.3.1, let $\Psi_2 f = \{\Psi_{2,j}f\}$, where

$$\Psi_{2,0}f = f(0), \quad \Psi_{2,j}f(t,s) = |I_j|^{-\frac{p}{2}} 1_{I_j}(t) 1_{I_j}(s)\big(f(e^{it}) - f(e^{is})\big), \quad j \in \mathbf{N}.$$

Then $\mathsf{L} \in \mathcal{Q}_{p,0}^*$ can be expressed as

$$\mathsf{L}(f) = c_0 f(0) + \sum_{j=1}^{\infty} c_j \int_{[0,2\pi]} \int_{[0,2\pi]} \overline{g_j(t,s)} \Psi_{2,j} f(t,s) |e^{it} - e^{is}|^{p-2} dt ds$$

$$= c_0 f(0) + \sum_{j=1}^{\infty} c_j |I_j|^{-\frac{p}{2}} \int_{I_j} \int_{I_j} \overline{g_j(t,s)} \big(f(e^{it}) - f(e^{is})\big)|e^{it} - e^{is}|^{p-2} dt ds,$$

where $\{c_j\} \in \ell^1$, and $g_j(\cdot,\cdot) \in \mathcal{L}^2([0,2\pi] \times [0,2\pi], |e^{it} - e^{is}|^{p-2} dt ds)$ is of the unit norm. As in (i), we can get such $\{c_j\}$ and $\{g_j\}$ that $\|\{c_j\}\|_{\ell^1} \precsim \|\mathsf{L}\|_{\mathcal{Q}_{p,0}^*}$. Since the class of all functions $v(t) - v(s)$ (where v produces an atom-like function of the second type) is closed in $\mathcal{L}^2([0,2\pi] \times [0,2\pi], |e^{it} - e^{is}|^{p-2} dt ds)$, we conclude from the Riesz Representation Theorem that for each j there is a function v_j such that

$$|I_j|^p \int_{I_j} \int_{I_j} |v_j(t) - v_j(s)|^2 |e^{it} - e^{is}|^{p-2} dt ds \leq 1$$

and

$$|I_j|^p \int_{I_j} \int_{I_j} \overline{g_j(t,s)} \big(f(e^{it}) - f(e^{is})\big)|e^{it} - e^{is}|^{p-2} dt ds$$

$$= \int_{I_j} \int_{I_j} \overline{(v_j(t) - v_j(s))} \big(f(e^{it}) - f(e^{is})\big)|e^{it} - e^{is}|^{p-2} dt ds.$$

Taking

$$b_j(z) = z \int_{I} \int_{I} \frac{(v_j(t) - v_j(s))(e^{-it} - e^{-is})}{(1 - ze^{-it})(1 - ze^{-is})} |e^{it} - e^{is}|^{p-2} dt ds,$$

and calculating $\overline{\mathsf{L}((1 - \bar{z}e^{it})^{-1})}$, we get the required assertion. $\qquad \square$

5.4 Extreme Points of Unit Balls

Given a Banach space $\mathcal{X}$ with norm $\|\cdot\|_{\mathcal{X}}$. The extreme points of the closed unit ball $\mathcal{B}_{\mathcal{X}} = \{f \in \mathcal{X} : \|f\|_{\mathcal{X}} \leq 1\}$ are the points which are not a proper convex combination of two different points of $\mathcal{B}_{\mathcal{X}}$.

From the previous discussions we have seen that the second dual space of $\mathcal{Q}_{p,0}$ is isomorphic to $\mathcal{Q}_p$ under the pairings $\langle \cdot, \cdot \rangle_{inv}$ and $\langle \cdot, \cdot \rangle$. So the well-known Krein–Milman Theorem then implies that the closed unit ball of $\mathcal{Q}_p$ has extreme points. We show that the closed unit ball of $\mathcal{Q}_{p,0}$ also has extreme points. The main result of this section is two characterizations of the extreme points of the closed unit ball of $\mathcal{Q}_{p,0}$.

Theorem 5.4.1. *Given $p \in (0, \infty)$ and $f \in \mathcal{Q}_p$.*

(i) *f is an extreme point of $\mathcal{B}_{\mathcal{Q}_{p,0}}$ with the norm $|f(0)| + \|f\|_{\mathcal{Q}_{p,1}}$ if and only if f is a constant function of module 1 or $f(0) = 0$ and $\|f\|_{\mathcal{Q}_{p,1}} = 1$.*

(ii) *If f is an extreme point of $\mathcal{B}_{\mathcal{Q}_p}$ with the norm $|f(0)| + \| \cdot \|_{\mathcal{Q}_{p,1}}$, then $|f(0| + \|f\|_{\mathcal{Q}_{p,1}} = 1$. Conversely, if f is a constant function of module 1 or there exists a point $w_0 \in \mathbf{D}$ such that $E_p(f, w_0) = 1$, then f is an extreme point of $\mathcal{B}_{\mathcal{Q}_p}$ with the norm $|f(0)| + \|f\|_{\mathcal{Q}_{p,1}}$.*

Proof. Defining

$$\mathcal{Q}_p^0 = \{ f \in \mathcal{Q}_p : \ f(0) = 0 \} \quad \text{and} \quad \mathcal{Q}_{p,0}^0 = \{ f \in \mathcal{Q}_{p,0} : \ f(0) = 0 \},$$

we find

$$\mathcal{Q}_p = \mathbf{C} \oplus \mathcal{Q}_p^0 \quad \text{and} \quad \mathcal{Q}_{p,0} = \mathbf{C} \oplus \mathcal{Q}_{p,0}^0.$$

Thus a routine argument tells us that $(|f(0)|, \|f\|_{\mathcal{Q}_{p,1}})$ is an extreme point for the closed unit ball of $\mathbf{R} \times \mathbf{R}$ in the Euclidean 1-norm if and only if either $f(0) = 0$ and $\|f\|_{\mathcal{Q}_{p,1}} = 1$ or $|f(0)| = 1$ and $\|f\|_{\mathcal{Q}_p} = 1$. With this equivalence, we make only the following consideration.

(i) Suppose f is an extreme point of $\mathcal{B}_{\mathcal{Q}_{p,0}^0}$ with respect to the norm $\|f\|_{\mathcal{Q}_{p,1}}$. If $\|f\|_{\mathcal{Q}_{p,1}} < 1$, then for

$$0 < \epsilon < \min\{1, \|f\|_{\mathcal{Q}_{p,1}}^{-1} - 1\}$$

let

$$f_1 = (1 - \epsilon)f \quad \text{and} \quad f_2 = (1 + \epsilon)f.$$

This choice gives $f_1, f_2 \in \mathcal{B}_{\mathcal{Q}_{p,0}}$ but $f = 2^{-1}(f_1 + f_2)$, contradicting the assumption that f is extreme. In order to verify the converse, we may assume $f \in \mathcal{Q}_{p,0}^0$ with $\|f\|_{\mathcal{Q}_{p,1}} = 1$. Then we have $\lim_{|w| \to 1} E_p(f, w) = 0$ and thus there is an $r \in (1/2, 1)$ such that $\sup_{|w| > r} (E_p(f, w)) \leq 1/2$. Accordingly, we get

$$1 = \|f\|_{\mathcal{Q}_{p,1}} = \sup_{w \in \mathbf{D}} E_p(f, w) = \max_{|w| \leq r} E_p(f, w).$$

Since $E_p(f, \cdot)$ is continuous on $\mathbf{D}$, it is uniformly continuous on the compact set $\{ w \in \mathbf{D} : |w| \leq r \}$. This yields that $E_p(f, w_0) = 1$ is valid for some $|w_0| \leq r$. Let now $g \in \mathcal{Q}_{p,0}^0$ be such that $\|f + g\|_{\mathcal{Q}_{p,1}} \leq 1$ and $\|f - g\|_{\mathcal{Q}_{p,1}} \leq 1$. Then

$$
\begin{aligned}
1 + \big(E_p(g, w_0)\big)^2 &= \big(E_p(f, w_0)\big)^2 + \big(E_p(g, w_0)\big)^2 \\
&= 2^{-1}\big(E_p(f + g, w_0)\big)^2 + \big(E_p(f - g, w_0)\big)^2 \\
&\leq 1.
\end{aligned}
$$

This forces $E_p(g, w_0) = 0$, and so $g = 0$ on $\mathbf{D}$. In other words, f is extreme.

(ii) The argument is essentially included in (i). □

Here it is worth pointing out that different norms produce usually different classes of the extreme points. To understand this principle, let us take $\| \cdot \|_{\mathcal{Q}_p, 2}$ into account. From now on, for $p \in (0, \infty)$ and a Lebesgue measurable function $f : \mathbf{T} \mapsto \mathbf{C}$ put

$$S_p(f, I) = (2\pi)^{-2} \int_I \int_I \frac{|f(\zeta) - f(\eta)|^2}{|\zeta - \eta|^{2-p}} |d\zeta||d\eta|.$$

We say that $f \in \mathcal{Q}_p(\mathbf{T})$ provided

$$\|f\|_{\mathcal{Q}_p(\mathbf{T})} = (2\pi)^{-1} \left| \int_{\mathbf{T}} f(\zeta)|d\zeta| \right| + \sup_{I \subseteq \mathbf{T}} |I|^{-p} S_p(f, I) < \infty$$

where the supremum is taken over all subarcs I of $\mathbf{T}$. Moreover, by $f \in \mathcal{Q}_{p,0}$ we mean

$$\lim_{\epsilon \to 0} \sup_{I \subset \mathbf{T}, |I| \le \epsilon} |I|^{-p} S_p(f, I) = 0.$$

Clearly, if $p \in (0, 1)$, then the nontangential boundary value function of a function f in $\mathcal{Q}_p$ or $\mathcal{Q}_{p,0}$ belongs to $\mathcal{Q}_p(\mathbf{T})$ or $\mathcal{Q}_{p,0}(\mathbf{T})$. Quite interesting is the following assertion.

Proposition 5.4.2. (i) *Let* $\mathcal{BMO}(\mathbf{T})$, *respectively* $\mathcal{VMO}(\mathbf{T})$, *be the class of all Lebesgue measurable functions* $f : \mathbf{T} \mapsto \mathbf{C}$ *with*

$$\int_I \left| f(\zeta) - (2\pi|I|)^{-1} \int_I f(\eta)|d\eta| \right|^2 |d\zeta| = O(|I|), \quad \text{respectively} \quad o(|I|),$$

as subarc $I \subseteq \mathbf{T}$ *varies. Then*

$$\mathcal{Q}_{p_1}(\mathbf{T}) \subseteq \mathcal{Q}_{p_2}(\mathbf{T}) \quad \text{and} \quad \mathcal{Q}_{p_1,0}(\mathbf{T}) \subseteq \mathcal{Q}_{p_2,0}(\mathbf{T}) \quad \text{for} \ \ 0 < p_1 < p_2 < \infty.$$

In particular,

$$\mathcal{Q}_p(\mathbf{T}) = \mathcal{BMO}(\mathbf{T}) \quad \text{and} \quad \mathcal{Q}_{p,0}(\mathbf{T}) = \mathcal{VMO}(\mathbf{T}) \quad \text{for} \ \ p \in (1, \infty).$$

(ii) *Let* $p \in (0, \infty)$ *and* $f \in \mathcal{Q}_p(\mathbf{T})$ *with*

$$d_{\mathcal{Q}_p(\mathbf{T})}(f, \mathcal{Q}_{p,0}(\mathbf{T})) = \inf\{\|f - g\|_{\mathcal{Q}_p(\mathbf{T})} : \ \ g \in \mathcal{Q}_{p,0}(\mathbf{T})\}.$$

Then

$$d_{\mathcal{Q}_p(\mathbf{T})}(f, \mathcal{Q}_{p,0}(\mathbf{T})) \approx \lim_{\delta \to 0} \sup_{I \subset \mathbf{T}, |I| < \delta} \left(\frac{S_p(f, I)}{|I|^p} \right)^{1/2}.$$

Proof. (i) The first result is trivial. To prove the special case $p > 1$, it suffices to show that each $\mathcal{Q}_{p,0}(\mathbf{T})$ coincides with $\mathcal{VMO}(\mathbf{T})$ whenever $p > 1$. First, we verify $\mathcal{VMO}(\mathbf{T}) \subseteq \mathcal{Q}_{p,0}(\mathbf{T})$. To do so, we observe that $\mathcal{Q}_{p,0}(\mathbf{T})$ has an integrated

Lip-characteristic which says that $f \in \mathcal{Q}_{p,0}(\mathbf{T})$ if and only if $\lim_{\delta \to 0} I_p(f, \delta) = 0$, where for $\delta \in (0, 1)$,

$$I_p(f, \delta) = \sup_{|I| < \delta} \int_0^{|I|} \sin^{p-2} \frac{\pi t}{2} dt \int_I |f(e^{i(s+t)}) - f(e^{is})|^2 ds.$$

Again, we write rI $(r > 0)$ for the arc with length $r|I|$ and the same center as I, and $f_J = (2\pi|J|)^{-1} \int_J f(\zeta)|d\zeta|$ for any subarc J of $\mathbf{T}$. Now if $f \in \mathcal{VMO}(\mathbf{T})$ then for any small $\epsilon > 0$ there is a $\delta \in (0, 1/3)$ such that as $|I| < \delta$,

$$\int_{3I} |f(e^{is}) - f_{3I}|^2 ds < 2\pi\epsilon|I|,$$

and hence

$$\int_0^{|I|} \sin^{p-2} \frac{\pi t}{2} dt \int_I |f(e^{is}) - f_{3I}|^2 ds \preceq \epsilon|I|^p,$$

which gives

$$\int_0^{|I|} \sin^{p-2} \frac{\pi t}{2} dt \int_I |f(e^{i(t+s)}) - f_{3I}|^2 ds \preceq \epsilon|I|^p.$$

Thus, $\lim_{\delta \to 0} I_p(f, \delta) = 0$, i.e., $f \in \mathcal{Q}_{p,0}(\mathbf{T})$.

Next, we show $\mathcal{Q}_{p,0}(\mathbf{T}) \subseteq \mathcal{VMO}(\mathbf{T})$. As for the case $p \in (1, 2]$, the result follows immediately from the definition. It remains to deal with the case $p \in (2, \infty)$. Let $f \in \mathcal{Q}_{p,0}(\mathbf{T})$. Then for arbitrarily small $\epsilon > 0$ there exists a $\delta \in (1, 1/2)$ such that $|J|^{-p} S_p(f, J) < \epsilon$ as $|J| < \delta$. Thus for $I \subseteq \mathbf{T}$ with $|I| < \delta$, we have

$$\int_I \int_I |f(e^{is}) - f(e^{it})|^2 ds dt$$

$$\leq \sum_{k=1}^{\infty} \int \int_{2^{-k} < \frac{|s-t|}{|T|} \leq 2^{1-k}} \frac{|e^{is} - e^{it}|^{2-p}|f(e^{is}) - f(e^{it})|^2}{|e^{is} - e^{it}|^{2-p}} ds dt$$

$$\preceq \sum_{k=1}^{\infty} \left(\frac{|I|}{2^k}\right)^{2-p} \int \int_{\frac{|s-t|}{|T|} \leq 2^{1-k}} \frac{|f(e^{is}) - f(e^{it})|^2}{|e^{is} - e^{it}|^{2-p}} ds dt$$

$$\preceq \sum_{k=1}^{\infty} \left(\frac{|I|}{2^k}\right)^{2-p} \int_{2^{2-k}I} \int_{2^{2-k}I} \frac{|f(e^{is}) - f(e^{it})|^2}{|e^{is} - e^{it}|^{2-p}} ds dt$$

$$\preceq \epsilon|I|^2,$$

which yields $f \in \mathcal{VMO}(\mathbf{T})$.

(ii) Since $d_{\mathcal{Q}_p(\mathbf{T})}(f, \mathcal{Q}_{p,0}(\mathbf{T}))$ equals 0 when $f \in \mathcal{Q}_{p,0}(\mathbf{T})$, it is easy to obtain

$$\lim_{\delta \to 0} \sup_{I \subset \mathbf{T}, |I| < \delta} \left(\frac{S_p(f, I)}{|I|^p}\right)^{1/2} \leq d_{\mathcal{Q}_p(\mathbf{T})}(f, \mathcal{Q}_{p,0}(\mathbf{T})), \quad f \in \mathcal{Q}_p(\mathbf{T}).$$

To establish the reversed estimate, we recall the Poisson integral:

$$f_r(e^{is}) = \frac{1}{2\pi} \int_{\mathbf{T}} f(\eta) \frac{1-r^2}{|\eta - re^{is}|^2} |d\eta|, \quad r \in (0,1), \quad f \in \mathcal{Q}_p(\mathbf{T}).$$

It is evident that $f_r \in \mathcal{Q}_{p,0}(\mathbf{T})$ and

$$f(\zeta) - f_r(\zeta) = (2\pi)^{-1} \int_{\mathbf{T}} \left(f(\zeta) - f(\zeta\bar{\lambda}) \right) \frac{1-r^2}{|1 - r\bar{\lambda}|^2} |d\lambda|, \quad \zeta = e^{is}, \quad \lambda = \zeta\bar{\eta}.$$

Letting $T_\lambda f(\zeta) = f(\zeta\bar{\lambda})$ and using Minkowski's inequality, we derive that for any small $\epsilon > 0$,

$$\|f - f_r\|_{\mathcal{Q}_p(\mathbf{T})}$$
$$\preceq \int_{\mathbf{T}} \|f - T_\lambda f\|_{\mathcal{Q}_p(\mathbf{T})} \frac{1-r^2}{|1 - r\bar{\lambda}|^2} |d\lambda|$$
$$\preceq \int_{|\lambda|<\epsilon} \|f - T_\lambda f\|_{\mathcal{Q}_p(\mathbf{T})} \frac{1-r^2}{|1 - r\bar{\lambda}|^2} |d\lambda| + \|f\|_{\mathcal{Q}_p(\mathbf{T})} \int_{\epsilon \leq |\lambda| \leq \pi} \frac{1-r^2}{|1 - r\bar{\lambda}|^2} |d\lambda|$$
$$= \operatorname{Term}_1(r) + \operatorname{Term}_2(r).$$

Given $\delta \in (0,1)$. The Lebesgue Dominated Convergence Theorem yields

$$\lim_{\lambda \to 0} \sup_{|I| > \delta} |I|^{-p} S_p(f - T_\lambda f, I) = 0.$$

In the meantime, the triangle inequality gives

$$\sup_{|I| \leq \delta} |I|^{-p} S_p(f - T_\lambda f, I) \preceq \sup_{|I| \leq \delta} \frac{S_p(f, I)}{|I|^p}.$$

Therefore, if $\epsilon \to 0$ then

$$\operatorname{Term}_1(r) \preceq \sup_{|I| < \delta} \left(\frac{S_p(f, I)}{|I|^p} \right)^{1/2}.$$

And, since $\lim_{r \to 1} \operatorname{Term}_2(r) = 0$, we conclude

$$d_{\mathcal{Q}_p(\mathbf{T})}(f, \mathcal{Q}_{p,0}(\mathbf{T})) \leq \lim_{r \to 1} \|f - f_r\|_{\mathcal{Q}_p(\mathbf{T})} \preceq \lim_{\delta \to 0} \sup_{I \subset \mathbf{T}, |I| < \delta} \left(|I|^{-p} S_p(f, I) \right)^{1/2},$$

as desired. $\square$

Let us come back to our topic on the extreme points.

Theorem 5.4.3. *For $p \in (0, \infty)$, let*

$$\mathcal{Q}_p^0(\mathbf{T}) = \left\{ f \in \mathcal{Q}_p(\mathbf{T}) : \int_{\mathbf{T}} f(\zeta) |d\zeta| = 0 \right\}$$

and

$$\mathcal{Q}_{p,0}^0(\mathbf{T}) = \left\{ f \in \mathcal{Q}_{p,0}(\mathbf{T}) : \int_{\mathbf{T}} f(\zeta) |d\zeta| = 0 \right\}.$$

(i) *If $f \in \mathcal{Q}_{p,0}^0(\mathbf{T})$, then f is an extreme point of $\mathcal{B}_{\mathcal{Q}_{p,0}^0(\mathbf{T})}$ with the norm $\|f\|_{\mathcal{Q}_p(\mathbf{T})}$ when and only when*

$$T_p(f, \zeta) = \sup_{\zeta \in I} |I|^{-p} S_p(f, I) = 1, \quad \zeta \in \mathbf{T},$$

where the supremum is taken over all arcs $I \subseteq \mathbf{T}$ containing ζ.

(ii) *If $f \in \mathcal{Q}_p^0(\mathbf{T})$ is an extreme point of $\mathcal{B}_{\mathcal{Q}_p^0(\mathbf{T})}$ with the norm $\| \cdot \|_{\mathcal{Q}_p(\mathbf{T})}$, then there are no two distinct points ζ_1, ζ_2 in $\mathbf{T}$ such that $T_p(f, \zeta_1) < 1$ and $T_p(f, \zeta_2) < 1$. Conversely, if $f \in \mathcal{Q}_p^0(\mathbf{T})$ is a function such that for all $z \in \mathbf{T}$ with one possible exception there are $T_p(f, \zeta) = 1$ and $|I_\zeta|^{-p} S_p(f, I_\zeta) = 1$ on some open subarc I_ζ containing ζ, then f is an extreme point of $\mathcal{B}_{\mathcal{Q}_p^0(\mathbf{T})}$ with the norm $\| \cdot \|_{\mathcal{Q}_p(\mathbf{T})}$.*

Proof. (i) Note that if $f \in \mathcal{Q}_p^0(\mathbf{T})$, then $T_p(f, \cdot)$ is lower semicontinuous; that is, $\{\zeta \in \mathbf{T} : T_p(f, \zeta) > t\}$ is an open set for every $t > 0$. So we have

$$\|f\|_{\mathcal{Q}_p(\mathbf{T})} = \left(\sup_{\zeta \in \mathbf{T}} T_p(f, \zeta) \right)^{\frac{1}{2}}.$$

Recalling that $\mathcal{C}(\mathbf{T})$ is the class of all continuous functions $f : \mathbf{T} \mapsto \mathbf{C}$, we can obtain $T_p(f, \cdot) \in \mathcal{C}(\mathbf{T})$. To see this, we suffice to show that $T_p(f, \cdot)$ is also upper semicontinuous, namely, $\{\zeta \in \mathbf{T} : T_p(f, \zeta) < t\}$ is an open subset of $\mathbf{T}$ for every $t > 0$. In so doing, we fix $t > 0$ and let $\zeta \in \mathbf{T}$ obey $T_p(f, \zeta) < t$. If $T_p(f, \zeta) \neq 0$, then by $f \in \mathcal{Q}_{p,0}^0(\mathbf{T})$, there is a $\delta > 0$ such that

$$\sup_{I \subseteq \mathbf{T}, |I| < \delta} \frac{S_p(f, I)}{|I|^p} \le \left(T_p(f, \zeta) \right)^2.$$

Now let $J \subseteq \mathbf{T}$ be an open subarc centered at ζ whose arclength $|J| = \epsilon$ is very small compared to δ. And, for $w \in J$ let $I \subset \mathbf{T}$ contain w. When $|I| < \delta$, we have $S_p(f, I)|I|^{-p} < t^2$. In the case $|I| \ge \delta$ we take $K = J \cup I$. Then K is an open subarc containing ζ and hence

$$\frac{S_p(f, I)}{|I|^p} \le \left(\frac{|K|}{|I|} \right)^p \frac{S_p(f, K)}{|K|^p} \le \left(\frac{|K|}{|I|} \right)^p \left(T_p(f, \zeta) \right)^2.$$

Further, putting $T_p(f, \zeta) = t - \tau$, $\tau \in (0, t)$, we derive $|I|^{-p} S_p(f, I) < t^2$ and thus

$$T_p(f, \eta) < t \quad \text{when} \quad \eta \in J \quad \text{and} \quad |J| = \epsilon < \delta\big(t/(t - \tau)\big)^{2/p} - 1\big).$$

On the other hand, if $T_p(f, \zeta) = 0$, then via the analysis on the open subarc K, we can select an open subarc J centered at ζ such that $T_p(f, \eta) = 0 < t$ as $\eta \in J$.

For the necessity, suppose $T_p(f, \cdot) \not\equiv 1$. Then by $T_p(f, \cdot) \in \mathcal{C}(\mathbf{T})$, there is an open subarc $J \subset \mathbf{T}$ such that $\sup_{\zeta \in J} T_p(f, \zeta) < 1$. Choose a function $g \in \mathcal{Q}_{p,0}^0(\mathbf{T})$ such that

$$\|g\|_{\mathcal{Q}_p(\mathbf{T})} \leq 1 - \sup_{\zeta \in J} T_p(f, \zeta),$$

$g = 0$ outside J, and $g \not\equiv 0$. When $I \subset \mathbf{T}$ is such that $I \cap J = \emptyset$, we evidently obtain

$$\frac{S_p(f + g, I)}{|I|^p} = \frac{S_p(f, I)}{|I|^p} \leq 1.$$

If $I \subset \mathbf{T}$ ensures $I \cap J \neq \emptyset$, then

$$\begin{aligned}
\left(\frac{S_p(f + g, I)}{|I|^p}\right)^{1/2} &\leq \left(\frac{S_p(f, I)}{|I|^p}\right)^{1/2} + \left(\frac{S_p(g, I)}{|I|^p}\right)^{1/2} \\
&\leq \sup_{\zeta \in J} T_p(f, \zeta) + \|g\|_{\mathcal{Q}_p(\mathbf{T})} \\
&\leq 1.
\end{aligned}$$

Thus $\|f + g\|_{\mathcal{Q}_p(\mathbf{T})} \leq 1$. Similarly, $\|f - g\|_{\mathcal{Q}_p(\mathbf{T})} \leq 1$. Due to $g \not\equiv 0$, those inequalities yield that f is not an extreme point of $\mathcal{B}_{\mathcal{Q}_{p,0}^0(\mathbf{T})}$, contradicting the assumption.

Assuming $T_p(f, \cdot) \equiv 1$, we derive that if $\zeta_0 \in \mathbf{T}$, then there exists a sequence of subarcs I_n containing ζ_0 such that $S_p(f, I_n)/|I_n|^p$ is convergent to 1. Without loss of generality, we may suppose $I_n = (a_n, b_n)$, intervals moving counterclockwise. Then, by passing a subsequence, we may also suppose $a_n \to a$ and $b_n \to b$. After that, let $I_{\zeta_0} \subset \mathbf{T}$ be the open subarc determined by the interval (a, b), i.e., $I_{\zeta_0} = (a, b)$. In this sense, I_{ζ_0} is viewed as the limiting open subarc of I_n. Accordingly, $S_p(f, I_{\zeta_0})/|I_{\zeta_0}|^p = 1$. Although I_{ζ_0} does not necessarily contain ζ_0, it is easy to see that ζ_0 belongs to $\bar{I}_{\zeta_0}$ — the closure of I_{ζ_0} — a closed subarc $[a, b]$ of $\mathbf{T}$. Observe that since $f \in \mathcal{Q}_{p,0}^0(\mathbf{T})$, I_n cannot get small and hence $I_{\zeta_0} \neq \emptyset$.

Now suppose

$$g \in \mathcal{Q}_{p,0}^0(\mathbf{T}) \quad \text{with} \quad \|f + g\|_{\mathcal{Q}_p(\mathbf{T})} \leq 1 \quad \text{and} \quad \|f - g\|_{\mathcal{Q}_p(\mathbf{T})} \leq 1.$$

In order to prove that f is an extreme point of $\mathcal{B}_{\mathcal{Q}_{p,0}^0(\mathbf{T})}$, we must show that $g \equiv 0$ on $\mathbf{T}$. Fix $\zeta_0 \in \mathbf{T}$ with the open subarc I_{ζ_0} constructed above. Thus,

$$\frac{f(z) - f(w)}{(|z - w|/|I_{\zeta_0}|)^{(2-p)/2}}\bigg|_{I_{\zeta_0} \times I_{\zeta_0}}$$

lies in the unit sphere of $\mathcal{L}^2(I_{\zeta_0} \times I_{\zeta_0}, (2\pi|I_{\zeta_0}|)^{-2}|dz||dw|)$. Using $\|f+g\|_{\mathcal{Q}_p(\mathbf{T})} \leq 1$, we also see that

$$\frac{f(z) - f(w) + g(z) - g(w)}{(|z-w|/|I_{\zeta_0}|)^{(2-p)/2}}\bigg|_{I_{\zeta_0} \times I_{\zeta_0}}$$

is a member of the closed unit ball of $\mathcal{L}^2(I_{\zeta_0} \times I_{\zeta_0}, (2\pi|I_{\zeta_0}|)^{-2}|dz||dw|)$, and similarly when g is replaced by $-g$. Notice that

$$\frac{f(z) - f(w)}{(|z-w|/|I_{\zeta_0}|)^{(2-p)/2}}\bigg|_{I_{\zeta_0} \times I_{\zeta_0}} = \frac{\big(f(z) - f(w)\big) + \big(g(z) - g(w)\big)}{2(|z-w|/|I_{\zeta_0}|)^{(2-p)/2}}\bigg|_{I_{\zeta_0} \times I_{\zeta_0}}$$
$$+ \frac{\big(f(z) - f(w)\big) - \big(g(z) - g(w)\big)}{2(|z-w|/|I_{\zeta_0}|)^{(2-p)/2}}\bigg|_{I_{\zeta_0} \times I_{\zeta_0}}$$

and even more importantly, that $\mathcal{L}^2(I_{\zeta_0} \times I_{\zeta_0}, (2\pi|I_{\zeta_0}|)^{-2}|dz||dw|)$ has the property: every point of the unit sphere is an extreme point of the closed unit ball [Rud, p.84, problem 16]. So, the last equation gives that $g(z) - g(w) = 0$ on $I_{\zeta_0} \times I_{\zeta_0}$; that is, g is a constant on I_{ζ_0}.

For each $\zeta_0 \in \mathbf{T}$ let J_{ζ_0} denote the largest open subarc covering I_{ζ_0} such that the restriction $g|_{J_{\zeta_0}}$ of g on J_{ζ_0} is constant. Obviously, as to $\zeta_0, \eta_0 \in \mathbf{T}$, $J_{\zeta_0} \cap J_{\eta_0} \neq \emptyset$ induces $J_{\zeta_0} = J_{\eta_0}$, and so the collection $\{J_{\zeta_0}\}$ can contain at most countably many different open subarcs, owing to $|\mathbf{T}| = 1$. Relabel these disjoint open subarcs $\{K_n\}$. The condition $\zeta_0 \in \bar{I}_{\zeta_0}$ implies $\mathbf{T} = \cup\bar{K}_n$, where $\bar{K}_n$ still stands for the closure of K_n. Thus $A = \mathbf{T} \setminus \cup K_n$ consists of all endpoints of all the subarcs K_n. In particular, A is a closed, countable set. If $A = \emptyset$, then $\{K_n\}$ contains only one element, namely $\mathbf{T}$, and thus g is a constant function on $\mathbf{T}$. Note that $g|_{\mathbf{T}} = 0$. Accordingly, $g \equiv 0$.

It remains to consider the case: $A \neq \emptyset$. Using the Baire Category Theorem, we see that the non-empty, countable, closed set A must have an isolated point $c \in \mathbf{T}$. Since $\mathbf{T} = \cup\bar{K}_n$, we can conclude that there must be two disjoint open subarcs K_n and K_m such that c is an endpoint for them. Recall that $g|_{K_n}$ and $g|_{K_m}$ are constant. If the two constants do not coincide, then g has a jump discontinuity at c, which certainly contradicts the condition $g \in \mathcal{Q}_{p,0}^0(\mathbf{T})$. If the two constants are the same, then g is constant on the open subarc $K_n \cup \{c\} \cup K_m$, which violates the maximality of the open subarcs $\{J_{\zeta_0}\}_{\zeta_0 \in \mathbf{T}}$. Thus the case $A \neq \emptyset$ cannot happen.

(ii) Let f be an extreme point of $\mathcal{B}_{\mathcal{Q}_p^0}(\mathbf{T})$. Of course, we must have $\|f\|_{\mathcal{Q}_p(\mathbf{T})} = 1$. In order to reach our purpose, suppose otherwise that there are two distinct points $\zeta_k \in \mathbf{T}$, $k = 1, 2$ such that $T_p(f, \zeta_k) < 1$, $k = 1, 2$; and use I_k, $k = 1, 2$ to denote the two open subarcs of $\mathbf{T}$ which have $\{\zeta_1, \zeta_2\}$ as endpoints. Define a function $g \in \mathcal{Q}_p^0(\mathbf{T})$ by $g|_{I_1} = \epsilon_1$, $g|_{I_2} = -\epsilon_2$, where $\epsilon_k > 0$, $k = 1, 2$ are chosen so that

$$g|_{\mathbf{T}} = 0 \quad \text{and} \quad \|g\|_{\mathcal{Q}_p(\mathbf{T})} \leq 1 - \max_{k=1,2} T_p(f, \zeta_k).$$

Now, let I be a subarc of $\mathbf{T}$. If both ζ_1 and ζ_2 are not in I, then g is constant on

I and consequently,

$$\frac{S_p(f+g, I)}{|I|^p} = \frac{S_p(f, I)}{|I|^p} \leq 1.$$

If one of ζ_k, for instance, ζ_1 is in I, then by the Minkowski inequality,

$$\left(\frac{S_p(f+g, I)}{|I|^p}\right)^{1/2} \leq T_p(f, \zeta_1) + \|g\|_{\mathcal{Q}_p(\mathbf{T})} \leq 1.$$

Accordingly, we get

$$\|f+g\|_{\mathcal{Q}_p(\mathbf{T})} \leq 1 \quad \text{and similarly} \quad \|f-g\|_{\mathcal{Q}_p(\mathbf{T})} \leq 1.$$

Because of $f = (f+g)/2 + (f-g)/2$, the function f is not an extreme point of $\mathcal{B}_{\mathcal{Q}_p^0(\mathbf{T})}$, contradicting the given condition.

On the other hand, suppose that $f \in \mathcal{Q}_p^0(\mathbf{T})$ is a function such that for all $\zeta \in \mathbf{T}$ with one possible exception $\eta \in \mathbf{T}$ there are

$$T_p(f, \zeta) = 1 \quad \text{and} \quad S_p(f, I_\zeta) = |I_\zeta|^p$$

on some open subarc I_ζ containing ζ. Now let $g \in \mathcal{Q}_p^0(\mathbf{T})$ obey $\|f \pm g\|_{\mathcal{Q}_p(\mathbf{T})} \leq 1$. To close the argument, we must show $g \equiv 0$ on $\mathbf{T}$. Applying the same reasoning as in the proof of the sufficiency of (i), we declare that $g|_{I_\zeta}$ is a constant. Since g is locally constant on the connected set $\mathbf{T} \setminus \{\eta\}$, $g \equiv 0$ by $g|_{\mathbf{T}} = 0$. □

We close this section by presenting some examples of either extreme points or nonextreme points.

Example 5.4.4. Given $p \in (0, \infty)$.

(i) For integers $n = \pm 1, \pm 2, \ldots$ let $f_n(z) = \lambda z^n$ where $z \in \mathbf{T}$ and $|\lambda| \equiv 1$. Then $g_n = \|f_n\|_{\mathcal{Q}_p(\mathbf{T})}^{-1} f_n$ are extreme points of $\mathcal{B}_{\mathcal{Q}_{p,0}^0(\mathbf{T})}$.

(ii) For $\delta \in (0, 1)$ let

$$f_\delta(e^{i\theta}) = \begin{cases} \frac{\theta}{\delta} & , \quad \theta \in [0, \delta], \\ 1 & , \quad \theta \in [\delta, 2\pi - \delta], \\ \frac{2\pi - \theta}{\delta} & , \quad \theta \in [2\pi - \delta, 2\pi). \end{cases}$$

Also put $g_\delta = f_\delta - 1 + (2\pi)^{-1}\delta$ on $[0, 2\pi)$ and extend it 2π-periodically. Then there exists a $\delta > 0$ such that $\|g_\delta\|_{\mathcal{Q}_p(\mathbf{T})}^{-1} g_\delta$ is a nonextreme point of $\mathcal{B}_{\mathcal{Q}_{p,0}^0(\mathbf{T})}$.

Proof. (i) By Theorem 5.4.3 we have to verify $T_p(g_n, \zeta) = 1$ for all $\zeta \in \mathbf{T}$. As a matter of fact, we use some elementary estimates to obtain

$$S_p(f_n, I) = 2^{p+1} \int_0^{|I|} (|I| - t) \sin^{p-2}(\pi t) \sin^2(n\pi t)\, dt$$

and so that

$$\left(T_p(f_n, \zeta)\right) = \sup_{|I| \in (0,1]} \frac{2^{p+1}}{|I|^p} \int_0^{|I|} (|I| - t) \sin^{p-2}(\pi t) \sin^2(n\pi t) dt.$$

Thus, $\|f_n\|_{\mathcal{Q}_p(\mathbf{T})} = T_p(f_n, \zeta)$ for each $\zeta \in \mathbf{T}$, and then $T_p(g_n, \cdot) \equiv 1$ follows.

(ii) A key observation is that g_δ is convergent to 0 as $\delta \to 0$. Since f_δ is a Lip1-function, we get $\|g_\delta\|_{\mathcal{Q}_p(\mathbf{T})}^{-1} g_\delta \in \mathcal{Q}_{p,0}^0(\mathbf{T})$. However, we are about to prove that it is not an extreme point of $\mathcal{B}_{\mathcal{Q}_{p,0}^0(\mathbf{T})}$. This will be done as long as $T_p(f_\delta, \cdot)$ is verified to be a nonconstant function for some δ. To this end, it suffices to verify $T_p(f_\delta, 0) \neq T_p(f_\delta, \pi)$ for some δ. First, for any subinterval $I = (a, b) \subset (0, \delta)$ we have

$$\frac{S_p(f_\delta, I)}{|I|^p} = \frac{2^p (2\pi)^{p-2}}{(b-a)^p \delta^2} \int_0^{b-a} \frac{(b - a - t)t^2}{\sin^{2-p} \frac{t}{2}} dt.$$

Thus

$$\sup_{I \subset (0,\delta)} \frac{S_p(f_\delta, I)}{|I|^p} \geq \frac{2^p (2\pi)^{p-2}}{\delta^{2+p}} \int_0^\delta \frac{(\delta - t)t^2}{\sin^{2-p} \frac{t}{2}} dt.$$

Moreover, using

$$\lim_{\delta \to 0} \frac{1}{\delta^{p+2}} \int_0^\delta \frac{(\delta - t)t^2}{\sin^{2-p} \frac{t}{2}} dt = \frac{2^{2-p}}{(p+1)(p+2)},$$

we can find a $\delta_1 \in (0, 1)$ such that

$$\delta \in (0, \delta_1) \Rightarrow T_p(f_\delta, 0) > 2^{-1} \left(\frac{2^p \pi^{p-2}}{(p+1)(p+2)} \right)^{1/2} = \frac{\mu_0}{2}.$$

Second, suppose $I \subseteq \mathbf{T}$ is any subarc containing π. We have

$$|I| \leq (\pi - \delta)/(2\pi) \Rightarrow S_p(f_\delta, I) = 0.$$

And, if $1 \geq |I| > (\pi - \delta)/(2\pi)$, then by the definition of f_δ,

$$S_p(f_\delta, I) \leq S_p(f_\delta, \mathbf{T}) = \frac{1}{(2\pi)^2} \int \int_\Omega \frac{|f_\delta(e^{is}) - f_\delta(e^{it})|^2}{|e^{is} - e^{it}|^{2-p}} ds dt,$$

where Ω is a domain defined by $\bigcup_{j=1}^4 \Omega_j$:

$$\Omega_j = \begin{cases} \{(s,t) : 0 \leq s \leq \delta, 0 \leq t \leq 2\pi\} & , \quad j = 1, \\ \{(s,t) : 2\pi - \delta \leq s \leq 2\pi, 0 \leq t \leq 2\pi\} & , \quad j = 2, \\ \{(s,t) : 0 \leq s \leq 2\pi, 0 \leq t \leq \delta\} & , \quad j = 3, \\ \{(s,t) : 0 \leq s \leq 2\pi, 2\pi - \delta \leq t \leq 2\pi\} & , \quad j = 4. \end{cases}$$

It is not hard to get a $\delta_2 \in (0, 1)$ such that as $\delta \in (0, \delta_2)$,

$$\int \int_{\Omega_j} \frac{|f_\delta(e^{is}) - f_\delta(e^{it})|^2}{|e^{is} - e^{it}|^{2-p}} ds dt \leq \left(\frac{\pi - \delta}{2\pi} \right)^p \left(\frac{2\pi\mu_0}{2} \right)^2, \quad j = 1, 2, 3, 4.$$

Hence

$$\frac{S_p(f_\delta, I)}{|I|^p} \leq \frac{1}{(2\pi)^2} \left(\frac{2\pi}{\pi - \delta}\right)^p \int\int_\Omega \frac{|f_\delta(e^{is}) - f_\delta(e^{it})|^2}{|e^{is} - e^{it}|^{2-p}} ds dt \leq \left(\frac{\mu_0}{2}\right)^2.$$

Consequently, $T_p(f_\delta, \pi) \leq \mu_0/2$ whenever $\delta \in (0, \delta_2)$. Therefore there exists a $\delta_3 \in (0, \min\{\delta_1, \delta_2\})$ such that $T_p(f_{\delta_3}, 0) > \mu_0/2$ and $T_p(f_{\delta_3}, \pi) \leq \mu_0/2$. This concludes the proof. $\square$

5.5 Notes

Note 5.5.1. Section 5.1 is mainly taken from [AlCaSi, Section 3]. For some related information on the Cauchy duality, see also [RosSh] and [DuRoSh]. In the proof of Example 5.1.2 (i) and (ii), we have actually used the concept of a sampling set for $\mathcal{A}^{-\alpha}$; see also [HedKZ, Chapter 5]. Moreover, the fact that there are sequences in $\mathbf{D}$ for which the representation in Example 5.1.2 (iii) holds is well known — see [Zhu2, Theorem 2.30].

Note 5.5.2. Section 5.2 is a slight modification of [AlCaSi, Section 4]. Corollary 5.2.4 (i) gives another description of the well-known fact that $\mathcal{H}^1 = \mathcal{H}^2 \odot \mathcal{H}^2$. Of course, this corollary also implies the following interesting factorization:

$$\mathcal{D}^{1,0} = \mathcal{D}^{1,-\frac{2-p}{2}} \odot \mathcal{D}^{2,2-p} \quad \text{for} \quad p \in (1, 2),$$

which has not been obtained until now. A more interesting reference on the dot product or weak factorization is [CohnVe].

Note 5.5.3. Section 5.3 is a sort of combination of Sections 2 and 6 in [AlCaSi]. Although the atom-like representations in Theorem 5.3.2 have no requirements on the compact support and the vanishing moment of either $a(z)$ or $b(z)$, they provide a good realization of Theorem 4.3.2 (i) on the Cauchy dual of $\mathcal{Q}_{p,0}$. Additionally, Theorem 5.3.2 may be viewed as a sort of dual form of Theorem 3.3.3.

Note 5.5.4. Section 5.4 consists of [WiXi2, Section 3] and [WiXi3]. There are several papers on the extreme points in the Bloch spaces — see e.g., [CiWo], [AnRo], [Wi1] [Wi2], [Col], and [CoheCol]. For a discussion on the extreme points in $\mathcal{VMO}$ and $\mathcal{BMO}$, see [AxSh]. And the notion of $\mathcal{VMO}$ is originally from [Sara].

Chapter 6

As Symbols of Hankel and Volterra Operators

The various holomorphic Q classes have good boundedness properties and consequently are often proper settings for the applications of complex analysis to, for instance, operator theory. In this chapter we will see that the $\mathcal{Q}_p$ spaces induce bounded holomorphic Hankel and Volterra operators acting between two Dirichlet-type spaces. This feature of function-theoretic operator theory will be shown naturally via the following components:

- Hankel and Volterra from Small to Large Spaces;
- Carleson Embeddings for Dirichlet Spaces;
- More on Carleson Embeddings for Dirichlet Spaces;
- Hankel and Volterra on Dirichlet Spaces.

6.1   Hankel and Volterra from Small to Large Spaces

For $f \in \mathcal{H}$ we consider two linear operators initially defined on $g \in \mathcal{H}(\bar{D})$ by

$$\mathsf{H}_f g(z) = \lim_{r \to 1} \int_{\mathbf{T}} \frac{f(r\zeta)\overline{g(r\zeta)}}{1 - zr\bar{\zeta}} |d\zeta|, \quad z \in \mathbf{D},$$

and

$$\mathsf{V}_f g(z) = \int_0^z g(w) f'(w) dw, \quad z \in \mathbf{D}.$$

The former is called a holomorphic Hankel operator and the latter is called a holomorphic Volterra operator. The name of Hankel is motivated by the fact that on the canonical basis $\{z^n\}$ one has:

$$\mathsf{H}_\phi(f, g) = \int_{\mathbf{T}} \phi(\zeta)\overline{f(\zeta)g(\zeta)}|d\zeta| \Rightarrow \mathsf{H}_\phi(z^m, z^n) = 2\pi \int_{\mathbf{T}} \phi(\zeta)\zeta^{-(m+n)}|d\zeta|,$$

which is associated with a Hankel matrix, while the name of Volterra is motivated by the Volterra integral $\int_0^z f(w)dw$. As a matter of fact, the Volterra operators may be regarded as Riemann–Stieltjes operators which generalize the classical Cesaro operators.

Note that if $f \in \mathcal{D}^{1,-(1-\frac{p}{2})}$, $p \in (0,2)$, then the mean-value-inequality for $|f'|$ produces

$$|f'(z)|(1-|z|)^{1+\frac{p}{2}} \preceq \|f\|_{\mathcal{D}^{1,-(1-\frac{p}{2})}} \quad \text{and so} \quad \mathcal{D}^{1,-(1-\frac{p}{2})} \subseteq \mathcal{D}^{2,p}.$$

Furthermore, the inclusion is proper. To see this, we first observe

$$f(z) = \sum_{j=0}^{\infty} a_j z^j \in \mathcal{D}^{2,p} \Leftrightarrow \sum_{j=0}^{\infty} |a_j|^2 (1+j)^{1-p} < \infty,$$

and next use Hardy's inequality (cf. [Du, p. 48])

$$\sum_{j=0}^{\infty} |a_j|(j+1)^{-1} \preceq \|f\|_{\mathcal{H}^1}, \quad f(z) = \sum_{j=0}^{\infty} a_j z^j \in \mathcal{H}^1$$

to yield that, for $f(z) = \sum_{j=0}^{\infty} a_j z^j \in \mathcal{D}^{1,-(1-\frac{p}{2})}$,

$$\infty > \int_{\mathbf{D}} |f'(z)|(1-|z|^2)^{\frac{p}{2}-1} dm(z)$$

$$= \int_0^1 (1-r^2)^{\frac{p}{2}-1} \left(\int_{\mathbf{T}} |f'(r\zeta)||d\zeta| \right) r dr$$

$$\succeq \sum_{j=0}^{\infty} |a_j| \int_0^1 r^n (1-r^2)^{\frac{p}{2}-1} r dr$$

$$\succeq \sum_{j=0}^{\infty} |a_j|(1+j)^{-\frac{p}{2}}.$$

Accordingly, it is quite natural to obtain the forthcoming characterization of $\mathcal{Q}_p$ in terms of the holomorphic Hankel and Volterra operators.

Theorem 6.1.1. *Let $p \in (0,2)$ and $f \in \mathcal{H}$. Then the following statements are equivalent:*

(i) H_f *extends to a continuous operator from $\mathcal{D}^{1,-(1-\frac{p}{2})}$ to $\mathcal{D}^{2,p}$.*

(ii) V_f *is a continuous operator from $\mathcal{D}^{1,-(1-\frac{p}{2})}$ to $\mathcal{D}^{2,p}$.*

(iii) $f \in \mathcal{Q}_p$.

Proof. (i)⇔(iii) If $f \in \mathcal{Q}_p$, then for $g \in \mathcal{D}^{1,-(1-\frac{p}{2})}$ and $h \in \mathcal{D}^{2,2-p}$ we use the density of polynomials in $\mathcal{D}^{1,-(1-\frac{p}{2})}$ and $\mathcal{D}^{2,2-p}$, plus Theorem 5.2.3, to get

$$|\langle \mathsf{H}_f g, h \rangle| = |\langle f, gh \rangle| \preceq (|f(0)| + \|f\|_{\mathcal{Q}_p,1}) \|g\|_{\mathcal{D}^{1,-(1-\frac{p}{2})}} \|h\|_{\mathcal{D}^{2,2-p}}.$$

Since the Cauchy dual of $\mathcal{D}^{2,2-p}$ is isomorphic to $\mathcal{D}^{2,p}$, we conclude from the just-established estimate that $\mathsf{H}_f g \in \mathcal{D}^{2,p}$ with

$$\|\mathsf{H}_f\|_{\mathcal{D}^{1,-(1-\frac{p}{2})}\mapsto\mathcal{D}^{2,p}} \precsim |f(0)| + \|f\|_{\mathcal{Q}_p,1}.$$

Conversely, if H_f extends to a continuous operator from $\mathcal{D}^{1,-(1-\frac{p}{2})}$ to $\mathcal{D}^{2,p}$, then the operator norm $\|\mathsf{H}_f\|_{\mathcal{D}^{1,-(1-\frac{p}{2})}\mapsto\mathcal{D}^{2,p}}$ is finite with

$$|\langle f, gh\rangle| = |\langle \mathsf{H}_f g, h\rangle| \precsim \|\mathsf{H}_f\|_{\mathcal{D}^{1,-(1-\frac{p}{2})}\mapsto\mathcal{D}^{2,p}}\|g\|_{\mathcal{D}^{1,-(1-\frac{p}{2})}}\|h\|_{\mathcal{D}^{2,2-p}}, \quad g, h \in \mathcal{H}(\bar{D}).$$

Of course, this assertion yields that f belongs to the Cauchy dual of $\mathcal{D}^{1,-(1-\frac{p}{2})} \odot \mathcal{D}^{2,2-p}$, i.e., $f \in \mathcal{Q}_p$ with

$$|f(0)| + \|f\|_{\mathcal{Q}_p,1} \precsim \|\mathsf{H}_f\|_{\mathcal{D}^{1,-(1-\frac{p}{2})}\mapsto\mathcal{D}^{2,p}},$$

owing to Theorem 5.2.3.

(ii)$\Leftrightarrow$(iii) If V_f sends $\mathcal{D}^{1,-(1-\frac{p}{2})}$ to $\mathcal{D}^{2,p}$ continuously, then for $a \in \mathbf{D}$ and its associated function $g_a(z) = (1 - |a|^2)^{1-\frac{p}{2}}(1 - \bar{a}z)^{-1}$, we have $\|g_a\|_{\mathcal{D}^{1,-(1-\frac{p}{2})}} \precsim 1$, and so that

$$\left(\int_{\mathbf{D}} \left(\frac{(1-|a|^2)^{2-p}}{|1-\bar{a}z|^2} \right) |f'(z)|^2 (1-|z|^2)^p dm(z) \right)^{\frac{1}{2}}$$
$$= \|\mathsf{V}_f g_a\|_{\mathcal{D}^{2,p}}$$
$$\leq \|\mathsf{V}_f\|_{\mathcal{D}^{1,-(1-\frac{p}{2})}\mapsto\mathcal{D}^{2,p}}\|g_a\|_{\mathcal{D}^{1,-(1-\frac{p}{2})}}$$
$$\precsim \|\mathsf{V}_f\|_{\mathcal{D}^{1,-(1-\frac{p}{2})}\mapsto\mathcal{D}^{2,p}},$$

implying $f \in \mathcal{Q}_p$ thanks to Lemma 3.1.1. Conversely, if $f \in \mathcal{Q}_p$ then for $g \in \mathcal{D}^{1,-(1-\frac{p}{2})}$ we use Example 5.1.2 to choose $\{w_j\} \subset \mathbf{D}$ and $\{c_j\} \in \ell^1$ such that

$$g(z) = \sum_{j=1}^{\infty} c_j (1 - |w_j|^2)^{1-\frac{p}{2}}(1 - \overline{w_j}z)^{-1} \quad \text{with} \quad \|\{c_j\}\|_{\ell^1} \precsim \|g\|_{\mathcal{D}^{1,-(1-\frac{p}{2})}}.$$

By Minkowski's inequality and Lemma 3.1.1, we derive

$$\|\mathsf{V}_f g\|_{\mathcal{D}^{2,p}} \leq \sum_{j=1}^{\infty} |c_j| \left(\int_{\mathbf{D}} \left(\frac{1-|w_j|^2)^{2-p}}{|1-\overline{w_j}z|^2} \right) |f'(z)|^2 (1-|z|^2)^p dm(z) \right)^{\frac{1}{2}}$$
$$\precsim \|\{c_j\}\|_{\ell^1}\|f\|_{\mathcal{Q}_p,1}$$
$$\precsim \|g\|_{\mathcal{D}^{1,-(1-\frac{p}{2})}}\|f\|_{\mathcal{Q}_p,1},$$

thereupon establishing the desired boundedness of V_f. $\qquad\square$

For an application of the previous theorem, we can characterize pointwise multipliers of $\mathcal{D}^{1,-(1-\frac{p}{2})}$ into $\mathcal{D}^{2,p}$. To this end, let $\mathsf{M}_f g = fg$ for any two functions f and g.

Corollary 6.1.2. *Let $p \in (0,2)$ and $f \in \mathcal{H}$. Then M_f is a continuous operator from $\mathcal{D}^{1,-(1-\frac{p}{2})}$ to $\mathcal{D}^{2,p}$ if and only if $f \in H^\infty \cap \mathcal{Q}_p$.*

Proof. Assuming

$$\mathsf{U}_f g(z) = \int_0^z f(w)g'(w)dw, \quad f,g \in \mathcal{H},$$

we have that U_f is a bounded operator from $\mathcal{D}^{1,-(1-\frac{p}{2})}$ to $\mathcal{D}^{2,p}$ if and only if $f \in \mathcal{H}^\infty$. In fact, if $f \in \mathcal{H}^\infty$, then using $\mathcal{D}^{1,-(1-\frac{p}{2})} \subset \mathcal{D}^{2,p}$ we derive

$$\|\mathsf{U}_f g\|_{\mathcal{D}^{2,p}} \leq \|f\|_{\mathcal{H}^\infty} \left(\int_{\mathbf{D}} |g'(z)|^2 (1-|z|^2)^p dm(z) \right)^{\frac{1}{2}} \preceq \|f\|_{\mathcal{H}^\infty} \|g\|_{\mathcal{D}^{1,-(1-\frac{p}{2})}}.$$

Conversely, if $\mathsf{U}_f : \mathcal{D}^{1,-(1-\frac{p}{2})} \mapsto \mathcal{D}^{2,p}$ is bounded, then

$$\left(\int_{\mathbf{D}} |f(z)|^2 |g'(z)|^2 (1-|z|^2)^p dm(z) \right)^{\frac{1}{2}}$$

$$\preceq \|\mathsf{U}_f\|_{\mathcal{D}^{1,-(1-\frac{p}{2})} \mapsto \mathcal{D}^{2,p}} \left(|g(0)| + \left(\int_{\mathbf{D}} |g'(z)|(1-|z|^2)^{\frac{p}{2}-1} dm(z) \right)^{\frac{1}{2}} \right).$$

If g takes g_a as in Theorem 6.1.1, then the mean-value-inequality for $|f|$ over the disk $\{z \in \mathbf{D} : |z-a| \leq \frac{1-|a|}{2}\}$, $a \in \mathbf{D}$, implies

$$|f(a)|^2 \preceq (1-|a|)^{-2} \int_{|z-a| \leq \frac{1-|a|}{2}} |f(z)|^2 dm(z)$$

$$\preceq \int_{\mathbf{D}} |f(z)|^2 |g_a'(z)|^2 (1-|z|^2)^p dm(z),$$

and hence

$$\|f\|_{\mathcal{H}^\infty} \preceq \|\mathsf{U}_f\|_{\mathcal{D}^{1,-(1-\frac{p}{2})} \mapsto \mathcal{D}^{2,p}} < \infty.$$

Upon noticing

$$(\mathsf{M}_f g)'(z) = (\mathsf{V}_f g)'(z) + (\mathsf{U}_f g)'(z), \quad z \in \mathbf{D},$$

we can use the previous result on U_f and Theorem 6.1.1 about V_f to get the desired assertion. $\qquad \square$

6.2   Carleson Embeddings for Dirichlet Spaces

Theorem 6.1.1 (ii) leads to the following Carleson embedding theorem.

Theorem 6.2.1. *Let $p \in (0,2)$ and μ be a nonnegative Radon measure on $\mathbf{D}$. Then*

$$\int_{\mathbf{D}} |g|^2 d\mu \preceq \|g\|^2_{\mathcal{D}^{1,-(1-\frac{p}{2})}}, \quad g \in \mathcal{D}^{1,-(1-\frac{p}{2})}$$

if and only if $\mu \in \mathcal{CM}_p$.

Proof. Suppose that μ is a measure satisfying the above integral inequality for $g \in \mathcal{D}^{1,-(1-\frac{p}{2})}$. Fix $a \in \mathbf{D}$. Then the function $g_a(z) = (1 - |a|^2)^{1-\frac{p}{2}}(1 - \bar{a}z)^{-1}$ belongs to $\mathcal{D}^{1,-(1-\frac{p}{2})}$ with $\|g_a\|_{\mathcal{D}^{1,-(1-\frac{p}{2})}} \preceq 1$. For any subarc $I \subset \mathbf{T}$ centered at ζ_I, set $a = (1 - |I|)\zeta_I$. Accordingly,

$$\frac{\mu(S(I))}{|I|^p} \preceq \int_{S(I)} \frac{(1 - |a|^2)^{2-p}}{|1 - \bar{a}z|^2} d\mu(z) \preceq 1,$$

namely, $\mu \in \mathcal{CM}_p$. Conversely, let $\mu \in \mathcal{CM}_p$. Then for $g \in \mathcal{D}^{1,-(1-\frac{p}{2})}$, using Example 5.1.2 we can find $\|\{\lambda_j\}\|_{\ell^1} \preceq \|g\|_{\mathcal{D}^{1,-(1-\frac{p}{2})}}$ and $\{a_j\} \subset \mathbf{D}$ such that $g(z) = \sum_j \lambda_j g_{a_j}$. Furthermore, we use the Minkowski inequality to conclude

$$\left(\int_{\mathbf{D}} |g|^2 d\mu \right)^{\frac{1}{2}} \leq \sum_j |\lambda_j| \left(\int_{\mathbf{D}} |g_{a_j}(z)|^2 d\mu(z) \right)^{\frac{1}{2}} \preceq \|\{\lambda_j\}\|_{\ell^1} \preceq \|g\|_{\mathcal{D}^{1,-(1-\frac{p}{2})}},$$

as desired. $\qquad\square$

Since $\mathcal{D}^{1,-(1-\frac{p}{2})} \subset \mathcal{D}^{2,p}$, a very natural question is: can one replace $\mathcal{D}^{1,-(1-\frac{p}{2})}$ by $\mathcal{D}^{2,p}$ in Theorem 6.2.1? To answer this, we introduce the Bessel capacity and its corresponding strong type inequality.

For $\alpha \in (0,1)$ let $K_\alpha(\theta) = |\theta|^{\alpha-1}$ for $\theta \in [-\pi,\pi]$ and be extended to be a 2π-periodical function on $\mathbf{R}$. Next, set

$$K_\alpha * g(e^{i\theta}) = \int_{-\pi}^{\pi} K_\alpha(\theta - t)g(e^{it})dt, \quad g \in \mathcal{L}^2(\mathbf{T}).$$

After that, for a subset E of $\mathbf{T}$, let $\tau(E)$ denote the corresponding subset in the interval $[-\pi, \pi]$ obtained by the natural identification of this interval with $\mathbf{R}$. With these notions, we define the $\mathcal{L}_\alpha^2$-Bessel capacity of a set $E \subseteq \mathbf{T}$ as

$$Cap_\alpha(E) = \inf \left\{ \|g\|_{\mathcal{L}^2(\mathbf{T})}^2 : \ g \in \mathcal{L}^2(\mathbf{T}), \ g \geq 0, \ K_\alpha * g \geq 1 \ \text{on} \ \tau(E) \right\}.$$

It is well known that $Cap_p(\cdot)$ is a subadditive set function and

$$Cap_\alpha(I) \approx |I|^{1-2\alpha}, \quad \text{whenever} \ \alpha \in (0,1/2)$$

holds for any subarc I of $\mathbf{T}$. Therefore

$$Cap_{\frac{1-p}{2}}(E) \preceq H_\infty^p(E), \quad E \subseteq \mathbf{T}, \ p \in (0,1).$$

Meanwhile, if $H_0^p(E) < \infty$ then $Cap_{\frac{1-p}{2}}(E) = 0$.

Below is the so-called strong type inequality for the Bessel capacity defined on $\mathbf{T}$.

Lemma 6.2.2. *Let* $\alpha \in (0, \frac{1}{2}]$ *and* $\mathcal{L}_+^2(\mathbf{T}) = \{f : \ f \geq 0, \ f \in \mathcal{L}^2(\mathbf{T})\}$. *Then*

$$\int_0^\infty Cap_\alpha(\{\zeta \in \mathbf{T} : \ K_\alpha * f(\zeta) > t\})dt^2 \preceq \|f\|_{\mathcal{L}^2(\mathbf{T})}^2, \quad f \in \mathcal{L}_+^2(\mathbf{T}).$$

Proof. To verify this assertion, we need the Bessel capacitary notion and its strong type estimate on $\mathbf{R}$. For $\alpha > 0$ and $x \in (-\infty, \infty)$, set

$$G_\alpha(x) = (4\pi)^{\frac{\alpha}{2}} \Gamma\left(\frac{\alpha}{2}\right) \int_0^\infty t^{\frac{\alpha-1}{2}} \exp\left(-\frac{\pi|x|^2}{t} - \frac{t}{4\pi}\right) \frac{dt}{t},$$

where $\Gamma(\cdot)$ is still the classical Gamma function. If x is near 0, then

$$G_\alpha(x) \approx \begin{cases} |x|^{\alpha-1} & , \quad \alpha \in (0,1), \\ \log|x|^{-1} & , \quad \alpha = 1, \\ 1 & , \quad \alpha > 1. \end{cases}$$

The Bessel capacity $Cap_{\alpha;\mathbf{R}}(E)$ of a set $E \subseteq \mathbf{R}$ is defined by

$$Cap_{\alpha;\mathbf{R}}(E) = \inf\{\|g\|^2_{\mathcal{L}^2(\mathbf{R})} : g \geq 0, \ G_\alpha * g \geq 1 \text{ on } E\},$$

where the infimum over the empty set is taken to be infinity, and $G_\alpha * g$ is the usual convolution of G_α and g. In particular, if E is an interval with sufficiently small length, then

$$Cap_{\alpha;\mathbf{R}}(E) \approx \begin{cases} |E|^{1-2\alpha} & , \quad \alpha \in (0, \frac{1}{2}), \\ (-\log|E|)^{-1} & , \quad \alpha = \frac{1}{2}. \end{cases}$$

Now, we make a claim: $Cap_\alpha(E) \approx Cap_{\alpha;\mathbf{R}}\big(\tau(E)\big)$ for any $E \subseteq \mathbf{T}$. To verify this claim, suppose $f \geq 0$ is in $\mathcal{L}^2(\mathbf{T})$ and set

$$h(x) = \begin{cases} f(x) & , \quad x \in [-2\pi, 2\pi], \\ 0 & , \quad x \in \mathbf{R} \setminus [-2\pi, 2\pi]. \end{cases}$$

Clearly, $h \geq 0$, $h \in \mathcal{L}^2(\mathbf{R})$ with $\|h\|_{\mathcal{L}^2(\mathbf{R})} \preceq \|f\|_{\mathcal{L}^2(\mathbf{T})}$ and

$$\int_{-\pi}^\pi f(x-y)K_\alpha(y)dy \preceq \int_{-\infty}^\infty h(x-y)G_\alpha(y)dy, \quad x \in [-\pi, \pi],$$

thanks to $G_\alpha \approx K_\alpha$ on $[-\pi, \pi]$. Accordingly, $Cap_{\alpha;\mathbf{R}}\big(\tau(E)\big) \preceq Cap_\alpha(E)$. To get the opposite estimate, it suffices to consider any set E for which $Cap_\alpha(E; \mathbf{R})$ is small. Accordingly, assume $Cap_\alpha(E; \mathbf{R}) < \epsilon$ for any small number $\epsilon > 0$. Then there is a function h obeying $\|h\|_{\mathcal{L}^2(\mathbf{R})} < \epsilon$, $h \geq 0$ and $G_\alpha * h \geq 1$ on E. Note that G_α decays exponentially at infinity. So the Cauchy–Schwarz inequality ensures that

$$\int_{|y| \geq \pi} h(x-y)G_\alpha(y)dy < \frac{1}{2} \quad \text{as} \quad \epsilon \to 0.$$

But

$$\frac{1}{2} \leq \int_{|y| < \pi} h(x-y)G_\alpha(y)dy \approx \int_{|y| < \pi} h(x-y)K_\alpha(y)dy \quad \text{when} \quad x \in \tau(E).$$

Now, let f_1 and f_2 be the periodic extensions of h on $[-2\pi, 0]$ and $[0, 2\pi]$ respectively. Then $f = c(f_1 + f_2)$ with an appropriate constant $c > 0$ will be a test function for $Cap_\alpha(E)$. Thus,

$$Cap_\alpha(E) \leq \|f\|^2_{\mathcal{L}^2(\mathbf{T})} \leq c^2 \|h\|^2_{\mathcal{L}^2(\mathbf{R})},$$

so that

$$Cap_\alpha(E) \leq c^2 Cap_{\alpha;\mathbf{R}}\big(\tau(E)\big).$$

By the preceding claim, we see that the desired strong type inequality for $Cap_\alpha(\cdot)$ is a consequence of the strong type inequality for $Cap_{\alpha;\mathbf{R}}(\cdot)$ below:

$$\int_0^\infty Cap_{\alpha;\mathbf{R}}\big(\{x \in \mathbf{R} : G_\alpha * h(x) > t\}\big) dt^2 \precsim \|h\|^2_{\mathcal{L}^2(\mathbf{R})}, \quad 0 \leq h \in \mathcal{L}^2(\mathbf{R}).$$

See also [AdHe, p. 189, Theorem 7.1.1]. □

Theorem 6.2.3. *Let μ be a nonnegative Radon measure on $\mathbf{D}$.*

(i) *If $p \in [1, 2)$, then*

$$\int_{\mathbf{D}} |g|^2 d\mu \precsim \|g\|^2_{\mathcal{D}^{2,p}}, \quad g \in \mathcal{D}^{2,p}$$

is equivalent to $\mu \in \mathcal{CM}_p$.

(ii) *If $p \in [0, 1)$, then*

$$\int_{\mathbf{D}} |g|^2 d\mu \precsim \|g\|^2_{\mathcal{D}^{2,p}}, \quad g \in \mathcal{D}^{2,p}$$

is equivalent to $\mu\big(\bigcup_{j=1}^n S(I_j)\big) \precsim Cap_{\frac{1-p}{2}}\big(\bigcup_{j=1}^n I_j\big)$ whenever $\{I_j\}_{j=1}^n$ are disjoint subarcs of $\mathbf{T}$.

Proof. (i) The necessity follows immediately from taking

$$g_w(z) = (1 - |w|^2)^{-\frac{p}{2}}(1 - \bar{w}z)^{-p}, \quad w \in \mathbf{D}.$$

As for the sufficiency, it suffices to consider $p \in (1, 2)$ since the case $p = 1$ is the well-known Carleson theorem — see also [Du, p. 157, Theorem 9.3] for a proof. In this case, if $\mu \in \mathcal{CM}_p$ then for fixed $r \in (0, 1)$,

$$\sup_{z \in \mathbf{D}} \frac{\mu\big(B(z, r)\big)}{(1 - |z|)^p} \precsim \|\mu\|_{\mathcal{CM}_p} < \infty,$$

where $B(z, r)$ is the hyperbolic ball with radius r and center z. Using Lemma 3.3.2

and its notation we get that if $g \in \mathcal{D}^{2,p}$, then $g \in \mathcal{A}^{2,p-2}$ and hence

$$
\begin{aligned}
\int_{\mathbf{D}} |g|^2 d\mu &\leq \sum_{j=1}^{\infty} \int_{B(z_j, \tau)} |g|^2 d\mu \\
&\preceq \sum_{j=1}^{\infty} \mu\big(B(z_j, \tau)\big) \sup_{z \in B(z_j, \tau)} |g(z)|^2 \\
&\preceq \sum_{j=1}^{\infty} \frac{\mu\big(B(z_j, \tau)\big)}{(1 - |z_j|)^p} \int_{B(z_j, 2\tau)} |g(z)|^2 (1 - |z|)^{p-2} dm(z) \\
&\preceq \|\mu\|_{\mathcal{CM}_p} \int_{\mathbf{D}} |g(z)|^2 (1 - |z|)^{p-2} dm(z),
\end{aligned}
$$

as desired.

(ii) Assume μ obeys the integral condition for $g \in \mathcal{D}^{2,p}$. Let $E = \bigcup_{j=1}^{n} I_j$ and f be a test function for E — namely — $f \geq 0$ and $f \in \mathcal{L}^2(\mathbf{T})$ with $K_{\frac{1-p}{2}} * f \geq 1$ on $\tau(E)$. Thus $K_{\frac{1-p}{2}} * f \geq 1_{I_j}$ for each $j = 1, 2, \ldots, n$. Note that $\widehat{1_{I_j}}$ — the Poisson extension of 1_{I_j} — is greater than or equal to $\frac{1}{4}$ on $S(I_j)$. So we find $\widehat{K_\alpha * f} \geq \frac{1}{4}$ on $\bigcup_{j=1}^{n} S(I_j)$. Accordingly,

$$
\mu\Big(\bigcup_{j=1}^{n} S(I_j)\Big) \leq 16 \int_{\mathbf{D}} |\widehat{K_\alpha * f}|^2 d\mu \preceq \|f\|_{\mathcal{L}^2(\mathbf{T})}^2.
$$

This clearly produces

$$
\mu\Big(\bigcup_{j=1}^{n} S(I_j)\Big) \preceq Cap_{\frac{1-p}{2}}\Big(\bigcup_{j=1}^{n} I_j\Big).
$$

Now for the converse, suppose that μ satisfies that geometric condition for all finite disjoint collections of subarcs of $\mathbf{T}$. Let $f \in \mathcal{L}^2(\mathbf{T})$ obey $f \geq 0$. Recall that $\mathrm{N}\big(\widehat{K_{\frac{1-p}{2}} * f}\big)(e^{i\theta})$ is the nontangential maximal function of $K_{\frac{1-p}{2}} * f$ at $e^{i\theta}$, and $\mathrm{M}(g)$ stands for the Hardy–Littlewood maximal function of g. Then it is well known that

$$
\mathrm{N}(\hat{g}) \preceq \mathrm{M}(g) \text{ and } \|\mathrm{M}(f)\|_{\mathcal{L}^2(\mathbf{T})} \preceq \|f\|_{\mathcal{L}^2(\mathbf{T})} \text{ hold for any } g \in \mathcal{L}^2(\mathbf{T}).
$$

For $t > 0$ let $\mathbf{K}$ be a compact subset of the level set

$$
E_t(f) = \{z \in \mathbf{D} : \widehat{K_{\frac{1-p}{2}} * f}(z) > t\}.
$$

Then there are finitely many points $\{w_j\}_{j=1}^{n}$ and subarcs $\{I_{w_j}\}$ centered at $w_j / |w_j|$ such that $\{S(2I_{w_j})\}$ covers $\mathbf{K}$. The union $\bigcup_{j=1}^{n} 2I_{w_j}$ can be written as the disjoint union of subarcs $\{J_j\}$. Clearly, $2I_{w_j}$ is contained in one of the subarcs J_j and so $\mathbf{K} \subseteq \bigcup_j S(J_j)$. Moreover, our assumption on the nontangential maximal function force

$$2I_{w_j} \subseteq \{e^{i\theta} : \mathsf{N}(\widehat{K_{\frac{1-p}{2}} * f})(e^{i\theta}) > t\}$$

and hence the same thing is valid for each J_j. Consequently,

$$\mu(\mathbf{K}) \preceq Cap_{\frac{1-p}{2}}\left(\{e^{i\theta} : \mathsf{N}(\widehat{K_{\frac{1-p}{2}} * f})(e^{i\theta}) > t\}\right)$$

and the regularity of μ yields

$$\mu\big(E_t(f)\big) \preceq Cap_{\frac{1-p}{2}}\left(\{e^{i\theta} : \mathsf{N}(\widehat{K_{\frac{1-p}{2}} * f})(e^{i\theta}) > t\}\right).$$

Using Lemma 6.2.2 and

$$\mathsf{M}\big(K_{\frac{1-p}{2}} * f\big) \leq K_{\frac{1-p}{2}} * \mathsf{M}(f),$$

we obtain

$$
\begin{aligned}
\int_{\mathbf{D}} |\widehat{K_{\frac{1-p}{2}} * f}|^2 d\mu &= \int_0^\infty \mu\big(E_t(f)\big)dt^2 \\
&\preceq \int_0^\infty Cap_{\frac{1-p}{2}}\left(\{e^{i\theta} : \mathsf{N}(\widehat{K_{\frac{1-p}{2}} * f})(e^{i\theta}) > t\}\right)dt^2 \\
&\preceq \int_0^\infty Cap_{\frac{1-p}{2}}\left(\{e^{i\theta} : \mathsf{M}(\widehat{K_{\frac{1-p}{2}} * f})(e^{i\theta}) > t\}\right)dt^2 \\
&\preceq \int_0^\infty Cap_{\frac{1-p}{2}}\left(\{e^{i\theta} : K_{\frac{1-p}{2}} * \mathsf{M}(f)(e^{i\theta}) > t\}\right)dt^2 \\
&\preceq \|\mathsf{M}(f)\|_{\mathcal{L}^2(\mathbf{T})}^2 \\
&\preceq \|f\|_{\mathcal{L}^2(\mathbf{T})}^2.
\end{aligned}
$$

Note that the Fourier coefficients of K_α are of the form $c_n(1 + n^2)^{-\frac{\alpha}{2}}$ where $a_n \approx 1$ for all $n = 0, \pm 1, \pm 2, \ldots$ and the Hardy space $\mathcal{H}^2$ is determined by its Taylor coefficients or as the space of harmonic extensions of functions $f \in \mathcal{L}^2(\mathbf{T})$ whose Fourier coefficients are zero for $n = -1, -2, \ldots$. So if $g \in \mathcal{D}^{2,p}$, then there exists a function $f \in \mathcal{H}^2$ such that

$$g(z) = \widehat{K_{\frac{1-p}{2}} * f}(z) \quad \text{and} \quad \|g\|_{\mathcal{D}^{2,p}} \approx \|f\|_{\mathcal{H}^2} \approx \|f\|_{\mathcal{L}^2(\mathbf{T})}.$$

Therefore, we finally derive

$$\int_{\mathbf{D}} |g|^2 d\mu \preceq \|f\|_{\mathcal{L}^2(\mathbf{T})}^2 \approx \|g\|_{\mathcal{D}^{2,p}}^2,$$

as desired. $\qquad\square$

6.3   More on Carleson Embeddings for Dirichlet Spaces

Motivated by Theorem 6.2.3, we give the following definition of Carleson measures involving the square Dirichlet spaces.

Definition 6.3.1. Let $p \in (0, \infty)$.

(i) A nonnegative Radon measure μ on $\mathbf{D}$ is called a $\mathcal{D}^{2,p}$-Carleson measure if

$$\|\mu\|_{CM\mathcal{D}_p} = \sup_{I \subseteq \mathbf{T}} \frac{\mu\big(S(I)\big)}{|I|^p} < \infty, \quad p \in [1, \infty),$$

where $S(I)$ still denotes the Carleson box based on the arc $I \subseteq \mathbf{T}$, and

$$\|\mu\|_{CM\mathcal{D}_p} = \sup \frac{\mu\Big(\bigcup_{j=1}^n S(I_j)\Big)}{Cap_{\frac{1-p}{2}}\Big(\bigcup_{j=1}^n I_j\Big)} < \infty, \quad p \in (0, 1),$$

where the supremum is taken over all finite sequences $\{I_j\}_{j=1}^n$ of disjoint subarcs on $\mathbf{T}$.

(ii) A function $f \in \mathcal{H}$ is of $\mathcal{W}_p$ class provided that $|f'(z)|^2(1 - |z|^2)^p dm(z)$ is a $\mathcal{D}^{2,p}$-Carleson measure on $\mathbf{D}$. The norm of $f \in \mathcal{W}_p$ is determined by

$$\|f\|_{\mathcal{W}_p} = \sup_{\|g\|_{\mathcal{D}^{2,p}} \leq 1} \left(\int_{\mathbf{D}} |g(z)|^2 |f'(z)|^2 (1 - |z|^2)^p dm(z) \right)^{\frac{1}{2}}.$$

Referring to the p-Carleson measure characterization of $\mathcal{Q}_p$, we read off $\mathcal{W}_p \subseteq \mathcal{Q}_p$ — but the problem is whether or not its equality can occur. Of course, it turns out from Theorem 6.2.3 that this is the case whenever $p \geq 1$. However, we will see that this is no longer true for $p \in (0, 1)$.

We start with the concept of an interpolating sequence for $\mathcal{D}^{2,2-p}$ which can be identified with $\mathcal{A}^{2,-p}$ whenever $p \in (0, 1)$. We say that a sequence $\{w_j\}$ in $\mathbf{D}$ is an interpolating sequence for $\mathcal{D}^{2,2-p}$ provided that the interpolation operator

$$I(f) = \{f(w_j)(1 - |w_j|)^{1-\frac{p}{2}}\} : \mathcal{D}^{2,2-p} \mapsto \ell^2$$

is surjective.

Lemma 6.3.2. *Given* $p \in (0, 1)$. *For* $j \in \mathbf{N} \cup \{0\}$ *and* $k = 0, 1, \ldots, 2^{n+4} - 1$, *set*

$$S_{j,k} = \{re^{i\theta} : \quad 1 - 2^{-j} < r < 1 - 2^{-j-1} \quad \& \quad \pi k 2^{-j-3} < \theta < \pi(k+1)2^{-j-3}\}$$

and

$$\zeta_{j,k} = (1 - 32^{-2-n}) \exp\big(2^{-j-4}(2k+1)\pi i\big).$$

Let $\Lambda = \{\zeta_1, \zeta_2, \cdots\}$ *be an enumeration of* $\{\zeta_{j,k}\}$. *Then:*

(i) Λ *is a finite union of interpolating sequences for* $\mathcal{D}^{2,2-p}$.

(ii) *There is a constant* $\kappa_p > 0$ *depending only on p such that for any* $g \in \mathcal{D}^{2,p}$ *and every nonnegative Radon measure* μ,

$$\int_{\mathbf{D}} |g|^2 d\mu \le \kappa_p \Big(\|g\|^2_{\mathcal{D}^{2,p}} \|\mu\|_{\mathcal{CM}_p} + \sum_{\zeta \in \Lambda} |g(\zeta)|^2 \mu(S_\zeta) \Big),$$

where S_ζ *denotes the unique set* $S_{j,k}$ *containing* ζ.

Proof. (i) Fix $l - 4 \in \mathbf{N}$ and put

$$\Lambda_j = \{\zeta_{n,k} : \quad k = m2^l + j \ \& \ m, n \in \mathbf{N} \cup \{0\}\}, \quad j = 0, 1, \ldots, 2^l - 1.$$

According to [DuSc, p.182], each Λ_j is a rotation of the so-called Luecking set with parameter $\gamma = 32^{2-l}$ and then [DuSc, p. 183, Corollary] gives that the so-called upper Seip density of $\{\Lambda_j\}$ is equal to $3(\log 2)^{-1} 2^{2-l}$. If $3/2^{l-3} < (1-p)\log 2$, then it follows from [HedKZ, p. 158, Theorem 5.22] that each Λ_j is an interpolating sequence for $\mathcal{D}^{2,2-p} = \mathcal{A}^{2,-p}$, as desired.

(ii) For $w \in \mathbf{D}$ and $\epsilon \in (0,1)$ let $\Delta(w, r) \subseteq \mathbf{D}$ be the disk with center w, radius $\epsilon(1 - |z|)$. If ϵ is small enough then $\mathbf{D} = \bigcup_{w \in \Lambda} \Delta(w, \epsilon)$ and any $z \in \mathbf{D}$ belongs to at most finite many $\Delta(w, \epsilon)$, $w \in \Lambda$. At the same time, we claim that if $0 < \epsilon_1 < \epsilon_2 < 1$ then there is a constant $\kappa(\epsilon_1, \epsilon_2) > 0$ such that for all $w \in \Lambda$ and any $g \in \mathcal{H}$ one has

$$\sup_{z \in \Delta(w, \epsilon_1)} |g(z) - g(w)|^2 \le \kappa(\epsilon_1, \epsilon_2) \int_{\Delta(w, \epsilon_2)} |g'|^2 dm.$$

To see this, we employ the Taylor expansion of g at w:

$$g(z) - g(w) = \sum_{j=1}^{\infty} c_j (z - w)^j, \quad |z - w| < 1 - |w|,$$

the Cauchy–Schwarz inequality and Parseval's identity to deduce

$$
\begin{aligned}
|g(z) - g(w)|^2 &\le \Big(\sum_{j=1}^{\infty} j |c_j|^2 (\epsilon_2(1 - |w|))^{2j} \Big) \Big(\sum_{j=1}^{\infty} j^{-1} (\epsilon_1 \epsilon_2^{-1})^{2j} \Big) \\
&= \pi^{-1} \Big(\int_{\Delta(w, \epsilon_2)} |g'|^2 dm \Big) \Big(\sum_{j=1}^{\infty} j^{-1} (\epsilon_1 \epsilon_2^{-1})^{2j} \Big),
\end{aligned}
$$

hence establishing the claim.

Taking $\epsilon \in (0,1)$ such that $\mathbf{D} = \bigcup_{w \in \Lambda} \Delta(w, \epsilon)$, we derive

$$
\int_{\mathbf{D}} |g|^2 d\mu \;\leq\; 2 \sum_{w \in \Lambda} \sup_{z \in S_w} |g(z) - g(w)|^2 \mu(S_w) + 2 \sum_{w \in \Lambda} \mu(S_w) |g(w)|^2
$$

$$
\preceq \sum_{w \in \Lambda} \mu(S_w) \left(\int_{\Delta(w,\epsilon_2)} |g'|^2 dm \right) + \sum_{w \in \Lambda} \mu(S_w) |g(w)|^2
$$

$$
\preceq \sum_{w \in \Lambda} \mu(S_w) \left((1 - |w|)^{-p} \int_{\Delta(w,\epsilon_2)} |g'(z)|^2 (1 - |z|)^p dm(z) \right)
$$

$$
+ \sum_{w \in \Lambda} \mu(S_w) |g(w)|^2
$$

$$
\preceq \|g\|_{\mathcal{D}^{2,p}}^2 \|\mu\|_{\mathcal{CM}_p} + \sum_{w \in \Lambda} \mu(S_w) |g(w)|^2,
$$

hence giving the desired estimate. $\qquad\qquad\qquad\qquad\qquad\qquad\qquad\square$

Theorem 6.3.3. *Let $p \in (0,1)$ and $g \in \mathcal{D}^{2,p}$. Then the following statements are equivalent:*

(i) $M_g \mathcal{D}^{2,2-p} \subseteq \mathcal{D}^{1,-(1-\frac{p}{2})} \odot \mathcal{D}^{2,2-p}$.

(ii) $U_g \mathcal{Q}_p \subseteq \mathcal{D}^{2,p}$.

(iii) $\mathcal{CM}_p \subseteq \{\mu : \ \mu \ \text{is a nonnegative Radon measure on } \mathbf{D} \text{ with } \int_{\mathbf{D}} |g|^2 d\mu < \infty\}$.

Proof. (iii)$\Rightarrow$(ii) This is evident because of

$$
f \in \mathcal{Q}_p \Leftrightarrow |f'(z)|^2 (1 - |z|^2)^p dm(z) \in \mathcal{CM}_p.
$$

(ii)$\Rightarrow$(i) By the Cauchy–Schwarz inequality and the Closed Graph Theorem, (ii) implies

$$
\left| \int_{\mathbf{D}} f'(z) \overline{\left(zg(z) h(z) \right)'} (1 - |z|^2) dm(z) \right|
$$

$$
= \left| \int_{\mathbf{D}} f' \overline{gh} dm \right|
$$

$$
\preceq \left(\int_{\mathbf{D}} |f'(z)|^2 |g(z)|^2 (1 - |z|^2)^p dm(z) \right)^{\frac{1}{2}} \left(\int_{\mathbf{D}} |h(z)|^2 (1 - |z|^2)^{-p} dm(z) \right)^{\frac{1}{2}}
$$

$$
\preceq (|f(0)| + \|f\|_{\mathcal{Q}_p,1}) \|h\|_{\mathcal{D}^{2,2-p}},
$$

hence giving $gh \in \mathcal{D}^{1,-(1-\frac{p}{2})} \odot \mathcal{D}^{2,2-p}$ owing to Theorem 5.2.3.

(i)$\Rightarrow$(iii) Suppose (i) holds. For any nonnegative Radon measure μ on $\mathbf{D}$, we take Lemma 6.3.2 into account. More precisely, given an interpolating sequence Γ for $\mathcal{D}^{2,2-p}$. Then, for any finite set $\Omega \subseteq \Gamma$ there exists a function $u_0 \in \mathcal{D}^{2,2-p}$ and a constant $C > 0$ depending only upon Γ such that

$$u_0(\zeta) = \begin{cases} (1 - |\zeta|)^{\frac{p}{2}-1}\overline{g(\zeta)}\sqrt{\mu(S_\zeta)} & , \quad \zeta \in \Omega, \\ 0 & , \quad \zeta \in \Gamma \setminus \Omega \end{cases}$$

and

$$\|u_0\|_{\mathcal{D}^{2,2-p}}^2 \leq C \sum_{\zeta \in \Omega} |g(\zeta)|^2 \mu(S_\zeta).$$

Consequently, if $u_0 \neq 0$ then $u = u_0\|u_0\|_{\mathcal{D}^{2,2-p}}^{-1}$ produces

$$\sqrt{\sum_{\zeta \in \Omega} |g(\zeta)|^2 \mu(S_\zeta)} \leq \sqrt{C} \sum_{\zeta \in \Omega} g(\zeta)u(\zeta)(1 - |\zeta|)^{1-\frac{p}{2}}\sqrt{\mu(S_\zeta)}.$$

Now (i) implies $gu \in \mathcal{D}^{1,-(1-\frac{p}{2})} \odot \mathcal{D}^{2,2-p}$. So it follows from the Closed Graph Theorem again that

$$\|gu\|_{\mathcal{D}^{1,-(1-\frac{p}{2})}\odot\mathcal{D}^{2,2-p}} \leq \sup_{\|v\|_{\mathcal{D}^{2,2-p}}=1} \|gv\|_{\mathcal{D}^{1,-(1-\frac{p}{2})}\odot\mathcal{D}^{2,2-p}} = \kappa(g) < \infty.$$

This means that gu can be expressed as

$$g(z)u(z) = \sum_{j=1}^{\infty} c_j(1 - |\lambda_j|^2)^{1-\frac{p}{2}}(1 - \overline{\lambda_j}z)^{-1}h_j(z)$$

where

$$\lambda_j \in \mathbf{D}, \quad h_j \in \mathcal{D}^{2,2-p}, \quad \|h_j\|_{\mathcal{D}^{2,2-p}} = 1 \quad \text{and} \quad \|\{c_j\}\|_{\ell^1} \leq 2\kappa(g).$$

Meanwhile, since Γ is an interpolating sequence for $\mathcal{D}^{2,2-p}$, we have

$$\sum_{\zeta \in \Omega} |v(\zeta)|^2 (1 - |\zeta|)^{2-p} \preceq 1, \quad \|v\|_{\mathcal{D}^{2,2-p}} \leq 1.$$

Putting all the above estimates together and using the Cauchy–Schwarz inequality, the implication

$$z \in S_\zeta \Rightarrow \left| \frac{1 - \overline{\lambda}z}{1 - \overline{\lambda}\zeta} \right| \leq 1 + \frac{|z - \zeta|}{1 - |\zeta|} \preceq 1, \quad \lambda \in \mathbf{D},$$

and Lemma 3.1.1, we derive

$$\sqrt{\sum_{\zeta \in \Gamma} |g(\zeta)|^2 \mu(S_\zeta)}$$

$$= \sqrt{\sum_{\zeta \in \Omega} |g(\zeta)|^2 \mu(S_\zeta)}$$

$$\leq 2\sqrt{C} \sup_{\lambda \in \mathbf{D}, \|v\|_{\mathcal{D}^{2,2-p}}} \sum_{\zeta \in \Omega} \frac{(1-|\lambda|^2)^{1-\frac{p}{2}}}{|1-\bar{\lambda}\zeta|} |v(\zeta)|(1-|\zeta|)^{1-\frac{p}{2}} \sqrt{\mu(S_\zeta)}$$

$$\preceq \kappa(g) \left(\sup_{\lambda \in \mathbf{D}} \frac{(1-|\lambda|^2)^{2-p}}{|1-\bar{\lambda}\zeta|^2} \mu(S_\zeta) \right)^{\frac{1}{2}}$$

$$\preceq \kappa(g) \left(\int_{\mathbf{D}} \frac{(1-|\lambda|^2)^{2-p}}{|1-\bar{\lambda}z|^2} d\mu(z) \right)^{\frac{1}{2}}$$

$$\preceq \kappa(g) \|\mu\|_{\mathcal{CM}_p}^{\frac{1}{2}}.$$

With the help of this treatment, we can handle the general case. Since the sequence Λ in Lemma 6.3.2 is a union of, say N_p, interpolating sequences Γ_j for $\mathcal{D}^{2,2-p}$, we can use Lemma 6.3.2 to get

$$\int_{\mathbf{D}} |g|^2 d\mu \leq \kappa_p \left(\|g\|_{\mathcal{D}^{2,p}}^2 \|\mu\|_{\mathcal{CM}_p} + \sum_{j=1}^{N_p} \sum_{\zeta \in \Gamma_j} |g(\zeta)|^2 \mu(S_\zeta) \right)$$

$$\preceq \|\mu\|_{\mathcal{CM}_p} \left(\|g\|_{\mathcal{D}^{2,p}}^2 + (\kappa(g))^2 \right),$$

thereupon establishing (iii). $\square$

We next verify $\mathcal{W}_p \neq \mathcal{Q}_p$ for $p \in (0,1)$.

Theorem 6.3.4. *Let $p \in (0,1)$. Then $\mathcal{W}_p$ is a proper subspace of $\mathcal{Q}_p$. Equivalently, $\mathcal{D}^{1,-(1-\frac{p}{2})} \odot \mathcal{D}^{2,2-p}$ is a proper subspace of $\mathcal{D}^{2,p} \odot \mathcal{D}^{2,2-p}$.*

Proof. Theorem 6.3.3 suggests that we show only that there are $f \in \mathcal{Q}_p, g \in \mathcal{D}^{2,p}$ satisfying $\|V_f g\|_{\mathcal{D}^{2,p}} = \infty$. To that end, it is enough to show that the conditions in Theorem 6.3.3 cannot hold for all $g \in \mathcal{D}^{2,p}$. By Theorem 6.2.3, we have

$$\sup_{\|g\|_{\mathcal{D}^{2,p}} \leq 1} \int_{\mathbf{D}} |g|^2 d\mu \approx \sup \frac{\mu\left(\bigcup_j S(I_j)\right)}{Cap_{\frac{1-p}{2}}\left(\bigcup_j I_j\right)},$$

where the supremum on the right-hand-side ranges over all finite disjoint unions of subarcs I_j of $\mathbf{T}$. Suppose that Theorem 6.3.3 is valid for all $g \in \mathcal{D}^{2,p}$. Then by Theorem 6.3.3(i) and the Closed Graph Theorem we derive

$$\sup_{\|v\|_{\mathcal{D}^{2,2-p}}=1} \|gv\|_{\mathcal{D}^{1,-(1-\frac{p}{2})} \odot \mathcal{D}^{2,2-p}} \preceq \|g\|_{\mathcal{D}^{2,p}}, \quad g \in \mathcal{D}^{2,p}.$$

Accordingly,

$$\sup \frac{\mu(\bigcup_j S(I_j))}{Cap_{\frac{1-p}{2}}(\bigcup_j I_j)} \preceq \|\mu\|_{\mathcal{CM}_p}$$

holds for all $\mu \in \mathcal{CM}_p$. For every finite union $E = \bigcup_{l=1}^m I_l$ of disjoint subarcs $I_l \subset \mathbf{T}$ with midpoint $e^{i\theta_l}$, let $\mu_E = \sum_{l=1}^m |I_l|^p \delta_l$ where δ_l is the Dirac measure at $(1 - |I_l|)e^{i\theta_l}$. Then

$$\mu_E\left(\bigcup_{l=1}^m S(I_l)\right) = \sum_{l=1}^m |I_l|^p \preceq Cap_{\frac{1-p}{2}}(E)\|\mu\|_{\mathcal{CM}_p}.$$

Note that $\mu_E(S(I)) = 0$ whenever $|I| < \min_{1 \le l \le m} |I_l|$. So, we will construct a Cantor-like set having positive p-dimensional Hausdorff measure. To begin with, let $\phi_n = 2^{-n/p}$ and note that $\phi_n - 2\phi_{n+1}$ are positive for all $n \in \mathbf{N}$. Remove from $[0, 1]$ the open middle interval of length $1 - 2\phi_1$, denoted by $I(2^{-1})$, obtaining a set K_1 which is the union of two closed intervals $[0, \phi_1]$ and $[1 - \phi_1, 1]$. Remove from each of these intervals the open middle intervals of length $\phi_1 - 2\phi_2$, denoted by $I(2^{-2})$ and $I(3 \cdot 2^{-2})$, obtaining a set K_2 which is the union of four closed intervals of length ϕ_2. Now proceed by induction. Assume that K_{n-1} is constructed. Remove from the different intervals making up this set the open middle intervals of length $\phi_{n-1} - 2\phi_n$, denoted by $I(2^{-n}), \ldots, I((2n - 1)2^{-n})$, to obtain K_n. Let $K = \bigcap_{n=1}^{\infty} K_n$. Then

$$\mathbf{K} = \{e^{2\pi i\theta} : \theta \in K \subset [0, 1]\}$$

is the natural identification of K on $\mathbf{T}$. Now $E_n = \bigcup_j I_{j,n}$ is the union of subarcs containing $\mathbf{K}$ obtained at the n-th stage of the construction of $\mathbf{K}$. More precisely, the 2^n subarcs $I_{j,n}$ of arclength r^n are obtained by removing from the middle of each $I_{k,n-1}$ a subarc of arclength $(1 - 2r)|I_{k,n-1}|$. In particular, if $I_{j,n}$ and $I_{j+1,n}$ are obtained from $I_{k,n-1}$ in this way, then by $2 = r^{-p}$,

$$|I_{j,n}|^p + |I_{j+1,n}|^p = 2r^{np} = r^{(n-1)p} = |I_{k,n-1}|.$$

To estimate $\|\mu\|_{\mathcal{CM}_p}$, recall first that if $\mu_{E_n}(S(I)) > 0$, then $|I| \ge r^n$. Denote by $\kappa(I, n)$ the number of subarcs $I_{j,n}$ that intersect the arc $2I$ and notice that if $\kappa(I, n) \le 2$, then $|I|^{-p}\mu_{E_n}(S(I)) \le 2$, and if $\kappa(I, n) > 2$, then

$$|I|^{-p}\mu_{E_n}(S(I)) \le 2\sum_{l=0}^{m-1} r^{(n-l)p} + |I|^{-p}\mu_{E_{n-m}}(S(I)) \quad \text{as} \quad \kappa(I, n - m + 1) > 2.$$

The process stops when m is the smallest integer with $\kappa(I, n - m) \le 2$. In this case we get $|I| > r^{n-m+1}$, producing

$$\frac{\mu_{E_n}(S(I))}{|I|^p} \le 2\left(1 + (1 - r^p)^{-1}\right).$$

On the other hand, since

$$H_0^p(\mathbf{K}) \leq \mu_{E_n}\left(\bigcup_j S(I_{j,n})\right) = 1,$$

we conclude

$$0 = Cap_{\frac{1-p}{2}}(\mathbf{K}) = \lim_{n \to \infty} Cap_{\frac{1-p}{2}}(E_n),$$

which however yields $\lim_{n \to \infty} \mu_{E_n}\left(\bigcup_j S(I_{j,n})\right) = 0$, a contradiction. $\qquad \square$

6.4   Hankel and Volterra on Dirichlet Spaces

In this section, we show that $\mathcal{W}_p$ behaves very much like $\mathcal{Q}_p$ with respect to the Hankel and Volterra operators acting on $\mathcal{D}^{2,p}$.

Lemma 6.4.1. *Let*

$$\mathsf{T}f(z) = \int_{\mathbf{D}} \frac{f(w)(1 - |w|^2)^{b-2}}{|1 - z\bar{w}|^b} dm(w).$$

(i) *If $\beta \in (-\infty, -1)$ and $b > 1 + (1 + \beta)/2$, then T is a continuous operator on $\mathcal{L}^2(\mathbf{D}, (1 - |z|^2)^\beta dm(z))$.*

(ii) *If*

$$\alpha \in (-\infty, 1/2], \ \beta > \max\{-1, -1 - 2\alpha\} \text{ and } b > \max\left\{\frac{\beta+3}{2}, \frac{\beta+3}{2} - \alpha\right\},$$

then $|\mathsf{T}f(z)|^2(1 - |z|^2)^\beta dm(z)$ is a $\mathcal{D}^{2,1-2\alpha}$-Carleson measure on $\mathbf{D}$ whenever $|f(z)|^2(1 - |z|^2)^\beta dm(z)$ is a $\mathcal{D}^{2,1-2\alpha}$-Carleson measure on $\mathbf{D}$.

Proof. (i) Rewriting

$$\mathsf{T}f(z) = \int_{\mathbf{D}} f(z)k(z,w)(1 - |z|^2)^\beta dm(z) \ \text{ where } \ k(z,w) = \frac{(1 - |z|^2)^{b-\beta-2}}{|1 - \bar{z}w|^b},$$

and noticing

$$\int_{\mathbf{D}} k(z,w)(1 - |z|^2)^{\gamma+\beta} dm(z) \preceq (1 - |w|^2)^\gamma$$

and

$$\int_{\mathbf{D}} k(z,w)(1 - |w|^2)^{\gamma+\beta} dm(z) \preceq (1 - |z|^2)^\gamma$$

provided that γ satisfies

$$\max\{1 - b, -1 - \beta\} < \gamma < \min\{0, b - 2 - \beta\}$$

which is valid under the assumption on β and b, we can use the Schur Lemma to derive the required continuity of the operator.

(ii) Thanks to (i), it suffices to prove that if $d\mu(z) = |f(z)|^2(1 - |z|^2)^\beta dm(z)$ is $\mathcal{D}^{2,1-2\alpha}$-Carleson measure, then $\mathsf{M}_{\mathsf{T}_f} - \mathsf{T} \circ \mathsf{M}_f$ is a continuous operator from $\mathcal{D}^{2,1-2\alpha}$ to $\mathcal{L}^2\big(\mathbf{D}, (1 - |z|^2)^\beta dm(z)\big)$. To do so, note that the formula

$$\left|\mathsf{M}_{\mathsf{T}f}g(w) - \mathsf{T} \circ \mathsf{M}_f g(w)\right|^2 = \left|\int_{\mathbf{D}} \frac{f(z)\big(g(w) - g(z)\big)}{|1 - \bar{z}w|^b(1 - |z|^2)^{2-b}} dm(z)\right|^2$$

holds for any $g \in \mathcal{D}^{2,1-2\alpha}$. So we consider two cases as follows.

Case 1: $\alpha = 1/2$. Since Lemma 2.5.1 is valid for this α, we apply this case to g, and employ the Cauchy–Schwarz inequality to get

$$\int_{\mathbf{D}} |\mathsf{M}_{\mathsf{T}f}g(w) - \mathsf{T} \circ \mathsf{M}_f g(w)|^2 (1 - |z|^2)^\beta dm(z)$$

$$\preceq \left(\int_{\mathbf{D}} |f(z)|^2(1 - |z|^2)^\beta dm(z)\right)\left(\int_{\mathbf{D}} \frac{|g(w) - g(z)|^2}{|1 - \bar{z}w|^{2b}(1 - |z|^2)^{4-2b+\beta}} dm(z)\right)$$

$$\preceq \left(\int_{\mathbf{D}} |f(z)|^2(1 - |z|^2)^\beta dm(z)\right)\|g\|^2_{\mathcal{D}^{2,1-2\alpha}}.$$

Case 2: $\alpha < 1/2$. Since the above-defined measure μ is a $\mathcal{D}^{2,1-2\alpha}$-Carleson measure, from the fact that for $\epsilon > 0$ and $\alpha < 1/2$, the function

$$g_a(z) = (1 - |a|^2)^{\epsilon/2}(1 - \bar{a}z)^{\alpha-1/2-\epsilon/2}$$

belongs to $\mathcal{D}^{2,1-2\alpha}$ we see

$$\sup_{a\in\mathbf{D}} \int_{\mathbf{D}} (1 - |a|^2)^\epsilon |1 - \bar{a}z|^{2\alpha-1-\epsilon} d\mu(z) \preceq \|\mu\|_{\mathcal{CMD}_{1-2\alpha}}.$$

Now, selecting such an $\epsilon > 0$ that those assumptions for Lemma 6.4.1 remain valid with β being replaced by $\beta - \epsilon$, we use Lemma 2.5.1 and the Cauchy–Schwarz inequality again to deduce

$$\int_{\mathbf{D}} |\mathsf{M}_{\mathsf{T}f}g(w) - \mathsf{T} \circ \mathsf{M}_f g(w)|^2(1 - |z|^2)^\beta dm(z)$$

$$\preceq \int_{\mathbf{D}} \left(\int_{\mathbf{D}} \frac{|f(z)|^2(1 - |z|^2)^\beta}{|1 - \bar{z}w|^{1+\epsilon-2\alpha}} dm(z)\right)$$

$$\times \left(\int_{\mathbf{D}} \frac{|g(w) - f(z)|^2}{|1 - \bar{z}w|^{2b-1-\epsilon+2\alpha}(1 - |z|^2)^{4-2b+\beta}} dm(z)\right)(1 - |w|^2)^\beta dm(w)$$

$$\preceq \|\mu\|_{\mathcal{CMD}_{1-2\alpha}} \int_{\mathbf{D}}\int_{\mathbf{D}} \frac{|g(w) - f(z)|^2(1 - |w|^2)^{\beta-\epsilon}}{|1 - \bar{z}w|^{2b-1-\epsilon+2\alpha}(1 - |z|^2)^{4-2b+\beta}} dm(z)dm(w)$$

$$\preceq \|\mu\|_{\mathcal{CMD}_{1-2\alpha}}\|g\|^2_{\mathcal{D}^{2,1-2\alpha}}. \qquad \square$$

For $\gamma > -1$, $z \in \mathbf{D}$ and $f, g \in \mathcal{H}(\bar{D})$, set

$$H_{f,\gamma}(g)(z) = \int_{\mathbf{D}} f'(w)\overline{g(w)} \left(\int_0^z \frac{1}{(1 - \bar{w}\zeta)^{2+\gamma}} d\zeta \right) (1 - |w|^2)^\gamma dm(w).$$

The forthcoming result, similar to Theorem 6.1.1, shows that each $\mathcal{W}_p$ exists as a symbol space of $H_{f,\gamma}$ acting continuously on $\mathcal{D}^{2,p}$ for a suitable γ.

Theorem 6.4.2. *Let $f \in \mathcal{H}$, $p \in [0, \infty)$ and $\gamma > (p-1)/2$. Then the following statements are equivalent:*

(i) *$H_{f,\gamma}$ exists as a continuous operator from $\mathcal{D}^{2,p}$ to $\mathcal{D}^{2,p}$.*

(ii) *V_f is a continuous operator from $\mathcal{D}^{2,p}$ to $\mathcal{D}^{2,p}$.*

(iii) *$f \in \mathcal{W}_p$.*

Proof. (i)$\Leftrightarrow$(iii) If $f \in \mathcal{W}_p$ then $g \in \mathcal{D}^{2,p}$ implies $\bar{f}'g \in \mathcal{L}^2\big(\mathbf{D}, (1 - |z|^2)^p dm(z)\big)$ and hence $H_{f,\gamma}g \in \mathcal{D}^{2,p}$ by Lemma 6.4.1 (i).

Conversely, suppose $H_{f,\gamma}$ extends to a continuous operator on $\mathcal{D}^{2,p}$. So its operator norm $\|H_{f,\gamma}\|_{\mathcal{D}^{2,p} \mapsto \mathcal{D}^{2,p}} < \infty$. Furthermore, if $g = 1$ then

$$H_{f,\gamma}(g)'(z) = \frac{\pi f'(z)}{1 + \gamma}, \quad f \in \mathcal{H}(\bar{\mathbf{D}}),$$

and hence

$$\|f\|_{\mathcal{D}^{2,p}} \preceq \|H_{f,\gamma}\|_{\mathcal{D}^{2,p} \mapsto \mathcal{D}^{2,p}} < \infty.$$

To prove $f \in \mathcal{W}_p$, we may in turn verify that

$$\int_{\mathbf{D}} \left| \frac{\pi}{\gamma + 1} f'(w)\overline{g(w)} - (H_{f,\gamma}g)'(w) \right|^2 (1 - |w|^2)^p dm(z) \preceq \|H_{f,\gamma}\|^2_{\mathcal{D}^{2,p} \mapsto \mathcal{D}^{2,p}} \|g\|^2_{\mathcal{D}^{2,p}}$$

holds for any $g \in \mathcal{D}^{2,p}$. Because of the formula

$$\frac{\pi}{\gamma + 1} f'(w)\overline{g(w)} - (H_{f,\gamma}g)'(w) = \int_{\mathbf{D}} \frac{f'(z)\big(\overline{g(w) - g(z)}\big)}{(1 - z\bar{w})^{2+\gamma}} (1 - |z|^2)^\gamma dm(z),$$

we can, in a similar manner to proving Lemma 6.4.1 (ii), derive that if $p = 0$, then

$$\int_{\mathbf{D}} \left| \frac{\pi}{\gamma + 1} f'(w)\overline{g(w)} - (H_{f,\gamma}g)'(w) \right|^2 (1 - |w|^2)^p dm(z)$$
$$\preceq \|f\|^2_{\mathcal{D}^{2,0}} \|g\|^2_{\mathcal{D}^{2,0}}$$
$$\preceq \|H_{f,\gamma}\|^2_{\mathcal{D}^{2,0} \mapsto \mathcal{D}^{2,0}} \|g\|^2_{\mathcal{D}^{2,0}},$$

but also that if $p > 0$, then

$$\int_{\mathbf{D}} \left| \frac{\pi}{\gamma + 1} f'(w)\overline{g(w)} - (H_{f,\gamma}g)'(w) \right|^2 (1 - |w|^2)^p dm(z) \preceq \|f\|^2_{\mathcal{B}} \|g\|^2_{\mathcal{D}^{2,p}}.$$

Of course, it remains to check that the boundedness of $\mathsf{H}_{f,\gamma}$ must yield

$$\|f\|_{\mathcal{B}} \preceq \|\mathsf{H}_{f,\gamma}\|_{\mathcal{D}^{2,p} \mapsto \mathcal{D}^{2,p}} < \infty \quad \text{for} \quad p > 0.$$

In fact, choose $-n = [(1-p)/2]$ — the greatest integer less than or equal to $(1-p)/2$ — and put

$$g_a(z) = \frac{z^{1+n}(1-|a|^2)^{1+n-p/2}}{(1-\bar{a}z)^{1+n}}, \quad h_a(z) = \frac{(1-|a|^2)^{1+p/2}(1-|z|^2)^{\gamma-p}}{(1-\bar{a}z)^{2+\gamma}}.$$

Then

$$\|g_a\|_{\mathcal{D}^{2,p}} \preceq 1, \quad \int_{\mathbf{D}} |h_a(z)|^2(1-|z|^2)^p dm(z) \approx 1.$$

Meanwhile, using the reproducing formula for $\mathcal{A}^{2,\gamma}$, we find

$$\int_{\mathbf{D}} \left(\mathsf{H}_{f,\gamma}g_a\right)'(w)\overline{h_a(w)}(1-|w|^2)^p dm(w)$$

$$= (1-|a|^2)^{1+p/2} \int_{\mathbf{D}} f'(z)\overline{g_a(z)} \left(\int_{\mathbf{D}} \frac{(1-|w|^2)^\gamma}{(1-a\bar{w})^{2+\gamma}(1-\bar{z}w)^{2+\gamma}} dm(w)\right) dm(z)$$

$$= \frac{\pi(1-|a|^2)^{1+p/2}}{\gamma+1} \int_{\mathbf{D}} f'(z)\overline{g_a(z)}(1-\bar{z}a)^{-(2+\gamma)} dm(z)$$

$$= \frac{\pi}{\gamma+1} \int_{\mathbf{D}} \frac{f'(z)\bar{z}^{1+n}(1-|a|^2)^{2+n}}{(1-\bar{z}a)^{3+\gamma+n}}$$

$$= \left(\frac{\pi}{\gamma+1}\right)^2 \frac{(1-|a|^2)^{n+2}f^{(n+2)}(a)}{(2+\gamma)(3+\gamma)\cdots(n+2+\gamma)},$$

so that

$$\sup_{a\in\mathbf{D}}(1-|a|^2)^{n+2}|f^{(n+2)}(a)|$$

$$\preceq \|\mathsf{H}_{f,\gamma}\|_{\mathcal{D}^{2,p}\mapsto\mathcal{D}^{2,p}} \sup_{a\in\mathbf{D}} \|g_a\|_{\mathcal{D}^{2,p}} \left(\int_{\mathbf{D}} |h_a(z)|^2(1-|z|^2)^p dm(z)\right)^{\frac{1}{2}},$$

hence implying the desired assertion thanks to the higher derivative characterization of $\mathcal{B}$ (cf. Theorem 1.1.3 (v) and Remark 1.1.4).

(ii)$\Leftrightarrow$(iii) This follows readily from the definitions of V_f, $\mathcal{W}_p$ and $\mathcal{D}^{2,p}$, and Theorem 6.2.3. $\qquad\square$

Corollary 6.4.3. *Let $p \in [0, \infty)$. Then $\mathcal{W}_p$ is isomorphic to the dual of $\mathcal{D}^{2,p} \odot \mathcal{D}^{2,2+p}$ with respect to the pairing*

$$\langle f, g\rangle_p = \int_{\mathbf{D}} f(z)\overline{g'(z)}(1-|z|^2)^p dm(z), \quad f \in \mathcal{D}^{2,p} \odot \mathcal{D}^{2,2+p}, \ g \in \mathcal{W}_p.$$

Proof. Suppose $f \in \mathcal{D}^{2,p} \odot \mathcal{D}^{2,2+p}$ and $g \in \mathcal{W}_p$. Then H_g is bounded on $\mathcal{D}^{2,p}$ and $f = \sum_{j=1}^{\infty} g_j h_j$ with $\sum_{j=1}^{\infty} \|g_j\|_{\mathcal{D}^{2,p}} \|h_j\|_{\mathcal{D}^{2,2+p}} < \infty$. Hence by the Cauchy–Schwarz inequality,

$$
\begin{aligned}
|\langle f, g \rangle_p| &= \left| \sum_{j=1}^{\infty} \langle g_j h_j, g \rangle_p \right| \\
&= \left| \sum_{j=1}^{\infty} \int_{\mathbf{D}} g_j(z) h_j(z) \overline{g'(z)} (1 - |z|^2)^p dm(z) \right| \\
&= \left| \sum_{j=1}^{\infty} \int_{\mathbf{D}} \mathsf{H}_{g,p}(g_j) \overline{h_j'(z)} (1 - |z|^2)^p dm(z) \right| \\
&\preceq \|\mathsf{H}_{g,p}\|_{\mathcal{D}^{2,p} \mapsto \mathcal{D}^{2,p}} \sum_{j=1}^{\infty} \|g_j\|_{\mathcal{D}^{2,p}} \|h_j\|_{\mathcal{D}^{2,p+2}},
\end{aligned}
$$

hence g induces a continuous linear functional on $\mathcal{D}^{2,p} \odot \mathcal{D}^{2,2+p}$ under the pairing $\langle \cdot, \cdot \rangle_p$ with

$$
|\langle f, g \rangle_p| \preceq \|\mathsf{H}_{g,p}\|_{\mathcal{D}^{2,p} \mapsto \mathcal{D}^{2,p}} \|f\|_{\mathcal{D}^{2,p} \odot \mathcal{D}^{2,2+p}}.
$$

Conversely, assume that L is a continuous linear functional on $\mathcal{D}^{2,p} \odot \mathcal{D}^{2,2+p}$. Then its norm $\|\mathsf{L}\| < \infty$ and hence for $g_j \in \mathcal{D}^{2,p}$ and $h_j \in \mathcal{D}^{2,2+p}$ we have

$$
|\mathsf{L}(g_j h_j)| \leq \|\mathsf{L}\| \|g_j h_j\|_{\mathcal{D}^{2,p} \odot \mathcal{D}^{2,p+2}} \leq \|g_j\|_{\mathcal{D}^{2,p}} \|h_j\|_{\mathcal{D}^{2,p+2}},
$$

thereby deriving that for fixed $g_j \in \mathcal{D}^{2,p}$, the mapping $h_j \mapsto \mathsf{L}(g_j h_j)$ is continuous from $\mathcal{D}^{2,p+2}$ to $\mathbf{C}$. Using the Riesz–Fischer Theorem, we find an element $G_j \in \mathcal{A}^{2,p}$ such that

$$
\mathsf{L}(g_j h_j) = \int_{\mathbf{D}} h_j(z) \overline{G_j(z)} (1 - |z|^2)^p dm(z).
$$

Consequently, g is uniquely determined by g_j and $g_j \mapsto G_j$ is bounded from $\mathcal{D}^{2,p}$ to $\mathcal{A}^{2,p}$ with $\|G_j\|_{\mathcal{A}^{2,p}} \preceq \|\mathsf{L}\| \|g_j\|_{\mathcal{D}^{2,p}}$. Using the reproducing formula for $\mathcal{A}^{2,p}$, we find

$$
G_j(w) = \frac{p+1}{\pi} \int_{\mathbf{D}} G_j(z) \overline{k_p(z, w)} (1 - |z|^2)^p dm(z) = \frac{p+1}{\pi} \overline{\mathsf{L}(g_j k_p(\cdot, w))},
$$

where $k_p(z, w) = (1 - \bar{w}z)^{-2-p}$ for $z, w \in \mathbf{D}$. Note that if w is fixed, then $g_j k_p(\cdot, w)$ belongs to $\mathcal{A}^{2,p}$. So there is $g \in \mathcal{D}^{2,p}$ such that

$$
\mathsf{L}(g_j k_p(\cdot, w)) = \int_{\mathbf{D}} g_j(z) k_p(z, w) \overline{g'(z)} (1 - |z|^2)^p dm(z).
$$

Consequently,

$$
\|\mathsf{H}_{g,p}(g_j)\|_{\mathcal{D}^{2,p}} \preceq \|G_j\|_{\mathcal{A}^{2,p}} \preceq \|\mathsf{L}\| \|g_j\|_{\mathcal{D}^{2,p}}.
$$

This gives that $\mathsf{H}_{g,p}$ is bounded on $\mathcal{D}^{2,p}$, and so that $g \in \mathcal{W}_p$ due to Theorem 6.4.2. Now, it is easy to deduce that if $f \in \mathcal{D}^{2,p} \odot \mathcal{D}^{2,p+2}$, then $f = \sum_{j=1}^{\infty} g_j h_j$ with $g_j \in \mathcal{D}^{2,p}$ and $h_j \in \mathcal{D}^{2,p+2}$ and hence

$$
\begin{aligned}
\mathsf{L}(f) &= \sum_{j=1}^{\infty} \mathsf{L}(g_j h_j) \\
&= \sum_{j=1}^{\infty} \int_{\mathbf{D}} g_j(z) h_j(z) \overline{G_j(z)} (1 - |z|^2)^p dm(z) \\
&= \sum_{j=1}^{\infty} \int_{\mathbf{D}} h_j(w) \left(\frac{p+1}{\pi} \int_{\mathbf{D}} \frac{g_j(z) k_p(z,w) \overline{g'(z)}}{(1-|z|^2)^{-p}} dm(z) \right) \frac{dm(w)}{(1-|w|^2)^{-p}} \\
&= \sum_{j=1}^{\infty} \int_{\mathbf{D}} \overline{g'(z)} g_j(z) \left(\frac{p+1}{\pi} \int_{\mathbf{D}} h_j(w) \frac{\overline{k_p(w,z)}}{(1-|w|^2)^{-p}} dm(w) \right) \frac{dm(z)}{(1-|z|^2)^{-p}} \\
&= \sum_{j=1}^{\infty} \int_{\mathbf{D}} g_j(z) h_j(z) \overline{g'(z)} (1 - |z|^2)^p dm(z) \\
&= \int_{\mathbf{D}} f(z) \overline{g'(z)} (1 - |z|^2)^p dm(z).
\end{aligned}
$$

This completes the proof.   □

Corollary 6.4.4. *Let $p \in [0,2)$ and $f \in \mathcal{H}$. Then M_f is a continuous operator from $\mathcal{D}^{2,p}$ to $\mathcal{D}^{2,p}$ if and only if $f \in H^{\infty} \cap \mathcal{W}_p$.*

Proof. By Theorem 6.4.2, it suffices to prove that the continuity of U_f on $\mathcal{D}^{2,p}$ implies $f \in \mathcal{H}^{\infty}$. To this end, given $a \in \mathbf{D}$, let

$$
g_a(z) = \begin{cases} \dfrac{(1-|a|^2)^{1-p/2}}{1-\bar{a}z} & , \quad p \in (0,2), \\ \dfrac{-\log(1-\bar{a}z)}{\sqrt{-\log(1-|a|^2)}} & , \quad p = 0, \end{cases}
$$

and evaluate

$$
\|g_a\|_{\mathcal{D}^{2,p}}^2 \preceq \begin{cases} 1 + \int_{\mathbf{D}} \dfrac{(1-|a|^2)^{2-p}(1-|z|^2)^p}{|1-\bar{a}z|^4} dm(z) \preceq 1 & , \quad p \in (0,2), \\ 1 + \int_{\mathbf{D}} \dfrac{\left(-\log(1-|a|^2)\right)^{-1}}{|1-\bar{a}z|^2} dm(z) \preceq 1 & , \quad p = 0. \end{cases}
$$

Using the proof of Corollary 6.1.2, we further derive

$$
\begin{aligned}
\|f\|_{\mathcal{H}^{\infty}} &\preceq \sup_{a \in \mathbf{D}} \left(\int_{\mathbf{D}} |f(z)|^2 |g_a'(z)|^2 (1 - |z|^2)^p dm(z) \right)^{\frac{1}{2}} \\
&\preceq \|\mathsf{U}_f\|_{\mathcal{D}^{2,p} \mapsto \mathcal{D}^{2,p}} \sup_{a \in \mathbf{D}} \|g_a\|_{\mathcal{D}^{2,p}} < \infty.
\end{aligned}
$$
                                                                                            □

A quite natural question is: what condition should f have so that M_f is surjective from $\mathcal{D}^{2,p}$ to $\mathcal{D}^{2,p}$? This question can be solved via the following corona type decomposition.

Theorem 6.4.5. *Let $p \in [0,1)$ and $n \in \mathbf{N}$. For $(f_1, f_2, \ldots, f_n) \in \mathcal{H} \times \mathcal{H} \times \cdots \times \mathcal{H}$, set*

$$\mathsf{M}_{(f_1, f_2, \ldots, f_n)}(g_1, g_2, \ldots, g_n) = \sum_{k=1}^{n} f_k g_k.$$

Then the following two statements are equivalent:

(i) $\mathsf{M}_{(f_1, f_2, \ldots, f_n)}$ *maps $\mathcal{D}^{2,p} \times \mathcal{D}^{2,p} \times \cdots \times \mathcal{D}^{2,p}$ onto $\mathcal{D}^{2,p}$.*

(ii)

$$(f_1, f_2, \ldots, f_n) \in (\mathcal{W}_p \cap \mathcal{H}^\infty) \times (\mathcal{W}_p \cap \mathcal{H}^\infty) \times \cdots \times (\mathcal{W}_p \cap \mathcal{H}^\infty)$$

with

$$\delta(f_1, f_2, \ldots, f_n) = \inf_{z \in \mathbf{D}} \sum_{k=1}^{n} |f_k(z)| > 0.$$

Proof. Suppose (i) holds. Then the first result in (ii) follows right away from Corollary 6.4.4. To see the second assertion of (ii), we use the Open Map Theorem to derive that for each $h \in \mathcal{D}^{2,p}$ there are $g_1, g_2, \ldots, g_n \in \mathcal{W}_p \cap \mathcal{H}^\infty$ such that

$$\|g_k\|_{\mathcal{D}^{2,p}} \preceq \|g_k\|_{\mathcal{H}^\infty} + \|g_k\|_{\mathcal{W}_p} \preceq \|h\|_{\mathcal{D}^{2,p}}, \quad k = 1, 2, \ldots, n,$$

and $h = \sum_{k=1}^{n} f_k g_k$. Note that every $f \in \mathcal{D}^{2,p}$ satisfies:

$$(1 - |z|^2)^{p/2}|f(z)| \preceq \|f\|_{\mathcal{D}^{2,p}}, \quad f \in \mathcal{D}^{2,p}, \ p \in (0,1)$$

and

$$\left(-\log(1 - |z|^2)\right)^{-\frac{1}{2}}|f(z)| \preceq \|f\|_{\mathcal{D}^{2,0}}, \quad f \in \mathcal{D}^{2,0}.$$

So, making use of these inequalities, and letting

$$h(z) = \begin{cases} \dfrac{(1-|w|^2)^{1-p/2}}{1-\bar{w}z} & , \quad p \in (0,1), \\ \log(1 - \bar{w}z)^{-1} & , \quad p = 0 \end{cases}$$

in $\sum_{k=1}^{n} f_k g_k = h$, we obtain that if $p \in (0,1)$, then

$$\frac{(1 - |w|^2)^{1-p/2}}{|1 - \bar{w}z|} \leq \sum_{k=1}^{n} |f_k(z)||g_k(z)| \preceq (1 - |z|^2)^{-p/2} \sum_{k=1}^{n} |f_k(z)|,$$

thus producing $\delta(f_1, f_2, \ldots, f_n) > 0$. Similarly, if $p = 0$, then

$$\log|1 - \bar{w}z|^{-1} \leq \sum_{k=1}^{n} |f_k(z)||g_k(z)|$$

$$\preceq \left(-\log(1 - |z|^2)\right)^{\frac{1}{2}} \|h\|_{\mathcal{D}^{2,0}} \sum_{k=1}^{n} |f_k(z)|$$

$$\preceq \left(-\log(1 - |z|^2)\right) \sum_{k=1}^{n} |f_k(z)|,$$

and hence $\delta(f_1, f_2, \ldots, f_n) > 0$.

Conversely, assume (ii) is true. If we can show

$$\mathcal{D}^{2,p} \subseteq M_{(f_1, f_2, \ldots, f_n)}(\mathcal{D}^{2,p} \times \cdots \times \mathcal{D}^{2,p}),$$

then we are done. For $f \in \mathcal{D}^{2,p}$ choose

$$h_k = \frac{f\bar{f}_k}{\sum_{j=1}^{n} |f_j|^2}, \quad k = 1, \ldots, n.$$

Then $(h_1, h_2, \ldots, h_n)$ is a solution to the equation $\sum_{k=1}^{n} f_k h_k = f$. But this solution is not holomorphic, and thus must be modified. Without loss of generality, by the normal family principle we may assume that each f_k is in $\mathcal{H}(\bar{D})$. Now, suppose we can find functions $b_{j,k}$, $1 \leq j, k \leq n$, defined on $\bar{\mathbf{D}}$ such that

$$\frac{\partial b_{j,k}(z)}{\partial \bar{z}} = f h_j \frac{\partial h_k(z)}{\partial \bar{z}}, \quad z \in \mathbf{D},$$

and such that the boundary value functions $b_{j,k}$ are of $\mathcal{D}^{2,p}(\mathbf{T})$, namely,

$$\int_{\mathbf{T}} \int_{\mathbf{T}} \frac{|b_{j,k}(\zeta) - b_{j,k}(\eta)|^2}{|\zeta - \eta|^{2-p}} |d\zeta||d\eta| < \infty.$$

Then

$$g_k = f h_k + \sum_{j=1}^{n} (b_{k,j} - b_{j,k}) f_j \in \mathcal{D}^{2,p} \quad \text{and} \quad \sum_{k=1}^{n} f_k g_k = f.$$

Accordingly, we have only to demonstrate that $\frac{\partial b}{\partial \bar{z}} = fh$ is solvable in $\mathcal{D}^{2,p}(\mathbf{T})$, where $b = b_{j,k}$ and $h = h_j \frac{\partial h_k}{\partial \bar{z}}$. Since each f_k is in $\mathcal{W}_p \cap \mathcal{H}^\infty$, $|f_k'(z)|^2(1-|z|^2)^p dm(z)$ is a $\mathcal{D}^{2,p}$-Carleson measure. Also since

$$|h(z)|^2 \leq \frac{2\sum_{k=1}^{n} |f_k'(z)|^2}{\left(\delta(f_1, f_2, \ldots, f_n)\right)^6},$$

[Ga, p. 326], $|h(z)|^2(1-|z|^2)^p dm(z)$ is a $\mathcal{D}^{2,p}$-Carleson measure. Note that a standard solution for the equation $\frac{\partial b}{\partial \bar{z}} = fh$ is given by

$$b(z) = \pi^{-1} \int_{\mathbf{D}} \frac{f(w)h(w)}{z-w} dm(z)$$

which is continuous on $\mathbf{C}$ and is $\mathcal{C}^2$ on $\mathbf{D}$. To prove that this function belongs to $\mathcal{D}^{2,p}(\mathbf{T})$, we will show

$$\int_{\mathbf{D}} |\nabla b(z)|^2 (1-|z|^2)^p dm(z) < \infty.$$

To this end, we use

$$|h(z)|^2 \preceq \sum_{k=1}^n |f_k'(z)|^2, \ f \in \mathcal{D}^{2,p} \ \text{ and } \ f_k \in \mathcal{W}_p \cap \mathcal{H}^\infty$$

to get

$$
\begin{aligned}
\int_{\mathbf{D}} \left| \frac{\partial b(z)}{\partial \bar{z}} \right|^2 (1-|z|^2)^p dm(z) &= \int_{\mathbf{D}} |f(z)h(z)|^2 (1-|z|^2)^p dm(z) \\
&\preceq \sum_{k=1}^n \int_{\mathbf{D}} |(ff_k)'(z)|^2 (1-|z|^2)^p dm(z) \\
&\quad + \sum_{k=1}^n \int_{\mathbf{D}} |f'(z)f_k(z)|^2 (1-|z|^2)^p dm(z) \\
&< \infty.
\end{aligned}
$$

Meanwhile, let $\mathsf{B}(F)$ denote the Beurling transform of a function $F \in \mathcal{L}^1_{loc}(\mathbf{C}, dm)$, that is, the following principal value integral:

$$\mathsf{B}(F)(z) = \text{p.v.} \int_{\mathbf{C}} \frac{F(w)}{(z-w)^2} dm(w), \quad z \in \mathbf{C}.$$

Then $\partial b / \partial z = \pi^{-1} \mathsf{B}(F)$ whenever

$$F(z) = \begin{cases} f(z)h(z) &, \quad z \in \mathbf{D}, \\ 0 &, \quad z \in \mathbf{C} \setminus \mathbf{D}. \end{cases}$$

Observe that $|1-|z|^2|^p$ is an A_2-weight [CuRu, p. 411] when $p \in [0,1)$, namely,

$$\sup_\Delta \left(\frac{1}{m(\Delta)} \int_\Delta |1-|z|^2|^p dm(z) \right) \left(\frac{1}{m(\Delta)} \int_\Delta |1-|z|^2|^{-p} dm(z) \right) < \infty,$$

where the supremum is taken over all Euclidean disks Δ in $\mathbf{C}$. Thus B is bounded on $\mathcal{L}^2(\mathbf{C}, |1-|z|^2|^p dm(z))$. Consequently,

$$
\begin{aligned}
\int_{\mathbf{D}} \left| \frac{\partial b(z)}{\partial z} \right|^2 (1-|z|^2)^p dm(z) &\preceq \int_{\mathbf{C}} |F(z)|^2 |1-|z|^2|^p dm(z) \\
&\approx \int_{\mathbf{D}} |f(z)h(z)|^2 (1-|z|^2)^p dm(z) < \infty.
\end{aligned}
$$

The above-established estimates, along with $|\nabla| \leq 2(|\partial/\partial\bar{z}| + |\partial/\partial z|)$, imply the required result. $\qquad\square$

6.5 Notes

Note 6.5.1. Section 6.1 is produced via modifying and improving [AlCaSi, Corollary 4.1 and Theorem 5.2].

Note 6.5.2. Section 6.2 consists of Theorem 6.2.1 (from [Xi5]), Lemma 6.2.2 and Theorem 6.2.3 (from [Ste1]). From Theorem 6.2.1 it turns out that

$$\int_{\mathbf{D}} |g| d\mu \preceq \|g\|_{\mathcal{D}^{1,-(1-\frac{p}{2})}}, \quad g \in \mathcal{D}^{1,-(1-\frac{p}{2})} \Leftrightarrow \mu \in \mathcal{C}\mathcal{M}_{p/2}.$$

An analog and generalization of the Carleson-type embedding in Theorem 6.2.3 (ii) can be established by introducing the Besov capacities and their strong type inequalities through the boundary value functions of the Dirichlet spaces. In fact, for $p \in (-1, 1)$ let $\mathcal{L}_p^2(\mathbf{T})$ be the space of all real-valued functions $f \in \mathcal{L}^2(\mathbf{T})$ with

$$\|f\|_{\mathcal{L}_p^2(\mathbf{T})} = \|f\|_{L^2(\mathbf{T})} + \left(\int_{\mathbf{T}} \int_{\mathbf{T}} \frac{|f(\zeta) - f(\eta)|^2}{|\zeta - \eta|^{2-p}} |d\zeta| |d\eta| \right)^{\frac{1}{2}} < \infty.$$

Given a compact subset K of $\mathbf{T}$, define

$$Cap(K; \mathcal{L}_p^2(\mathbf{T})) = \inf\{\|f\|_{\mathcal{L}_p^2(\mathbf{T})}^2 : f \geq 1_K\}$$

as the Besov $\mathcal{L}_p^2(\mathbf{T})$-capacity of K. This definition extends to any set $E \subseteq \mathbf{T}$ in terms of

$$Cap(E; \mathcal{L}_p^2(\mathbf{T})) = \sup\{Cap(K; \mathcal{L}_p^2(\mathbf{T})) : \text{ compact } K \subseteq E\}.$$

From [Wu3, Theorem 2.2] we see that the following capacitary strong type inequality holds for any $f \in \mathcal{L}_p^2(\mathbf{T})$,

$$\int_0^\infty Cap(\{\zeta \in \mathbf{T} : |f(\zeta)| > t\}; \mathcal{L}_p^2(\mathbf{T})) dt^2 \preceq \|f\|_{\mathcal{L}_p^2(\mathbf{T})}^2.$$

This estimate, plus the following two facts:

$$\|\mathsf{N}(f)\|_{\mathcal{L}_p^2(\mathbf{T})} \preceq \|f\|_{\mathcal{L}_p^2(\mathbf{T})} \text{ and } \{z \in \mathbf{D} : |f(z)| > t\} \subseteq T(\{\zeta \in \mathbf{T} : \mathsf{N}(f)(\zeta) > t\});$$

see also [Wu3, Lemmas 2.3 and 3.2], implies that a nonnegative Radon measure μ on $\mathbf{D}$ is $\mathcal{D}^{2,p}$-Carleson measure if and only if $\mu(T(O)) \preceq Cap(O; \mathcal{L}_p^2(\mathbf{T}))$ holds for any open set $O \subset \mathbf{T}$ – see also [Ve, Theorem D].

Note 6.5.3. Section 6.3 is taken from [AlCaSi, Section 5]. The extreme case

$$\mathcal{W}_0 \neq \mathcal{Q}_0 = \mathcal{D}^{2,0} = \mathcal{D}$$

is a consequence of [Ste1, Theorem 4.2]. In addition, the following inequality for $p \in (0,1)$ is of independent interest:

$$\|g\|_{\mathcal{D}^{q,p}}^q \|f\|_{\mathcal{Q}_{p,1}}^q \succeq \begin{cases} \int_{\mathbf{D}} |g(z)|^q |f'(z)|^q (1 - |z|^2)^p dm(z) \text{ , } q \in (1,2), \\ \int_{\mathbf{D}} |g(z)|^q |f'(z)|^q (1 - |z|^2)^p \left(\log \frac{2}{1-|z|}\right)^{-1} dm(z) \text{ , } q = 2. \end{cases}$$

See also [AlCaSi, Theorem 5.2 (iii)-(iv)] for more details.

Note 6.5.4. Section 6.4 is formed via modifying [RocWu, Lemmas A, 4 and 6; Theorem 2′], [Wu1, Theorem 1] and [Xi2, Theorems 3.1 and 3.4]. There are many works on the Volterra operators acting on holomorphic spaces — see also [Sis1] and [Sis2].

A follow-up problem of Corollary 6.4.3 is to prove or disprove: $\mathcal{W}_p$ is isomorphic to $\mathcal{D}^{2,p} \odot \mathcal{D}^{2,2-p}$, $p \in [0,2)$ with the Cauchy pairing — equivalently, $\mathsf{H}_f : \mathcal{D}^{2,p} \mapsto \mathcal{D}^{2,2-p}$ is continuous if and only if $f \in \mathcal{W}_p$, since the Cauchy dual of $\mathcal{D}^{2,2-p}$ is isomorphic to $\mathcal{D}^{2,p}$, but also since according to the well-known equivalent principle between the boundedness of the Hankel operator and the Cauchy duality of weak factorization (cf. [JaPeSe], [CohnVe]) one has

$$\mathsf{H}_f : \mathcal{D}^{2,p} \mapsto \mathcal{D}^{2,p} \Leftrightarrow f \in \Theta\big((\mathcal{D}^{2,p} \odot \mathcal{D}^{2,2-p})^*\big).$$

Of course, the key point is to consider whether any element in the Cauchy dual of $\mathcal{D}^{2,p} \odot \mathcal{D}^{2,2-p}$ induces an element of $\mathcal{W}_p$ — here [Wu2] is maybe helpful. This is because: if

$$f \in \mathcal{W}_p \quad \text{and} \quad g = \sum_{j=1}^{\infty} u_j v_j \in \mathcal{D}^{2,p} \odot \mathcal{D}^{2,2-p}$$

where

$$u_j \in \mathcal{D}^{2,p} \quad \text{and} \quad v_j \in \mathcal{D}^{2,2-p},$$

then

$$\langle f, g \rangle = 2\pi f(0)\overline{g(0)} + 2 \lim_{r \to 1} \int_{\mathbf{D}} f'(rz)\overline{g'(rz)}(-\log|z|^2)dm(z)$$

and

$$\left| \int_{\mathbf{D}} (u_j v_j)'(z) \overline{g'(z)} (-\log|z|^2) dm(z) \right|$$

$$\leq \int_{\mathbf{D}} |u_j'(z)||v_j(z)||g'(z)|(-\log|z|^2) dm(z)$$

$$+ \int_{\mathbf{D}} |u_j(z)||v_j'(z)||g'(z)|(-\log|z|^2) dm(z)$$

$$\preceq \|u_j\|_{\mathcal{D}^{2,p}} \left(\int_{\mathbf{D}} |v_j(z)|^2 |g'(z)|^2 (1-|z|^2)^{2-p} dm(z) \right)^{\frac{1}{2}}$$

$$+ \|v_j\|_{\mathcal{D}^{2,2-p}} \left(\int_{\mathbf{D}} |u_j(z)|^2 |g'(z)|^2 (1-|z|^2)^p dm(z) \right)^{\frac{1}{2}}$$

$$\preceq \|u_j\|_{\mathcal{D}^{2,p}} \|v_j\|_{\mathcal{D}^{2,2-p}} \|g\|_{\mathcal{B}} + \|u_j\|_{\mathcal{D}^{2,p}} \|v_j\|_{\mathcal{D}^{2,2-p}} \|g\|_{\mathcal{W}_p}$$

$$\preceq \|u_j\|_{\mathcal{D}^{2,p}} \|v_j\|_{\mathcal{D}^{2,2-p}} \|g\|_{\mathcal{W}_p}.$$

Here we have used the inclusion:

$$\mathcal{W}_p \subseteq \mathcal{Q}_p \subseteq \mathcal{B}, \quad p \in [0,2).$$

Moreover, Theorem 6.4.5 describes the corona structure of $\mathcal{W}_p \cap \mathcal{H}^\infty$, $p \in [0,1)$ which says:

$$\mathrm{M}_{f_1,f_2,\dots,f_n} : (\mathcal{W}_p \cap \mathcal{H}^\infty) \times (\mathcal{W}_p \cap \mathcal{H}^\infty) \times \cdots \times (\mathcal{W}_p \cap \mathcal{H}^\infty) \mapsto \mathcal{W}_p \cap \mathcal{H}^\infty$$

is onto if and only if

$$(f_1, f_2, \dots, f_n) \in (\mathcal{W}_p \cap \mathcal{H}^\infty) \times (\mathcal{W}_p \cap \mathcal{H}^\infty) \times \cdots \times (\mathcal{W}_p \cap \mathcal{H}^\infty)$$

with

$$\delta(f_1, f_2, \dots, f_n) = \inf_{z \in \mathbf{D}} \sum_{k=1}^{n} |f_k(z)| > 0.$$

In particular, the case $p = 0$ gives an affirmative answer to [BroSh, Question 18]. This assertion can be proved by using Lemma 6.4.1 (ii) and an explicit $\mathcal{L}^\infty$-solution to the $\bar{\partial}$ equation — see [Jo] and [Xi2, Theorem 3.4]. Additionally, for a discussion on the stable rank of $\mathcal{Q}_p \cap \mathcal{H}^\infty$, $p \in (0,1)$, see also [PauS].

Chapter 7

Estimates for Growth and Decay

We present in this chapter several of the size estimates that arise in the process of determining growth and decay of a Q_p-function. The details are included in the following four sections:

- Convexity Inequalities;
- Exponential Integrabilities;
- Hadamard Convolutions;
- Characteristic Bounds of Derivatives.

7.1 Convexity Inequalities

For convenience, we begin by recalling some basic concepts on Calderón's complex interpolation.

Suppose $(\mathcal{X}_0, \mathcal{X}_1)$ is a compatible couple of complex Banach spaces with norms $\|\cdot\|_{\mathcal{X}_0}$ and $\|\cdot\|_{\mathcal{X}_1}$, i.e., there is a Hausdorff topological vector space $\mathcal{V}$ containing both $\mathcal{X}_0$ and $\mathcal{X}_1$. Then $\mathcal{X}_0 \cap \mathcal{X}_1$ and $\mathcal{X}_0 + \mathcal{X}_1$ are two subspaces of $\mathcal{V}$, but also Banach spaces equipped with norms below:

$$\|f\|_{\mathcal{X}_0 \cap \mathcal{X}_1} = \max\left\{\|f\|_{\mathcal{X}_0}, \|f\|_{\mathcal{X}_1}\right\}$$

and

$$\|f\|_{\mathcal{X}_0 + \mathcal{X}_1} = \inf\left\{\|f_0\|_{\mathcal{X}_0} + \|f_1\|_{\mathcal{X}_1}, \ f = f_0 + f_1, \ f_0 \in \mathcal{X}_0, \ f_1 \in \mathcal{X}_1\right\}.$$

A Banach space $\mathcal{X}$ is called an intermediate space between $\mathcal{X}_0$ and $\mathcal{X}_1$ provided

$$\mathcal{X}_0 \cap \mathcal{X}_1 \subseteq \mathcal{X} \subseteq \mathcal{X}_0 + \mathcal{X}_1.$$

Given $S = \{z \in \mathbf{C} : 0 \leq \Re z \leq 1\}$ and $S^{\circ} = \{z \in \mathbf{C} : 0 < \Re z < 1\}$. Let $\mathcal{F}(\mathcal{X}_0, \mathcal{X}_1)$ be a family of mappings $F : S \mapsto \mathcal{X}_0 + \mathcal{X}_1$ such that

(i) F is holomorphic on S° and continuous on S;

(ii) $\sup_{z \in S} \|F(z)\|_{\mathcal{X}_0 + \mathcal{X}_1} < \infty$;

(iii) $F(iy) \in \mathcal{X}_0$ and $F(1 + iy) \in \mathcal{X}_1$ for all $y \in \mathbf{R}$;

(iv) $y \mapsto F(iy)$ and $y \mapsto F(1 + iy)$ are continuous and bounded on $\mathbf{R}$.

Then $\mathcal{F}(\mathcal{X}_0, \mathcal{X}_1)$ becomes a Banach space once it is equipped with the norm

$$\|F\|_{\mathcal{F}(\mathcal{X}_0, \mathcal{X}_1)} = \max \left\{ \sup_{y \in \mathbf{R}} \|F(iy)\|_{\mathcal{X}_0}, \ \sup_{y \in \mathbf{R}} \|F(iy)\|_{\mathcal{X}_1} \right\}.$$

And hence, for $s \in (0, 1)$, the complex interpolation space $[\mathcal{X}_0, \mathcal{X}_1]_s$ is defined by

$$[\mathcal{X}_0, \mathcal{X}_1]_s = \{f \in \mathcal{X}_0 + \mathcal{X}_1 : f = F(s), \ F \in \mathcal{F}(\mathcal{X}_0, \mathcal{X}_1)\}$$

with the norm

$$\|f\|_{[\mathcal{X}_0, \mathcal{X}_1]_s} = \inf \left\{ \|F\|_{\mathcal{F}(\mathcal{X}_0, \mathcal{X}_1)} : \ f = F(s) \right\}.$$

When regarding $\mathcal{D}$, $\mathcal{BMOA}$ and $\mathcal{B}$ as the endpoint spaces in the chain of Q_p, $p \in [0, \infty)$, we naturally see the following result.

Example 7.1.1. Let $s \in (0, 1)$. Then

(i) $[\mathcal{D}, \mathcal{B}]_s$ equals the holomorphic Besov space $\mathcal{B}_{\frac{2}{1-s}}$ consisting of all $f \in \mathcal{H}$ with

$$\|f\|_{[\mathcal{D}, \mathcal{B}]_s} = \left(\int_{\mathbf{D}} |f'(z)|^{\frac{2}{1-s}} (1 - |z|^2)^{\frac{2s}{1-s}} \, dm(z) \right)^{\frac{1-s}{2}} < \infty.$$

(ii) $[\mathcal{D}, \mathcal{BMOA}]_s$ comprises all $f \in \mathcal{H}$ with

$$\|f\|_{[\mathcal{D}, \mathcal{BMOA}]_s} = \left(\int_{\mathbf{T}} \left(\int_0^1 |f'(r\zeta)|^2 (1 - r^2)^s r \, dr \right)^{\frac{1}{1-s}} |d\zeta| \right)^{\frac{1-s}{2}} < \infty.$$

(iii) $[\mathcal{D}, \mathcal{BMOA}]_s \subset [\mathcal{D}, \mathcal{B}]_s \subset Q_s$.

Proof. (i) See [Zhu1, Theorem 5.3.8].

(ii) As for this assertion, we note the boundary behavior of functions in $\mathcal{D}$ and $\mathcal{BMOA}$ respectively, and use [Tr, p. 45, Theorem (i) (4) where $s_0 = \frac{1}{2}$, $p_0 = 2$, $p_1 = 2$, $p = 2$; $s_1 = 0$, $q_0 = 2$, $q_1 = \infty$, $q = \frac{2}{1-s}$] to derive that $f \in [\mathcal{D}, \mathcal{BMOA}]_s$ if and only if f is holomorphic on $\mathbf{D}$ and its boundary value function f on $\mathbf{T}$ obeys

$$\int_{\mathbf{T}} \left(\int_{\mathbf{T}} \frac{|f(\zeta) - f(\eta)|^2}{|\zeta - \eta|^{2-s}} |d\zeta| \right)^{\frac{1}{1-s}} |d\eta| < \infty.$$

This finiteness condition amounts to (cf. e.g. [Ve, Theorem F])

$$\int_{\mathbf{T}} \left(\int_0^1 |f'(r\zeta)|^2 (1-r)^s r \, dr \right)^{\frac{1}{1-s}} |d\eta| < \infty.$$

(iii) The first strict inclusion follows from (i) and (ii). Since $\mathcal{B}_{\frac{2}{1-s}} \subset \mathcal{Q}_s$, we conclude $[\mathcal{D}, \mathcal{B}]_s \subset \mathcal{Q}_s$. □

Motivated by Example 7.1.1, we can establish some convexity inequalities and their corresponding dual forms which can be viewed as the improved isoperimetric inequalities without sharp constants.

Theorem 7.1.2. *For $p_1, p_2 \in [0, \infty)$ and $s \in [0, 1]$ let $p = sp_1 + (1-s)p_2$. Then*

$$E_p(f, w) \le \left(E_{p_1}(f, w) \right)^s \left(E_{p_2}(f, w) \right)^{1-s}$$

and

$$F_p(f, w) \le \left(F_{p_1}(f, w) \right)^s \left(F_{p_2}(f, w) \right)^{1-s}$$

hold for $f \in \mathcal{H}$ and $w \in \mathbf{D}$. Consequently:

(i) $\|f\|_{\mathcal{Q}_p,1} \lesssim \|f\|_{\mathcal{D}}^{1-p} \|f\|_{BMOA}^p, \quad f \in \mathcal{D}, \quad p \in [0, 1].$

(ii) $\|f\|_{\mathcal{D}} \preceq \|f\|_{\mathcal{P}_p}^{1-p} \|f\|_{\mathcal{H}'}^p, \quad f \in \mathcal{H}', \quad p \in (0, 1].$

(iii) $\|f\|_{\mathcal{Q}_p,1} \preceq \|f\|_{\mathcal{D}}^s \|f\|_{\mathcal{B}}^{1-s}, \quad f \in \mathcal{D}, \quad s \in (0, 1], \quad 1-s < p \le 1.$

(iv) $\|f\|_{\mathcal{D}} \preceq \|f\|_{\mathcal{P}_p}^{1-p} \|f\|_{\mathcal{B}_1}^p, \quad f \in \mathcal{B}_1, \quad p \in (0, 1].$

Proof. When $p_1, p_2 \in [0, \infty)$ and $s \in (0, 1)$ satisfy $p = sp_1 + (1-s)p_2$, Hölder's inequality implies

$$E_p(f, w)$$

$$= \left(\int_{\mathbf{D}} |f'(z)|^{2s+2(1-s)} \left(-\log |\sigma_w(z)| \right)^{sp_1+(1-s)p_2} \right)^{\frac{1}{2}}$$

$$\le \left(\int_{\mathbf{D}} \frac{|f'(z)|^2}{\left(-\log |\sigma_w(z)| \right)^{-p_1}} dm(z) \right)^{\frac{s}{2}} \left(\int_{\mathbf{D}} \frac{|f'(z)|^2}{\left(-\log |\sigma_w(z)| \right)^{-p_2}} dm(z) \right)^{\frac{1-s}{2}}$$

$$= \left(E_{p_1}(f, w) \right)^s \left(E_{p_2}(f, w) \right)^{1-s}, \quad f \in \mathcal{H}, \quad w \in \mathbf{D}.$$

This argument also works for $F_p(\cdot, \cdot)$.

(i) and (iii) follow from the second convexity inequality above with $p = s$, $p_1 = 0$, $p_2 = 1$ and $p = (1-s)p_2$, $p_2 > 1$ (which ensures $\mathcal{Q}_{p_2} = \mathcal{B}$), $p_1 = 0$ respectively.

(ii) Suppose ω is a nonnegative function on $\mathbf{D}$ such that $\|\omega\|_{\mathcal{LN}^1(H_\infty^p)} \le 1$. Then $\omega(z) \preceq (1-|z|^2)^{-p}$. By Hölder's inequality, we have that if $f \in \mathcal{H}'$, then $f \in \mathcal{P}_p \subset \mathcal{D}$ with $\|f\|_{\mathcal{D}} \preceq \|f\|_{\mathcal{H}'}$ and hence

$$\|f\|_{\mathcal{D}}^2 \preceq \int_{\mathbf{D}} |(1-|z|^2)(zf(z))''|^2 dm(z)$$

$$\preceq \left(\int_{\mathbf{D}} |(1-|z|^2)(zf(z))''|^2 (\omega(z))^{-1} (1-|z|^2)^{-p} dm(z) \right)^{1-p}$$

$$\times \left(\int_{\mathbf{D}} |(1-|z|^2)(zf(z))''|^2 (\omega(z))^{\frac{1-p}{p}} (1-|z|^2)^{1-p} dm(z) \right)^{p}$$

$$\preceq \left(\int_{\mathbf{D}} |(1-|z|^2)(zf(z))''|^2 (\omega(z))^{-1} (1-|z|^2)^{-p} dm(z) \right)^{1-p}$$

$$\times \left(\int_{\mathbf{D}} |(1-|z|^2)(zf(z))''|^2 dm(z) \right)^{p}$$

$$\preceq \left(\int_{\mathbf{D}} |(1-|z|^2)(zf(z))''|^2 (\omega(z))^{-1} (1-|z|^2)^{-p} dm(z) \right)^{1-p} \|f\|_{\mathcal{D}}^{2p}$$

$$\preceq \left(\int_{\mathbf{D}} |(1-|z|^2)(zf(z))''|^2 (\omega(z))^{-1} (1-|z|^2)^{-p} dm(z) \right)^{1-p} \|f\|_{\mathcal{H}'}^{2p}.$$

By taking the infimum over all ω in the above estimates, we derive

$$\|f\|_{\mathcal{D}}^2 \preceq \|f\|_{\mathcal{P}_p}^{2(1-p)} \|f\|_{\mathcal{H}'}^{2p}.$$

(iv) Observe that (cf. Note 4.6.4) $\mathcal{B}_1 \subset \mathcal{H}' \subset \mathcal{P}_p \subset \mathcal{D}$, $\quad p \in (0,1)$. So the argument for (ii) gives (iv) right away. $\qquad \square$

In view of Theorem 7.1.2 we can next handle $\mathcal{Q}_p$-seminorms of the dilated functions $f_r(z) = f(rz)$, $r \in (0,1)$ via the product between an f-free quantity and a convex multiplication of both $\mathcal{Q}_q$-seminorm and $\mathcal{B}$-seminorm of $f \in \mathcal{H}$.

Theorem 7.1.3. *Let $p, q \in [0, \infty)$ and $s \in [1, \infty]$. If $f \in \mathcal{Q}_q$ and $r \in (0,1)$, then*

$$\|f_r\|_{\mathcal{Q}_p} \le \sqrt{C_1(p,q,r,s)} \|f\|_{\mathcal{Q}_q}^{1-\frac{1}{s}} \|f\|_{\mathcal{B}}^{\frac{1}{s}}$$

where

$$C_1(p,q,r,s) = \begin{cases} \left(\int_0^{\frac{2r}{1+r^2}} \left(\frac{(-\log t)^{ps}}{(-\log t)^{q(s-1)}} \right) \frac{dt^2}{(1-t^2)^2} \right)^{\frac{1}{s}} & , \quad s \in [1, \infty), \\ \sup \left\{ \frac{(-\log t)^p}{(-\log t)^q} : \ 0 < t < \frac{2r}{1+r^2} \right\} & , \quad s = \infty. \end{cases}$$

Proof. Case 1: $s \in (1, \infty)$. For $r \in (0,1)$ and $w \in \mathbf{D}$ let

$$\mathbf{D}_{rw} = \sigma_{rw}(\{z \in \mathbf{D} : \ |z| < r\}).$$

Then
$$\mathbf{D}_{rw} \subset \{z \in \mathbf{D} : |z| < 2r(1+r^2)^{-1}\}.$$

Applying the maximum principle to the function
$$z \mapsto \frac{z}{\sigma_w(r^{-1}\sigma_{rw}(z))} \quad \text{over} \quad \mathbf{D}_{rw},$$

we derive $|z| \leq |\sigma_w(r^{-1}\sigma_{rw}(z))|$ and consequently,

$$
\begin{aligned}
\|f_r\|_{\mathcal{Q}_p}^2 &= \sup_{w \in \mathbf{D}} \int_{|z| \leq r} |f'(z)|^2 \big(-\log|\sigma_w(r^{-1}z)| \big)^p dm(z) \\
&= \sup_{w \in \mathbf{D}} \int_{\mathbf{D}_{rw}} |(f \circ \sigma_{rw})'(z)|^2 \big(-\log|\sigma_w(r^{-1}\sigma_{rw}(z))| \big)^p dm(z) \\
&\leq \sup_{w \in \mathbf{D}} \int_{\mathbf{D}_{rw}} |(f \circ \sigma_{rw})'(z)|^2 \big(-\log|z| \big)^p dm(z).
\end{aligned}
$$

Next, writing
$$g(z) = (1-|z|^2)^{2/s}(-\log|z|)^{q(s-1)/s-p},$$

and using the Hölder inequality plus the estimate
$$(1-|z|^2)|(f \circ \sigma_{rw})'(z)| \leq \|f\|_{\mathcal{B}},$$

we compute

$$
\begin{aligned}
\|f_r\|_{\mathcal{Q}_p}^2 &\leq \sup_{w \in \mathbf{D}} \left(\int_{\mathbf{D}_{rw}} |g(z)|^{-s} dm(z) \right)^{\frac{1}{s}} \\
&\quad \times \sup_{w \in \mathbf{D}} \left(\int_{\mathbf{D}_{rw}} \frac{|(f \circ \sigma_{rw})'(z)|^2 \big(-\log|z| \big)^q}{\big((1-|z|^2)|(f \circ \sigma_{rw})'(z)|\big)^{\frac{2}{1-s}}} dm(z) \right)^{\frac{s-1}{s}} \\
&\preceq \left(\int_0^{\frac{2r}{1+r^2}} |g(t)|^{-s} dt^2 \right)^{\frac{1}{s}} \|f\|_{\mathcal{B}}^{2/s} \|f\|_{\mathcal{Q}_q}^{2(s-1)/s} \\
&\preceq C_1(p,q,r,s) \|f\|_{\mathcal{B}}^{2/s} \|f\|_{\mathcal{Q}_q}^{2(s-1)/s},
\end{aligned}
$$

hence getting the desired inequality.

Case 2: $s = 1$. This follows from the limit of Case 1 as $s \to 1$.

Case 3: $s = \infty$. This can be checked directly because the factor $\|f\|_{\mathcal{B}}^{2/s}$ is replaced by 1. $\qquad\square$

The following two extreme cases of Theorem 7.1.3 are quite interesting. First, if $s = \infty$ then

$$
\|f_r\|_{\mathcal{Q}_p} \leq \begin{cases} \|f\|_{\mathcal{Q}_p}, & p \in [0,\infty), \\ \left(\log \frac{1+r^2}{2r}\right)^{\frac{p-1}{2}} \|f\|_{\mathcal{Q}_1}, & p \in [0,1], \end{cases}
$$

and

$$\|f_r\|_{\mathcal{D}} \le \left(\log\frac{1+r^2}{2r}\right)^{-\frac{p}{2}}\|f\|_{\mathcal{Q}_p}, \quad p \in [0,\infty).$$

Second, if $s = 1$ then

$$\|f_r\|_{BMOA} \preceq \left(\int_0^{\frac{2r}{1+r^2}}\frac{-\log t}{(1-t^2)^2}dt^2\right)^{\frac{1}{2}}\|f\|_{\mathcal{B}} \preceq \left(\log(1-r^2)^{-1}\right)^{\frac{1}{2}}\|f\|_{\mathcal{B}}.$$

We demonstrate next that the last estimate can be improved under a superharmonic condition.

Theorem 7.1.4. *Let $p, q \in [0, \infty)$, $s \in [1, \infty)$ and $ps - q(s-1) > 0$. If $f \in \mathcal{Q}_q$ and $r \in (0, 1)$, then*

$$\|f_r\|_{\mathcal{Q}_p} \le \sqrt{C_2(p,q,r,s)}\|f\|_{\mathcal{Q}_q}^{1-\frac{1}{s}}\|f\|_{\mathcal{B}}^{\frac{1}{s}},$$

where

$$C_2(p,q,r,s) = \left(\pi\int_0^r\left(\log\frac{r}{t}\right)^{ps-q(s-1)}\frac{dt^2}{(1-t^2)^2}\right)^{\frac{1}{s}}.$$

Proof. Since $ps - q(s-1) > 0$ ensures that $z \mapsto (-\log|z|)^{ps-q(s-1)}$ is a superharmonic function on $\mathbf{D}$ and vanishes on $\mathbf{T}$, we conclude that the radial function

$$L(p,q,r,s;w) = \sup_{w\in\mathbf{D}}\int_{\mathbf{D}}\left(-\log|\sigma_w(z)|\right)^{ps-q(s-1)}\left(\frac{r^2}{(1-r^2|z|^2)^2}\right)dm(z)$$

is superharmonic on $\mathbf{D}$ and so assumes the maximum at the origin; this is,

$$\sup_{w\in\mathbf{D}}L(p,q,r,s;w) = \int_{\mathbf{D}}(-\log|z|)^{ps-q(s-1)}\left(\frac{r^2}{(1-r^2|z|^2)^2}\right)dm(z)$$
$$= \left(C_2(p,q,r,s)\right)^s.$$

Next we employ Hölder's inequality with $s \in (1, \infty)$ to derive

$$\left(E_p(f_r,w)\right)^2 \le \|f\|_{\mathcal{B}}^{\frac{2}{s}}\int_{\mathbf{D}}|f_r'(z)|^{\frac{2(s-1)}{s}}\left(\frac{r}{1-|rz|^2}\right)^{\frac{2}{s}}\left(-\log|\sigma_w(z)|\right)^p dm(z)$$

$$\le C_2(p,q,r,s)\|f\|_{\mathcal{B}}^{\frac{2}{s}}\left(\int_{\mathbf{D}}|f_r'(z)|^2\left(-\log|\sigma_w(z)|\right)^q dm(z)\right)^{\frac{s-1}{s}}$$

$$\le C_2(p,q,r,s)\|f\|_{\mathcal{B}}^{\frac{2}{s}}\|f\|_{\mathcal{Q}_q}^{\frac{2(s-1)}{s}},$$

hence establishing the desired inequality.

By a slight change of the preceding argument, we can readily verify that the case $s = 1$ is also true. □

Remark 7.1.5. Here it is worthy of special remark that Theorem 7.1.4 with $p = 1$, $q = 2$ and $s = 1$ goes back to the Korenblum inequality below:

$$\|f_r\|_{\mathcal{Q}_1} \le \|f\|_{\mathcal{B}}\left(2^{-1}\pi|\log(1-r^2)|\right)^{\frac{1}{2}}.$$

7.2 Exponential Integrabilities

In this section we concern ourselves with the majorization and length-area principle via the exponential growth of a holomorphic function.

First of all, we compare the quadratic Dirichlet spaces with the Bergman and Hardy spaces.

Theorem 7.2.1. *Let* $p \in [0, 2)$. *Then*

$$\mathcal{D}^{2,p} \subset \begin{cases} \mathcal{A}^{2,0} & , \quad p \in (1, 2), \\ \bigcap_{0 < q < \frac{2}{p}} \mathcal{H}^q & , \quad p \in [0, 1]. \end{cases}$$

Furthermore, if $\|f\|_{\mathcal{H}^2} > 0$, $p \in [0, 1]$ *and* $\epsilon \in (-p, 1 - p)$, *then*

$$\exp\left(\int_{\mathbf{T}} \left(\frac{|f(\zeta)|}{\|f\|_{\mathcal{H}^2}} \right)^2 \left(\log \frac{|f(\zeta)|}{\|f\|_{\mathcal{H}^2}} \right) |d\zeta| \right) \preceq \left(\frac{\|f\|_{\mathcal{D}^{2,p}}}{\|f\|_{\mathcal{H}^2}} \right)^{\frac{1}{1-p-\epsilon}}.$$

Proof. Case 1: $p \in (1, 2)$. If $f(z) = \sum_{j=0}^{\infty} a_j z^j$ lies in $\mathcal{D}^{2,p}$, then

$$\|f\|_{\mathcal{A}^{2,0}} \approx \left(\sum_{j=0}^{\infty} (1+j)^{-1} |a_j|^2 \right)^{\frac{1}{2}} \preceq \left(\sum_{j=0}^{\infty} (1+j^2)^{\frac{1-p}{2}} |a_j|^2 \right)^{\frac{1}{2}}.$$

To see the strictness of the inclusion, we take $f(z) = \sum_{j=0}^{\infty} 2^{j/2} z^{2^j}$. This function is in $\mathcal{A}^{2,0} \setminus \mathcal{D}^{2,p}$ for $p \in (1, 2)$.

Case 2: $p \in [0, 1]$. We make the following consideration.

If $p = 0$, then $\mathcal{D} \subset \mathcal{BMOA} \subset \cap_{q<\infty} \mathcal{H}^q$, as desired.

If $p = 1$, then $\mathcal{D}^{2,p} = \mathcal{H}^2$ which is contained properly in $\cap_{0<q<2} \mathcal{H}^q$. This is well-known result.

Assume now $p \in (0, 1)$. If $f(z) = \sum_{j=0}^{\infty} a_j z^j$ is in $\mathcal{D}^{2,p}$, then

$$\|f\|_{\mathcal{D}^{2,p}}^2 \approx \sum_{j=0}^{\infty} (1+j^2)^{\frac{1-p}{2}} |a_j|^2.$$

Accordingly, a combination of the Hausdorff–Young Theorem (cf. [Du, Theorem 6.1]) and the Cauchy–Schwarz inequality implies that if $q = \alpha/(\alpha - 1) \in (0, 2/p)$, $\alpha \in (1, 2]$ and $\alpha > 2p$, then

$$\|f\|_{\mathcal{H}^q}^q \preceq \sum_{j=0}^{\infty} |a_j|^\alpha \preceq \left(\sum_{j=0}^{\infty} |a_j|^2 (1+j^2)^{\frac{1-p}{2}} \right)^{\frac{\alpha}{2}} \approx \|f\|_{\mathcal{D}^{2,p}}^\alpha,$$

as required. Of course, the fact that

$$f(z) = \sum_{j=1}^{\infty} j^{-2} z^{2^j} \quad \text{belongs to} \quad \left(\bigcap_{0 < q < \frac{2}{p}} \mathcal{H}^q \right) \setminus \mathcal{D}^{2,p}$$

shows the strict inclusion.

To verify the second assertion, we use the Hölder inequality and the first inclusion, and we derive that for $2 < q < 2/(p + \epsilon)$ and $0 < \epsilon < 1 - p$,

$$
\int_{\mathbf{T}} |f(\zeta)|^q |d\zeta| = \int_{\mathbf{T}} |f(\zeta)|^{\frac{q-2}{1-p-\epsilon}} |f(\zeta)|^{\frac{2-q(p+\epsilon)}{1-p-\epsilon}} |d\zeta|
$$

$$
\leq \left(\int_{\mathbf{T}} |f(\zeta)|^{\frac{2}{p+\epsilon}} |d\zeta| \right)^{\frac{(p+\epsilon)(q-2)}{2(1-p-\epsilon)}} \left(\int_{\mathbf{T}} |f(\zeta)|^2 |d\zeta| \right)^{1 - \frac{(p+\epsilon)(q-2)}{2(1-p-\epsilon)}}
$$

$$
\preceq \|f\|_{\mathcal{D}^{2,p}}^{\frac{q-2}{1-p-\epsilon}} \left(\int_{\mathbf{T}} |f(\zeta)|^2 |d\zeta| \right)^{1 - \frac{(p+\epsilon)(q-2)}{2(1-p-\epsilon)}}.
$$

Accordingly, if $\int_{\mathbf{T}} |f(\zeta)|^2 |d\zeta| = 1$, then $|f(\zeta)|^2 |d\zeta|$ can be treated as a probability measure on $\mathbf{T}$ and hence

$$
\left(\int_{\mathbf{T}} |f(\zeta)|^{q-2} |f(\zeta)|^2 |d\zeta| \right)^{\frac{1}{q-2}} = \left(\int_{\mathbf{T}} |f(\zeta)|^q |d\zeta| \right)^{\frac{1}{q-2}} \preceq \|f\|_{\mathcal{D}^{2,p}}^{\frac{1}{1-p-\epsilon}}.
$$

Letting $q \to 2$, we find

$$
\exp \left(\int_{\mathbf{T}} |f(\zeta)|^2 \log |f(\zeta)| |d\zeta| \right) \preceq \|f\|_{\mathcal{D}^{2,p}}^{\frac{1}{1-p-\epsilon}}.
$$

The general case follows from considering $f\|f\|_{\mathcal{H}^2}^{-1}$. $\square$

As a matter of fact, the first part of Theorem 7.2.1 is a sort of holomorphic Sobolev inequality — its conformally invariant form appears to be the well-known embedding $\mathcal{Q}_p \subset \mathcal{BMOA}$ for $p \in [0, 1)$. Meanwhile, the second part of Theorem 7.2.1 may be viewed as the logarithmic Sobolev inequality associated with the weighted Dirichlet space. When exploring its conformally invariant form, we feel that it is necessary to investigate certain exponential integrals of functions in $\mathcal{Q}_p$ through a dual form of $\mathcal{L}_\alpha^2$-Bessel capacity.

To be more specific, for a compact set $K \subset \mathbf{C}$ let μ be a positive measure on $\mathbf{C}$ that takes 1 on K and 0 on $\mathbf{C} \setminus K$. Then we introduce the potential

$$
\mathsf{P}(\mu, \alpha)(z) = \begin{cases} \int_K \log |z - \zeta|^{-1} d\mu(\zeta) & , \quad \alpha = 0, \\ \int_K |z - \zeta|^{-\alpha} d\mu(\zeta) & , \quad \alpha \in (0, 2), \end{cases}
$$

and then define the α-capacity of K as

$$
cap_\alpha(K) = \left(\inf_\mu \sup_z \mathsf{P}(\mu, \alpha)(z) \right)^{-1}.
$$

The compact sets K satisfying $cap_\alpha(K) = 0$ have the property that a countable union of them has α-capacity 0. Of course, if $\left(cap_\alpha(K) \right)^{-1} < \infty$, then there is a unique μ_α such that the corresponding potential $\mathsf{P}(\mu_\alpha, \alpha)(z)$ is not greater than

$(cap_\alpha(K))^{-1}$ for all $z \in \mathbf{C}$ but also is equal to $(cap_\alpha(K))^{-1}$ for all $z \in K$ except perhaps for a set with α-capacity 0. For the general set E, we define

$$cap_\alpha(E) = \sup_{K \subset E} cap_\alpha(K)$$

where the supremun is taken over all compact subsets K of E.

We need the following three auxiliary results. The first one is the Beurling weak-type estimate for the extreme case $cap_0(\cdot)$ — the logarithmic capacity.

Theorem 7.2.2. *Given* $f \in \mathcal{H}(\bar{\mathbf{D}})$. *For* $r \in (0, 1]$ *let* $M = \sup_{|z| \le r} |f(z)| < \infty$. *Then*

$$cap_0\left(\{\zeta \in \mathbf{T} : |f(\zeta)| > t\}\right) \le r^{-1/2} \exp\left(-\pi \int_M^t \frac{ds}{\int_{\{z \in \mathbf{D} : |f(z)| = s\}} |f'(z)||dz|}\right).$$

Proof. For a domain $\Omega \subset \mathbf{C}$ and two subsets E_1, E_2 of the boundary $\partial\Omega$, suppose $\rho_\Omega(E_1, E_2)$ is the extremal distance between E_1 and E_2 with respect to Ω. For example, if $\mathbf{T}_{r_1}$ and $\mathbf{T}_{r_2}$ are the circles centered at the origin with radii r_1 and r_2 and if $0 < r_1 < r_2$ and $\mathbf{A}_{r_1, r_2}$ denotes the annulus bounded by $\mathbf{T}_{r_1}$ and $\mathbf{T}_{r_2}$, then

$$\rho_{\mathbf{A}_{r_1, r_2}}(\mathbf{T}_{r_1}, \mathbf{T}_{r_2}) = (2\pi)^{-1} \log \frac{r_2}{r_1}.$$

For the sake of simplicity, we will remove the subscript Ω in the sequel. According to [Ahl, Theorems 2-4 and 4-9], we have that if $\gamma_s = \{z \in \mathbf{D} : |f(z)| = s\}$, then using the hypothesis $M = \sup_{|z| \le r} |f(z)| < \infty$ for $r \in (0, 1]$,

$$\log\left(cap_0(\{\zeta \in \mathbf{T} : |f(\zeta)| > t\})\right)$$
$$\le -\pi\left(\lim_{s \to 0}\left(\rho(\mathbf{T}_s, \gamma_t) + (2\pi)^{-1} \log s\right)\right)$$
$$\le -\pi\left(\lim_{s \to 0}\left(\rho(\mathbf{T}_s, \mathbf{T}_r) + \rho(\mathbf{T}_r, \gamma_M) + \rho(\gamma_M, \gamma_t) + (2\pi)^{-1} \log s\right)\right)$$
$$\le -\left(2^{-1} \log r + \pi\rho(\gamma_M, \gamma_t)\right).$$

To complete the argument, we next define a metric on $\mathbf{D}$ by

$$\rho(z) = |f'(z)|\left(\int_{\gamma_s} |f'(z)||dz|\right)^{-1}, \quad z \in \gamma_s$$

and let $\Omega_{M,t} = \{z \in \mathbf{D} : M < |f(z)| < t\}$. Then

$$\int_{\Omega_{M,t}} (\rho(z))^2 dtdy = \int_M^t \int_{\gamma_s} |f'(z)|\left(\int_{\gamma_s} |f'(w)||dw|\right)^{-2} |dz|ds$$
$$= \int_M^t \left(\int_{\gamma_s} |f'(z)||dz|\right)^{-1} ds.$$

Consequently, if γ is a curve in $\Omega_{M,t}$ such that $\gamma(0) \in \gamma_M$ and $\gamma(1) \in \gamma_t$, then

$$\int_{\gamma} \rho(z)|dz| = \int_{f(\gamma)} \left(\int_{\gamma_s} |f'(z)||dz| \right)^{-1} |dw| \geq \int_{M}^{t} \left(\int_{\gamma_s} |f'(z)||dz| \right)^{-1} ds.$$

Taking the infimum in the last estimate over all the possible curves γ, we derive

$$\rho(\gamma_M, \gamma_t) \geq \int_{M}^{t} \left(\int_{\gamma_s} |f'(z)||dz| \right)^{-1} ds,$$

hence deducing the desired inequality. $\square$

The second one is Marshall's Lemma as follows.

Lemma 7.2.3. *There exists a numerical constant $r > 0$ with the following property: if $f \in \mathcal{D}$ is normalized by $f(0) = 0$ and $\|f\|_{\mathcal{D}}^2 \leq \pi$, then there exists a constant M depending only on f, with*

$$M \in (0, 1]; \quad \{z \in \mathbf{D} : |z| < r\} \subseteq \{z \in \mathbf{D} : |f(z)| < M\};$$

and

$$\int_{0}^{M} \left(\int_{\{z \in \mathbf{D}: \ |f(z)|=s\}} |f'(z)||dz| \right) ds \geq \frac{\pi M^2}{3}.$$

Proof. Since $f \in \mathcal{D}$ has the growth

$$|f(z)| \leq \pi^{-\frac{1}{2}} \|f\|_{\mathcal{D}} \left(- \log(1 - |z|^2) \right)^{\frac{1}{2}},$$

we conclude from the hypothesis that $|z| < \sqrt{1 - e^{-M^2}}$ implies $|f(z)| < M$. In the meantime,

$$\pi t^2 \leq \int_{0}^{t} \left(\int_{\gamma_s} |f'(z)||dz| \right) ds \quad \text{as} \quad t \in (0,1) \quad \text{is small.}$$

Clearly, if

$$2^{-1}\pi t^2 \leq \int_{0}^{t} \left(\int_{\gamma_s} |f'(z)||dz| \right) ds \quad \text{for all} \quad t \in (0,1),$$

then we choose $M = 1$ and $r = \sqrt{1 - e^{-1}}$. If this is not the case, then we may suppose that there is an $M \in (0, 1]$ obeying

$$3^{-1}\pi M^2 \leq \int_{0}^{M} \left(\int_{\gamma_s} |f'(z)||dz| \right) ds \leq 2^{-1}\pi M^2.$$

Consequently, $f(\mathbf{D})$ omits at least two points $w_1, w_2 \in \mathbf{C}$ with

$$2^{-1}M \leq |w_1| \leq |w_2| \leq M \quad \text{and} \quad |w_2 - w_1| \geq 2^{-1}M.$$

Now, let h be the covering map from $\mathbf{D}$ onto $\mathbf{C} \setminus \{0, 1\}$ and h^{-1} its inverse. Then

$$g(z) = h^{-1}\left(\frac{f(z) - w_1}{w_2 - w_1}\right)$$

defines an element of $\mathcal{H}^\infty$ with $\|g\|_{\mathcal{H}^\infty} \leq 1$. Note that

$$2^{-2} \leq \left|\frac{w_1}{w_2 - w_1}\right| \leq 2 \quad \text{and} \quad 2^{-2} \leq \left|\frac{w_2}{w_2 - w_1}\right|.$$

So, it follows that $|g(0)| < c < 1$ for a numerical constant c. An application of Schwarz's Lemma to g yields that if $r > 0$ (independent of f and M) is sufficiently small, then $(f(z) - w_1)/(w_2 - w_1)$ must be in a disk of radius at most 2^{-3} and hence $|f(z)| < M$ whenever $|z| < r$. We are done. $\qquad\square$

The third one is the well-known Moser's Theorem.

Theorem 7.2.4. *There exists a constant $\kappa > 0$ such that*

$$\int_0^\infty \left(\phi(t)\right)^2 dt \leq 1 \Rightarrow \int_0^\infty \exp\left(\left(\int_0^t \phi(s)ds\right)^2 - t\right) dt \leq \kappa.$$

Proof. Assume

$$E_s = \left\{t \geq 0 : t - \left(\int_0^t \phi(s)ds\right)^2 \leq s\right\}, \quad s > 0.$$

To check the theorem, we just verify that $t_1, t_2 \in E_s$ and $2s \leq t_1 < t_2$ imply $t_2 - t_1 \leq 20s$. For then we have

$$|E_s| \leq 22s \quad \text{and} \quad \int_0^\infty \exp\left(\left(\int_0^t \phi(s)ds\right)^2 - t\right) dt = \int_0^\infty |E_s|e^{-s}ds \leq 22.$$

Needless to say, in the above and below, $|E|$ stands for the Lebesgue measure of a set $E \subseteq (-\infty, \infty)$. Now the Cauchy–Schwarz inequality gives:

$$t \in E_s \Rightarrow t - s \leq \left(\int_0^t \phi(x)dx\right)^2 \leq t\int_0^t \left(\phi(x)\right)^2 dx \leq t - t\int_t^\infty \left(\phi(x)\right)^2 dx,$$

and so $\int_t^\infty \left(\phi(x)\right)^2 dx \leq s/t$. Using this and the Cauchy-Schwarz inequality again, we further derive

$$
\begin{aligned}
t_2 - s &\leq \left(\int_0^{t_1} \phi(x)dx + \int_{t_1}^{t_2} \phi(x)dx\right)^2 \\
&\leq \left(t_1^{\frac{1}{2}} + (t_2 - t_1)^{\frac{1}{2}}\left(\int_{t_1}^\infty \left(\phi(x)\right)^2 dx\right)^{\frac{1}{2}}\right)^2 \\
&\leq t_1 + 2\sqrt{(t_2 - t_1)s} + 2^{-1}(t_2 - t_1),
\end{aligned}
$$

hence getting $t_2 - t_1 < 20s$. $\qquad\square$

Now, we are at a position to state the Chang–Marshall Theorem for the exponential growth of the Dirichlet space.

Theorem 7.2.5. *There exists a constant $\kappa > 0$ such that if $f \in \mathcal{D}$ with $f(0) = 0$ and $\|f\|_{\mathcal{D}}^2 \leq \pi$, then $\int_{\mathbf{T}} e^{|f(\zeta)|^2} |d\zeta| \leq \kappa$.*

Proof. Without loss of generality, we may assume that $f \in \mathcal{H}(\bar{D})$ and set

$$L(f,s) = \int_{\{z \in D: \, |f(z)| = s\}} |f'(z)| |dz|.$$

Using [Ahl, Theorem 2-7], Theorem 7.2.2 and Lemma 7.2.3 we find

$$\sin \left(2^{-2} |\{z \in \mathbf{T} : |f(\zeta)| > t\}| \right) \leq r^{-1/2} \exp \left(-\pi \int_M^t \left(L(f,s) \right)^{-1} ds \right).$$

This yields

$$
\begin{aligned}
\int_{\mathbf{T}} e^{|f(\zeta)|^2} |d\zeta| &= 2\pi + \int_0^\infty |\{z \in \mathbf{T} : |f(\zeta)| > t\}| e^{t^2} dt^2 \\
&\leq 2\pi \left(1 + \int_0^M e^{t^2} dt^2 \right) \\
&\quad + 2^{-1} \pi r^{-1/2} \int_M^{\|f\|_{\mathcal{H}^\infty}} \exp \left(t^2 - \pi \int_M^t \left(L(f,s) \right)^{-1} ds \right) dt^2.
\end{aligned}
$$

For our purpose, set

$$\psi(x) = \begin{cases} \pi M x \left(\int_0^M L(f,s) ds \right)^{-1} & , \quad x \in [0, M], \\ \pi \int_M^x \left(L(f,s) \right)^{-1} ds + \pi M^2 \left(\int_0^M L(f,s) ds \right)^{-1} & , \quad x \in [M, \|f\|_{\mathcal{H}^\infty}). \end{cases}$$

Then Lemma 7.2.3 tells us that

$$\psi(M) \leq 3, \quad \|\psi\|_\infty = \sup_{x \in [0, \|f\|_{\mathcal{H}^\infty})} |\psi(x)| < \infty \quad \text{and} \quad M \leq 1.$$

Accordingly, it is enough to control the integral

$$\int_0^{\|f\|_{\mathcal{H}^\infty}} \exp \left(x^2 - \psi(x) \right) dx^2$$

from above. In so doing, we apply Theorem 7.2.4 to the function

$$\phi(y) = \begin{cases} x & , \quad y = \psi(x), \\ \|f\|_{\mathcal{H}^\infty} & , \quad y > \|\psi\|_\infty. \end{cases}$$

Of course, this function ϕ satisfies

$$\phi(0) = 0 \quad \text{and} \quad \int_0^\infty \left(\phi'(y)\right)^2 dy = \pi^{-1} \int_0^{\|f\|_{\mathcal{H}\infty}} L(f, s) ds = \pi^{-1} \|f\|_{\mathcal{D}}^2 \leq 1.$$

Therefore

$$\int_0^{\|f\|_{\mathcal{H}\infty}} \exp\left(x^2 - \psi(x)\right) dx^2 = 2 \int_0^\infty \exp\left(\left(\phi(t)\right)^2 - t\right) \phi'(t) dx.$$

From this formula and an approximation we may assume that ϕ' is continuous and has compact support in $(0, \infty)$. Integrating by parts, we know that it is enough to prove that there exists a constant $\kappa > 0$ such that if ϕ is absolutely continuous with $\phi(0) = 0$ and $\int_0^\infty \left(\phi'(t)\right)^2 dt \leq 1$, then

$$\int_0^\infty \exp\left(\left(\phi(t)\right)^2 - t\right) dt = \int_0^\infty \exp\left(\left(\int_0^t \phi'(s) ds\right)^2 - t\right) dt \leq \kappa,$$

but this has been verified via Theorem 7.2.4. So we are done. $\square$

The foregoing discussion leads to the following exponential integral estimates.

Theorem 7.2.6. *Let* $p \in [0, \infty)$. *Then:*

(i)
$$\sup_{r \in (0,1),\ \|f\|_{\mathcal{D}^{2,p}} > 0} \int_{\mathbf{T}} \exp\left(\frac{\pi(1-r^2)^p |f(r\zeta) - f(0)|^2}{\|f\|_{\mathcal{D}^{2,p}}^2}\right) |d\zeta| < \infty.$$

(ii)
$$\sup_{r \in (0,1),\ w \in \mathbf{D},\ \|f\|_{\mathcal{Q}_p} > 0} \int_{\mathbf{T}} \exp\left(\frac{\pi(-\log r)^p |f \circ \sigma_w(r\zeta) - f(w)|^2}{\|f\|_{\mathcal{Q}_p}^2}\right) |d\zeta| < \infty.$$

Proof. Note that for any $r \in (0,1)$ and $w \in \mathbf{D}$,

$$\left\|\left(f - f(0)\right)_r\right\|_{\mathcal{D}}^2 \leq (1 - r^2)^{-p} \int_{\mathbf{D}} |f'(z)|^2 (1 - |z|^2)^p dm(z)$$

and

$$\left\|\left(f \circ \sigma_w - f(w)\right)_r\right\|_{\mathcal{D}}^2 \leq (-\log r)^{-p} \int_{\mathbf{D}} |(f \circ \sigma_w)'(z)|^2 (-\log|z|)^p dm(z).$$

So, the desired assertions follow from Theorem 7.2.5. $\square$

As an immediate consequence of Theorem 7.2.6, we obtain an interesting exp-characterization of the Bloch space.

Corollary 7.2.7. *Let* $p \in (1, \infty)$ *and* $f \in \mathcal{H}$. *Then* $f \in \mathcal{B}$ *if and only if there is a constant* $c(f) > 0$ *depending only on* f *such that*

$$\sup_{r \in (0,1), w \in \mathbf{D}} \int_{\mathbf{T}} \exp\left(c(f)(-\log r)^p |f \circ \sigma_w(r\zeta) - f(w)|^2\right) |d\zeta| < \infty.$$

Proof. Suppose $f \in \mathcal{B}$. If $\|f\|_{\mathcal{B}} = 0$, then there is nothing to argue. Hence, assume $\|f\|_{\mathcal{B}} > 0$, i.e., $\|f\|_{\mathcal{Q}_p} > 0$ due to $p > 1$. Taking $c(f) = \pi \|f\|_{\mathcal{Q}_p}^{-2}$ and employing Theorem 7.2.6, we derive the desired finiteness. Conversely, if a constant $c(f) > 0$ (depending only on f) obeys

$$C(f) = \sup_{r \in (0,1), w \in \mathbf{D}} \int_{\mathbf{T}} \exp\left(c(f)(-\log r)^p |f \circ \sigma_w(r\zeta) - f(w)|^2\right) |d\zeta| < \infty,$$

then

$$\int_{\mathbf{T}} (1-r)^p |f \circ \sigma_w(r\zeta) - f(w)|^2 |d\zeta| \le \frac{C(f)}{c(f)}, \quad r \in (0,1), \quad w \in \mathbf{D}.$$

Multiplying the last inequality by $r dr$ and integrating over $(0,1)$, we find

$$\int_{\mathbf{D}} |f \circ \sigma_w(z) - f(w)|^2 (1 - |z|)^p dm(z) \le \frac{C(f)}{c(f)},$$

and consequently,

$$\sup_{w \in \mathbf{D}} \int_{\mathbf{D}} |(f \circ \sigma_w)'(z)|^2 (1 - |z|^2)^{p+2} dm(z) \preceq \frac{C(f)}{c(f)};$$

that is, $f \in \mathcal{Q}_{p+2} = \mathcal{B}$ thanks to $p + 2 > 1$. $\qquad\square$

In view of Theorem 7.2.6, we can get the following exponential estimate of Korenblum type regarding the Bloch space.

Theorem 7.2.8. *There exist numerical constants $\kappa > 0$ such that*

$$\sup_{r \in (0,1),\ \|f\|_{\mathcal{B}} > 0} \int_{\mathbf{T}} \exp\left(\frac{\kappa |f(r\zeta) - f(0)|}{\|f\|_{\mathcal{B}} \sqrt{|\log(1 - r^2)|}}\right) |d\zeta| < \infty.$$

Moreover

$$\limsup_{r \to 1} \sup_{\|f\|_{\mathcal{B}} > 0} \frac{|f(r\zeta) - f(0)|}{\|f\|_{\mathcal{B}}\big(\log|\log(1 - r)|\big)\sqrt{|\log(1 - r^2)|}} < \infty.$$

Proof. Because the well-known John–Nirenberg distribution theorem on $\mathcal{BMOA}$ implies that there is a numerical constant $\kappa > 0$ such that

$$\sup_{\|f\|_{\mathcal{BMOA}} > 0} \int_{\mathbf{T}} \exp\left(\frac{\kappa |f(\zeta) - f(0)|}{\|f\|_{\mathcal{BMOA}}}\right) |d\zeta| < \infty,$$

the first assertion of the theorem follows from an application of the just-mentioned exponential estimate for $\mathcal{BMOA}$ and Remark 7.1.5 to $(f - f(0))_r$, $r \in (0,1)$.

To see the second assertion, we use the first part to get another numerical constant $\kappa_1 > 0$ such that for any $f \in \mathcal{B}$ with $\|f\|_\mathcal{B} > 0$,

$$\int_0^1 \left(\int_\mathbf{T} \exp\left(\frac{\kappa|f(r\zeta) - f(0)|}{\|f\|_\mathcal{B}\sqrt{|\log(1-r^2)|}} \right) |d\zeta| \right) \frac{dr}{(1-r)\left(1 - \log(1-r)\right)^2} < \kappa_1.$$

Consequently, for almost all $\zeta \in \mathbf{T}$, we have

$$\int_0^1 \left(\exp\left(\frac{\kappa|f(r\zeta) - f(0)|}{\|f\|_\mathcal{B}\sqrt{|\log(1-r^2)|}} \right) \right) \frac{dr}{(1-r)\left(1 - \log(1-r)\right)^2} < \infty$$

thereupon producing

$$\lim_{r \to 1} \int_r^{\frac{1+r}{2}} \left(\exp\left(\frac{\kappa|f(t\zeta) - f(0)|}{\|f\|_\mathcal{B}\sqrt{|\log(1-t^2)|}} \right) \right) \frac{dt}{(1-t)\left(1 - \log(1-t)\right)^2} < \infty.$$

Writing

$$\varpi(f, r, \zeta) = \min\left\{ |f(r\zeta) - f(0)| : r \le t \le \frac{1+r}{2} \right\},$$

we derive

$$\lim_{r \to 1} \left(\frac{\kappa\varpi(f, r, \zeta)}{\|f\|_\mathcal{B}\sqrt{|\log(1-r)|}} - 2\log|\log(1-r)| \right) = -\infty.$$

Accordingly, for almost all $\zeta \in \mathbf{T}$,

$$\varpi(f, r, \zeta) \preceq \|f\|_\mathcal{B}\sqrt{|\log(1-r)|}\log|\log(1-r)| \quad \text{as} \quad r \to 1.$$

This estimate, together with the inequality

$$|f(t_2\zeta) - f(t_1\zeta)| \le (\log 2)\|f\|_\mathcal{B} \quad \text{for all} \quad t_1, t_2 \in [r, \frac{1+r}{2}],$$

yields the above-desired limit. $\qquad\square$

7.3 Hadamard Convolutions

We have seen from the last section that the dilation f_r plays an important role in dealing with the exponential decay of $f \in \mathcal{Q}_p$. Therefore, another look at f_r would reveal a few more properties on $\mathcal{Q}_p$. Let now $f(z) = \sum_{j=0}^\infty a_j z^j$ be in $\mathcal{H}$. Then for $r \in (0,1)$ and $z \in \bar{\mathbf{D}}$ one has $f_r(z) = \sum_{j=0}^\infty a_j r^j z^j$. If $g(z) = \sum_{j=0}^\infty z^n = (1-z)^{-1}$ for $z \in \mathbf{D}$, then $g_r(z) = (1-rz)^{-1}$ and hence $f_r(z) = f \star g_r(z)$. Here and hereafter, $\star$ denotes the Hadamard convolution; that is, if $f, g \in \mathcal{H}$ with $f(z) = \sum_{j=0}^\infty a_j z^j$ and $g(z) = \sum_{j=0}^\infty b_j z^j$, then

$$f \star g(s\zeta) = \sum_{j=0}^\infty a_j b_j (s\zeta)^j = (2\pi)^{-1} \int_\mathbf{T} f(\sqrt{s}\eta) g(\sqrt{s}\zeta\bar{\eta}) |d\eta|, \quad s \in (0,1), \ \zeta \in \mathbf{T}.$$

Recall that for $q \in (0, \infty]$ and $\alpha \in (0, 1)$, $\Lambda(q, \alpha)$ stands for the mean Lipschitz class of $f \in \mathcal{H}$ with

$$\|f\|_{\Lambda(q,\alpha)} = |f(0)| + \sup_{r \in (0,1)} (1 - r)^{1-\alpha} \|(f')_r\|_{\mathcal{H}^q} < \infty.$$

It is not hard to see that $f \in \Lambda(q, \alpha)$ when and only when

$$\sup_{r \in (0,1)} (1 - r)^{2-\alpha} \|(f'')_r\|_{\mathcal{H}^q} < \infty.$$

Using this fact and the monotonic property of $\|(f')_r\|_{\mathcal{H}^q}$ in $r \in (0, 1)$, we can easily check that if $\mathsf{D}f(z) = (zf'(z))'$, then

$$\|f\|_{\Lambda(q,\alpha)} \approx |f(0)| + \sup_{r \in (0,1)} (1 - r)^{2-\alpha} \|(\mathsf{D}f)_r\|_{\mathcal{H}^q}.$$

At the same time, denote by $\Lambda(q, p, \alpha)$, where $q, p \in (0, \infty]$ and $\alpha \in (0, 1)$, the integrated Lipschitz class of $f \in \mathcal{H}$ obeying

$$\|f\|_{\Lambda(q,p,\alpha)} = |f(0)| + \left(\int_0^1 \left((1 - r)^{1-\alpha} \|(f')_r\|_{\mathcal{H}^q} \right)^p (1 - r)^{-1} dr \right)^{\frac{1}{p}} < \infty.$$

Similarly, we have that $f \in \Lambda(q, p, \alpha)$ if and only if

$$\int_0^1 \left((1 - r)^{2-\alpha} \|(f'')_r\|_{\mathcal{H}^q} \right)^p (1 - r)^{-1} dr < \infty.$$

Accordingly,

$$\|f\|_{\Lambda(q,p,\alpha)} \approx |f(0)| + \left(\int_0^1 \left((1 - r)^{2-\alpha} \|(\mathsf{D}f)_r\|_{\mathcal{H}^q} \right)^p (1 - r)^{-1} dr \right)^{\frac{1}{p}}.$$

Clearly, $\Lambda(q, \alpha)$ can be treated as the limit space of $\Lambda(q, p, \alpha)$ as $p \to \infty$. Moreover, if $q = \infty$ then $\Lambda(\infty, q^{-1}) = \Lambda(\infty, 0) = \mathcal{B}$. Taking its $\mathcal{Q}_p$-setting into account, we find the following assertion.

Lemma 7.3.1. *Let* $p \in (0, 1]$.

(i) *If* $2 < q < \frac{2}{1-p}$, *then* $f \in \Lambda(q, q^{-1})$ *implies* $f \in \mathcal{Q}_p$ *with*

$$\|f\|_{\mathcal{Q}_p,2} \preceq \sup_{r \in (0,1)} (1 - r)^{2-q^{-1}} \|(\mathsf{D}f)_r\|_{\mathcal{H}^q}.$$

(ii) *If* $q = \frac{2}{1-p}$, *then* $f \in \Lambda(q, 2, q^{-1})$ *implies* $f \in \mathcal{Q}_p$ *with*

$$\|f\|_{\mathcal{Q}_p,2} \preceq \left(\int_0^1 (1 - r)^{-1} \left((1 - r)^{2-q^{-1}} \|(\mathsf{D}f)_r\|_{\mathcal{H}^q} \right)^2 dr \right)^{\frac{1}{2}}.$$

Proof. To see $f \in \mathcal{Q}_p$ provided that f obeys the conditions in (i) and (ii), we fix a Carleson box $S(I)$ based on a subarc I of $\mathbf{T}$, and then use Hölder's inequality with $2 < q \le 2/(1-p)$ to obtain

$$\int_{S(I)} |f'(z)|^2 (1-|z|)^p dm(z) \preceq |I|^{1-\frac{2}{q}} \int_{1-|I|}^1 \left(\int_{\mathbf{T}} |f'(r\zeta)|^q |d\zeta| \right)^{\frac{2}{q}} (1-r)^p dr$$

$$\preceq |I|^{1-\frac{2}{q}} \int_0^1 \left(\int_{\mathbf{T}} |f'(r\zeta)|^q |d\zeta| \right)^{\frac{2}{q}} (1-r)^p dr.$$

(i) Let $f \in \Lambda(q, q^{-1})$. If $2 < q < \frac{2}{1-p}$, the last estimates give

$$\int_{S(I)} |f'(z)|^2 (1-|z|)^p dm(z) \preceq |I|^p \left(\sup_{r \in (0,1)} (1-r)^{1-q^{-1}} \|(f')_r\|_{\mathcal{H}^q} \right)^2$$

$$\approx |I|^p \left(\sup_{r \in (0,1)} (1-r)^{2-q^{-1}} \|(Df)_r\|_{\mathcal{H}^q} \right)^2,$$

hence giving $f \in \mathcal{Q}_p$ with the required seminorm inequality.

(ii) Let $f \in \Lambda(q, 2, q^{-1})$. If $2 < q = \frac{2}{1-p}$, we similarly get

$$\int_{S(I)} |f'(z)|^2 (1-|z|)^p dm(z) \preceq |I|^p \int_0^1 (1-r)^{-1} \left((1-r)^{1-q^{-1}} \|(f')_r\|_{\mathcal{H}^q} \right)^2 dr$$

$$\approx |I|^p \int_0^1 (1-r)^{-1} \left((1-r)^{2-q^{-1}} \|(Df)_r\|_{\mathcal{H}^q} \right)^2 dr,$$

as required. □

The above lemma founds a basis of the Hadamard convolution inequalities below.

Theorem 7.3.2. *Let* $p \in (0, 1]$.

(i) *If* $2 < q < \frac{2}{1-p}$, *then* $f \in \Lambda(q, q^{-1})$ *and* $g \in \Lambda(1, 0)$ *imply* $f \star g \in \mathcal{Q}_p$ *with*

$$\|f \star g\|_{\mathcal{Q}_p, 2} \preceq \|f\|_{\Lambda(q, q^{-1})} \|g\|_{\Lambda(1, 0)}.$$

(ii) *If* $q = \frac{2}{1-p}$, *then* $f \in \Lambda(q, q^{-1})$ *and* $g \in \mathcal{H}^1$ *imply* $f \star g \in \mathcal{Q}_p$ *with*

$$\|f \star g\|_{\mathcal{Q}_p, 2} \preceq \|f\|_{\Lambda(q, q^{-1})} \|g\|_{\mathcal{H}^1}.$$

(iii) *If* $q = \frac{2}{2-p}$, *then* $f \in \mathcal{D}^{2, p}$ *and* $g \in \Lambda(q, 0)$ *imply* $f \star g \in \mathcal{Q}_p$ *with*

$$\|f \star g\|_{\mathcal{Q}_p, 2} \preceq \|f\|_{\mathcal{D}^{2, p}} \|g\|_{\Lambda(q, 0)}.$$

Proof. The starting point is the identity

$$\mathsf{D}(f \star g) = f' \star g' \quad \text{for} \quad f, g \in \mathcal{H}.$$

(i) By Lemma 7.3.1 (i) and the Minkowski inequality with $q > 2$, we get that if $f \in \Lambda(q, q^{-1})$ and $g \in \Lambda(1, 0)$, then

$$
\begin{aligned}
\|f \star g\|_{\mathcal{Q}_p, 2} &\preceq \sup_{r \in (0,1)} (1-r)^{2-q^{-1}} \|(\mathsf{D}(f \star g))_r\|_{\mathcal{H}^q} \\
&\preceq \sup_{r \in (0,1)} (1-r)^{2-q^{-1}} \|(f' \star g')_r\|_{\mathcal{H}^q} \\
&\preceq \sup_{r \in (0,1)} (1-r)^{2-q^{-1}} \|(f')_{\sqrt{r}}\|_{\mathcal{H}^q} \|(g')_{\sqrt{r}}\|_{\mathcal{H}^1} \\
&\preceq \|f\|_{\Lambda(q, q^{-1})} \|g\|_{\Lambda(1, 0)}.
\end{aligned}
$$

(ii) When $f \in \Lambda(q, q^{-1})$ and $g \in \mathcal{H}^1$, we use the following Hardy–Littlewood inequality

$$\int_0^1 (1-r) \left(\int_{\mathbf{T}} |g'(r\zeta)| |d\zeta| \right)^2 dr \preceq \|g\|_{\mathcal{H}^1}^2$$

and Lemma 7.3.1 (ii) to get

$$
\begin{aligned}
\|f \star g\|_{\mathcal{Q}_p, 2}^2 &\preceq \|f\|_{\Lambda(q, q^{-1})}^2 \int_0^1 (1-r) \|(g')_r\|_{\mathcal{H}^1}^2 dr \\
&\preceq \|f\|_{\Lambda(q, q^{-1})}^2 \|g\|_{\mathcal{H}^1}^2,
\end{aligned}
$$

hence producing the desired inequality.

(iii) Suppose $f \in \mathcal{D}^{2,p}$ and $g \in \Lambda(q, 0)$. Then these two functions enjoy Young's inequality (more general than Minkowski's inequality):

$$\|(f \star g)_r\|_{\mathcal{H}^{q_3}} \leq \|f_{\sqrt{r}}\|_{\mathcal{H}^{q_1}} \|f_{\sqrt{r}}\|_{\mathcal{H}^{q_2}} \quad \text{where} \quad \frac{1}{q_1} + \frac{1}{q_2} - \frac{1}{q_3} = 1 \text{ and } q_1, q_2, q_3 \geq 1.$$

Using this inequality with $q_1 = 2$, $q_2 = \frac{2}{2-p}$, $q_3 = \frac{2}{1-p}$, plus Lemma 7.2.8 (ii), we get

$$
\begin{aligned}
\|f \star g\|_{\mathcal{Q}_p, 2}^2 &\preceq \int_0^1 (1-r)^{p+2} \left(\|(f')_{\sqrt{r}}\|_{\mathcal{H}^2} \|(g')_{\sqrt{r}}\|_{\mathcal{H}^{\frac{2}{2-p}}} \right)^2 dr \\
&\preceq \|g\|_{\Lambda(\frac{2}{2-p}, 0)}^2 \int_0^1 (1-r)^p \|(f')_{\sqrt{r}}\|_{\mathcal{H}^2}^2 dr \\
&\preceq \|g\|_{\Lambda(\frac{2}{2-p}, 0)}^2 \|f\|_{\mathcal{D}^{2,p}}^2,
\end{aligned}
$$

thereupon establishing the required inequality.   □

Corollary 7.3.3. *Let $p \in [0, \infty)$. Then $\mathcal{Q}_p \star \mathcal{Q}_p \subseteq \mathcal{Q}_p$.*

Proof. If $p \in (0,1]$, then the assertion follows immediately from Theorem 7.3.2 (iii) and two inclusions

$$\mathcal{Q}_p \subset \mathcal{D}^{2,p} \quad \text{and} \quad \mathcal{B} \subseteq \Lambda(q,0), \quad q > 0.$$

In the case of $p \in (1,\infty)$, we have $\mathcal{Q}_p = \mathcal{B}$. So if $f, g \in \mathcal{B}$ then

$$\|f \star g\|_{\mathcal{B}} \preceq \sup_{r \in [0,1), \, \zeta \in \mathbf{T}} (1-r)^2 |D(f \star g)(r\zeta)|$$

$$\preceq \sup_{r \in [0,1), \, \zeta \in \mathbf{T}} \int_{\mathbf{T}} (1-r)^2 |f'(\sqrt{r}\eta)||g'(\sqrt{r}\zeta\bar{\eta})||d\eta|$$

$$\preceq \|f\|_{\mathcal{B}} \|g\|_{\mathcal{B}},$$

as required.

In the case of $p = 0$, we get that if $f, g \in \mathcal{D}$, then by $\mathcal{D} \subset \mathcal{B}$ and the Cauchy–Schwarz inequality,

$$\|f \star g\|_{\mathcal{D}}^2 \preceq \int_0^1 \left(\int_{\mathbf{T}} |D(f \star g)(r\zeta)|^2 |d\zeta| \right) (1-r)^2 dr$$

$$\preceq \int_0^1 \int_{\mathbf{T}} \left(\int_{\mathbf{T}} |f'(\sqrt{r}\eta)||g'(\sqrt{r}\zeta\bar{\eta})||d\eta| \right)^2 |d\zeta|(1-r)^2 dr$$

$$\preceq \|g\|_{\mathcal{B}}^2 \int_0^1 \int_{\mathbf{T}} |f'(\sqrt{r}\eta)|^2 |d\eta| dr$$

$$\preceq \|g\|_{\mathcal{D}}^2 \|f\|_{\mathcal{D}}^2,$$

as desired. □

In fact, Lemma 7.3.1 (ii) can tell us a forward and backward estimate for the $\mathcal{Q}_p$ seminorm.

Theorem 7.3.4. *Given $p \in [0,1]$ and $q = \frac{2}{1-p}$. For $n \in \mathbf{N}$ and $f(z) = \sum_{j=0}^{\infty} a_j z^j \in \mathcal{H}$ let $I_0 = \{0\}$, $I_n = \{k \in \mathbf{N} : 2^{n-1} \le k < 2^n\}$ and $(\Delta_n f)(z) = \sum_{k \in I_n} a_k z^k$. Then*

$$\left(\sum_{n=1}^{\infty} \frac{\|\Delta_n f\|_{\mathcal{L}^2(\mathbf{T})}^2}{2^{(p-1)n}} \right)^{\frac{1}{2}} \preceq \|f\|_{\mathcal{D}^{2,p}} \preceq \|f\|_{\mathcal{Q}_{p,2}} \preceq \left(\sum_{n=1}^{\infty} \frac{\|\Delta_n f\|_{\mathcal{L}^q(\mathbf{T})}^2}{2^{(p-1)n}} \right)^{\frac{1}{2}}.$$

Proof. Note that $f \in \Lambda(q, 2, q^{-1})$ amounts to $\sum_{n=1}^{\infty} 2^{(1-p)n} \|\Delta_n f\|_{\mathcal{L}^q(\mathbf{T})}^2 < \infty$. So, from Lemma 7.3.1 (ii) it turns out that we only need to check the left-hand-side of the above-desired inequalities. Since

$$f \in \mathcal{D}^{2,p} \Leftrightarrow \int_0^1 (1-r)^p \|(f')_r\|_{\mathcal{H}^2}^2 dr < \infty,$$

we conclude that $f \in \mathcal{Q}_p \subseteq \mathcal{D}^{2,p}$ gives $f \in \Lambda(2,2,\frac{1-p}{2})$ with

$$\|f\|_{\Lambda(2,2,\frac{1-p}{2})} - |f(0)| \preceq \|f\|_{\mathcal{D}^{2,p}} \preceq \|f\|_{\mathcal{Q}_p}.$$

But, the most left quantity of the last inequalities dominates

$$\sum_{n=1}^{\infty} 2^{(1-p)n} \|\Delta_n f\|_{\mathcal{L}^2(\mathbf{T})}^2,$$

and thus this completes the proof. $\square$

Corollary 7.3.5. *Given $p \in [0,1]$. Suppose $\{n_k\}$ is a sequence of natural numbers with $\inf_{k \in \mathbf{N}} n_{k+1}/n_k > 1$. Then $f(z) = \sum_{k=0}^{\infty} a_k z^{n_k}$ is in $\mathcal{Q}_p$ if and only if $\sum_{k=1}^{\infty} \sum_{2^{k-1} \leq n_j < 2^k} |a_j|^2$ is convergent.*

Proof. Theorem 7.3.4, along with

$$\|\Delta_n f\|_{\mathcal{L}^q(\mathbf{T})} \leq \|\Delta_n f\|_{\mathcal{L}^\infty(\mathbf{T})} \leq \sum_{2^{n-1} \leq n_j < 2^n} |a_j|$$

and the fact that the cardinal of $\{j \in \mathbf{N} : 2^{k-1} \leq n_j < 2^k\}$ is not greater than

$$1 + \frac{\log 2}{\log\left(\inf_{k \in \mathbf{N}} \frac{n_{k+1}}{n_k}\right)},$$

yields the desired assertion. $\square$

7.4 Characteristic Bounds of Derivatives

For $f \in \mathcal{H}$ and $r \in [0,1)$, let

$$T(r,f) = (2\pi)^{-1} \int_{\mathbf{T}} \max\left\{\log|f(r\zeta)|, 0\right\} |d\zeta|$$

be the Nevanlinna characteristic of f. It is easy to see that if $f \in \mathcal{B}$ then

$$T(r,f') \leq -\log(1-r) + O(1) \quad \text{as} \quad r \to 1.$$

The estimate is sharp in the sense that there exists an $f \in \mathcal{B}$ such that

$$-\log(1-r) - T(r,f') = O(1) \quad \text{as} \quad r \to 1,$$

which gives

$$\int_0^1 (1-r)\exp\left(2T(r,f')\right) dr = \infty.$$

Since $\mathcal{B} = \mathcal{Q}_p$ for $p > 1$, a natural question is: how about $\mathcal{Q}_p$, $p \in [0,1]$? Below is the answer.

Theorem 7.4.1. *Let $p \in [0,1]$. Then $f \in \mathcal{Q}_p$ implies*

$$\int_0^1 (1-r)^p \exp\left(2T(r,f')\right) dr < \infty.$$

Furthermore, if $\phi : (0,1) \mapsto (0,\infty)$ is an increasing function and satisfies the following three conditions:

(i) *$r \mapsto (1-r)^{\frac{1+p}{2}} \exp\left(\phi(r)\right)$ is a decreasing function on $(0,1)$;*

(ii) *$\phi(r) - \phi(s) \to \infty$ as $(1-r)/(1-s) \to 0$;*

(iii) *$\int_0^1 (1-r)^p \exp\left(2\phi(r)\right) dr < \infty$,*

then there is an $f \in \mathcal{Q}_p$ such that $T(r,f') > \phi(r)$ as $r \to 1$.

Proof. Suppose $f \in \mathcal{Q}_p$ and $r \in (0,1)$. Using Jensen's inequality, we derive

$$\int_0^1 \exp\left(2T(r,f')\right)(1-r)^p dr$$

$$\leq \int_0^1 \left(\exp\left((2\pi)^{-1} \int_{\mathbf{T}} \log\left(1+|f'(r\zeta)|^2\right)|d\zeta|\right)\right)(1-r)^p dr$$

$$\leq (2\pi)^{-1} \int_0^1 \left(\int_{\mathbf{T}} \left(1+|f'(r\zeta)|^2\right)|d\zeta|\right)(1-r)^p dr$$

$$\leq 1 + \int_0^1 \left(\int_{\mathbf{T}} |f'(r\zeta)|^2 |d\zeta|\right)(1-r)^p dr$$

$$\precsim 1 + \|f\|_{\mathcal{D}^{2,p}}^2$$

$$\precsim 1 + \|f\|_{\mathcal{Q}_p,1}^2.$$

To check the second part of the theorem, we assume that ϕ obeys those hypotheses. Since ϕ is increasing on $(0,1)$, the condition (iii) implies

$$\infty > \int_0^1 (1-r)^p \exp\left(2\phi(r)\right) dr$$

$$\geq \sum_{k=1}^{\infty} \int_{1-2^{-k}}^{1-2^{-k-1}} (1-r)^p \exp\left(2\phi(r)\right) dr$$

$$\geq 2^{-1-p} \sum_{k=1}^{\infty} 2^{-k(1+p)} \exp\left(2\phi(1-2^{-k})\right).$$

This finiteness, along with the Dini Theorem in [Kno, p. 297], generates an increasing sequence $\{c_k\}$ of natural numbers greater than 2 such that $\lim_{k\to\infty} c_k = \infty$, $\lim_{k\to\infty} c_{k+1}/c_k = 1$ and

$$\sum_{k=1}^{\infty} c_k^{1+p} 2^{-k(1+p)} \exp\left(2\phi(1-2^{-k})\right) \leq \sum_{k=1}^{\infty} c_k^2 2^{-k(1+p)} \exp\left(2\phi(1-2^{-k})\right) < \infty.$$

Choosing now $n_1 = 1$ and $n_{k+1} = c_k n_k$ for $k \in \mathbf{N}$, we have $n_{k+1} > 2^k$ for $k \in \mathbf{N}$, and

$$\sum_{k=1}^{\infty} c_k^{1+p} n_{k+1}^{-1-p} \exp\left(2\phi(1 - n_{k+1}^{-1})\right) \le \sum_{k=1}^{\infty} c_k^{1+p} 2^{-k(1+p)} \exp(2\phi(1 - 2^{-k})) < \infty,$$

thanks to the condition (i). If

$$f(z) = \sum_{k=1}^{\infty} a_k z^{n_k} \quad \text{with} \quad a_k = 10 n_k^{-1} \exp\left(2\phi(1 - n_{k+1}^{-1})\right),$$

then it is easy to find $f \in \mathcal{Q}_p$ due to Corollary 7.3.5.

The argument will end up with verifying $T(r, f') \ge \phi(r)$. In so doing, we note that for $k - 1 \in \mathbf{N}$ and $|z| = 1 - n_k^{-1}$,

$$
\begin{aligned}
|f'(z)| \;\ge\; |zf'(z)| &= \left| \sum_{j=1}^{\infty} a_j n_j z^{n_j} \right| \\
&\ge\; a_k |z|^{n_k} - \sum_{j=1}^{k-1} a_j n_j |z|^{n_j} - \sum_{j=1}^{k-1} a_j n_j \left(1 - \frac{1}{n_k}\right)^{n_j} \\
&=\; (I) - (II) - (III).
\end{aligned}
$$

For the first term, we trivially have $(I) \ge 2^{-2} a_k n_k$ for $n_k \ge 2$.

For the second term, we use the condition (ii) to get

$$\lim_{k \to \infty} \frac{a_k n_k}{a_{k+1} n_{k+1}} = \lim_{k \to \infty} \exp\left(\phi(1 - n_{k+1}^{-1}) - \phi(1 - n_{k+2}^{-1})\right) = 0$$

which readily deduces $\lim_{k \to \infty}(II)/(a_k n_k) = 0$.

For the third term, we employ the elementary inequality

$$(1 - t)^n < 2(nt)^{-2} \quad \text{for} \quad t \in (0, 1) \quad \text{and} \quad n \in \mathbf{N}$$

to derive $(III) \le 2 n_k^2 \sum_{j=k+1}^{\infty} a_j n_j^{-1}$. By the condition (i), we find

$$\lim_{k \to \infty} \frac{n_k a_k^{-1}}{n_{k+1} a_{k+1}^{-1}} = \lim_{k \to \infty} \frac{\exp\left(\phi(1 - n_{k+2}^{-1})\right)}{c_k^2 \exp\left(\phi(1 - n_{k+1}^{-1})\right)} \le \lim_{k \to \infty} c_k^{\frac{p-3}{2}} \left(\frac{c_{k+1}}{c_k}\right)^{\frac{1+p}{2}} = 0,$$

hence we get $\lim_{k \to \infty}(III)/(a_k n_k) = 0$ via the technique in the ratio test for positive series.

All together, we can select a natural number k_0 so that $k \ge k_0$ implies

$$|f'(z)| > 2^{-3} a_k n_k > \exp\left(\phi(1 - n_{k+1}^{-1})\right) \quad \text{for} \quad |z| = 1 - n_k^{-1}.$$

Consequently,

$$T(1 - n_k^{-1}, f') = (2\pi)^{-1} \int_{\mathbf{T}} \max\left\{ |f'((1 - n_k^{-1})\zeta)|, 0 \right\} |d\zeta| > \phi(1 - n_{k+1}^{-1}).$$

As for $r \geq 1 - n_{k_0}^{-1}$, we take $k \geq k_0$ to ensure $1 - n_k^{-1} \leq r < 1 - n_{k+1}^{-1}$. Because $T(r, f)$ and $\phi(r)$ increase with r, we finally reach

$$T(r, f') \geq T(1 - n_k^{-1}, f') > \phi(1 - n_{k+1}^{-1}) \geq \phi(r),$$

as required. $\qquad \square$

On the basis of the radial growth of f' for $f \in \mathcal{B}$: $(1 - |z|)|f'(z)| \leq \|f\|_{\mathcal{B}}$ which is sharp in the sense that if $f(z) = \sum_{j=1}^{\infty} z^{n^j}$ with $n \in \mathbf{N}$ being big enough, then $f \in \mathcal{B}$ with

$$1 \leq (1 - |z|^2)|f'(z)| \quad \text{for} \quad 1 - n^{-k} \leq |z| \leq 1 - n^{-k-1/2}$$

and so

$$1 \precsim \limsup_{r \to 1}(1 - r)|f'(r\zeta)| \quad \text{for all } \zeta \in \mathbf{T}.$$

We next deal with the radial growth of the derivative of a function in $\mathcal{Q}_p$ for $p \in [0, 1]$.

Theorem 7.4.2. *Let $p \in [0, 1]$. If $f \in \mathcal{Q}_p$, then*

$$\lim_{r \to 1}(1 - r)^{\frac{1+p}{2}}|f'(r\zeta)| = 0 \quad \text{for a.e. } \zeta \in \mathbf{T}.$$

Furthermore, if $\phi : (0, 1) \mapsto (0, \infty)$ increases and satisfies $\int_0^1 (1 - r)^p (\phi(r))^2 dr < \infty$, then there exists $f \in \mathcal{Q}_p$ such that

$$\limsup_{r \to 1} (\phi(r))^{-1}|f'(r\zeta)| = \infty \quad \text{for all } \zeta \in \mathbf{T}.$$

Proof. Assume $f \in \mathcal{Q}_p$. From the Hardy–Littlewood maximal estimate for $\mathcal{H}^2$ it follows that

$$\int_0^1 \left(\int_{\mathbf{T}} \max_{t \in [0, r]} |f'(t\zeta)|^2 |d\zeta| \right) (1 - r^2)^p dr^2$$

$$\precsim \int_0^1 \left(\int_{\mathbf{T}} |f'(r\zeta)|^2 |d\zeta| \right) (1 - r^2)^p dr^2$$

$$\precsim \|f\|_{\mathcal{Q}_p, 1}^2.$$

Accordingly, for a.e. $\zeta \in \mathbf{T}$, we have

$$\int_0^1 \max_{t \in [0, r]} |f'(t\zeta)|^2 (1 - r^2)^p dr^2 < \infty,$$

hence we get

$$\lim_{r \to 1}(1 - r)^{1+p}|f'(r\zeta)|^2 \leq (1 + p) \lim_{r \to 1} \max_{t \in [0, r]} |f'(t\zeta)|^2 \int_r^1 (1 - s)^p ds$$

$$\leq (1 + p) \lim_{r \to 1} \int_r^1 \max_{t \in [0, s]} |f'(t\zeta)|^2 (1 - s)^p ds$$

$$= 0.$$

This proves the radial growth of f'.

As for the second part of the theorem, we may assume without loss of generality that $\lim_{r \to 1} \phi(r) = \infty$ and prove

$$1 \preceq \limsup_{r \to 1} \left(\phi(r)\right)^{-1} |f'(r\zeta)| \quad \text{for every} \quad \zeta \in \mathbf{T}$$

since otherwise it is possible to choose an increasing function $\phi_1 : (0,1) \mapsto (0,\infty)$ with

$$\lim_{r \to 1} \phi_1(r) = \infty \quad \text{and} \quad \int_0^1 (1-r)^p \left(\phi(r)\phi_1(r)\right)^2 dr < \infty,$$

thereby yielding that for all $\zeta \in \mathbf{T}$,

$$\infty = \limsup_{r \to 1} \phi_1(r) \left(\phi(r)\phi_1(r)\right)^{-1} |f'(r\zeta)| = \limsup_{r \to 1} \left(\phi(r)\right)^{-1} |f'(r\zeta)|.$$

Under this assumption, we choose an increasing sequence $\{r_k\}$ such that $\lim_{k \to \infty} r_k = 1$ and

$$r_1 > 2^{-2}; \; \frac{r_{k+1} - r_k}{1 - r_k} > 2^{-1}; \; \lim_{k \to \infty} \frac{\phi(r_{k+1})}{\phi(r_k)} = \infty; \; \limsup_{k \to \infty} \frac{(1 - r_{k+1})^{\frac{3-p}{2}}}{(1 - r_k)^2} < \infty.$$

With these conditions, we make a simple calculation to get

$$\sum_{k=1}^{\infty} (1 - r_k)^{1+p} \left(\phi(r_k)\right)^2 \preceq \sum_{k=1}^{\infty} \int_{r_k}^{r_{k+1}} (1-r)^p \left(\phi(r_k)\right)^2 dr$$

$$\preceq \sum_{k=1}^{\infty} \int_{r_k}^{r_{k+1}} (1-r)^p \left(\phi(r)\right)^2 dr$$

$$\preceq \int_0^1 (1-r)^p \left(\phi(r)\right)^2 dr$$

$$< \infty.$$

Now, for $k \in \mathbf{N}$, suppose that n_k is the unique natural number ensuring

$$n_k \leq (1 - r_k)^{-1} < n_k + 1.$$

So we have

$$1 - n_k^{-1} \leq r_k < 1 - (1 + n_k)^{-1}, \quad 2^{-2} < n_k(1 - r_k) \leq 1 \quad \text{and} \quad n_{k+1}/n_k > 5/4.$$

If $f(z) = \sum_{k=1}^{\infty} (1 - r_k)\phi(r_k)z^{n_k}$, then $f \in \mathcal{H}$. Our aim is to verify $f \in \mathcal{Q}_p$. In so doing, for $j \in \mathbf{N}$ let $k(j) \in \mathbf{N}$ be the unique number obeying $2^{k(j)} \leq n_j < 2^{1+k(j)}$.

Then the requirements on $\{r_k\}$ imply

$$\sum_{k=0}^{\infty} 2^{k(1-p)} \sum_{2^k \leq n_j < 2^{k+1}} (1 - r_j)^2 \big(\phi(r_j)\big)^2 \leq \sum_{j=1}^{\infty} n_j^{1-p} (1 - r_j)^2 \big(\phi(r_j)\big)^2$$

$$\leq \sum_{j=1}^{\infty} (1 - r_j)^{1+p} \big(\phi(r_j)\big)^2$$

$$< \infty,$$

hence giving $f \in \mathcal{Q}_p$.

Finally, we are about to demonstrate

$$1 \preceq \limsup_{r \to 1} \big(\phi(r)\big)^{-1} |f'(r\zeta)| \quad \text{for all} \quad \zeta \in \mathbf{T}.$$

If $k - 1 \in \mathbf{N}$ and $|z| = 1 - n_k^{-1}$, then

$$|f'(z)| \geq |zf'(z)| = \left| \sum_{j=1}^{\infty} n_j (1 - r_j)\phi(r_j) z^{n_j} \right|$$

$$\geq n_k (1 - r_k)\phi(r_k) r_k^{n_k} - \sum_{j \neq k} n_j (1 - r_j)\phi(r_j) r_k^{n_j}$$

$$\geq 2^{-2}\phi(r_k)\Big(1 - \frac{1}{n_k}\Big)^{n_k} - \sum_{j=1}^{k-1} \phi(r_j) - \sum_{j=k+1}^{\infty} \phi(r_j)\Big(1 - \frac{1}{1 + n_k}\Big)^{n_j}$$

$$= (IV) - (V) - (VI).$$

In a similar manner to controlling (I) and (II) in the argument for Theorem 7.4.1, we easily obtain $\phi(r_k) \preceq (IV)$ and $\lim_{k\to\infty}(V)/\phi(r_k) = 0$. Moreover, the definitions of r_k and n_k yield

$$\frac{n_j^2/\phi(r_j)}{n_{j+1}^2/\phi(r_{j+1})} \leq \Big(\frac{2^4}{\phi(2^{-2})}\Big)\Big(\frac{(1 - r_{j+1})^2 \phi(r_{j+1})}{(1 - r_j)^2}\Big)$$

$$= \Big(\frac{2^4}{\phi(2^{-2})}\Big)\Big(\frac{(1 - r_{j+1})^{\frac{3-p}{2}}}{(1 - r_j)^2}\Big)\Big((1 - r_{j+1})^{\frac{1+p}{2}} \phi(r_{j+1})\Big)$$

$$\to 0 \quad \text{as} \quad j \to \infty.$$

Accordingly, we achieve

$$\limsup_{k\to\infty} \frac{(VI)}{\phi(r_k)} \leq 2 \limsup_{k\to\infty} \frac{(n_k + 1)^2}{\phi(r_k)} \sum_{j=k+1}^{\infty} \frac{\phi(r_j)}{n_j^2} = 0,$$

hence getting a constant $c > 0$ and a natural number k_0 such that

$$\big(\phi(r_k)\big)^{-1} |f'(r_k\zeta)| \geq c \quad \text{whenever} \quad k \geq k_0 \quad \text{and} \quad \zeta \in \mathbf{T}.$$

Clearly, the desired estimate for the limit superior follows. □

7.5   Notes

Note 7.5.1. Section 7.1 is taken from [Xi4, Section 4] and [AlSi, Section 3]. An interesting problem related to Example 7.1.1 is to give certain function-theoretic characterizations of $[\mathcal{BMOA}, \mathcal{B}]_s$ and even $[\mathcal{D}, \mathcal{Q}_p]_s$ for $p \in (0,1)$.

Observe that $\| \prod_{k=1}^{n} \sigma_{w_k} \|_{\mathcal{D}} \approx n^{\frac{1}{2}}$ for any finite subset $\{w_k\}_{k=1}^{n}$ of $\mathbf{D}$ (see e.g. [ArFiPe3, Theorem]). Thus, Theorem 7.1.2 (i) yields

$$\left\| \prod_{k=1}^{n} \sigma_{w_k} \right\|_{\mathcal{Q}_p,1} \leq \left\| \prod_{k=1}^{n} \sigma_{w_k} \right\|_{\mathcal{D}}^{1-p} \left\| \prod_{k=1}^{n} \sigma_{w_k} \right\|_{\mathcal{BMOA}}^{p} \preceq n^{\frac{1-p}{2}}.$$

On the other hand, we know from [Bo, Theorem 4.1 with $s = \frac{1}{2}$] that

$$\|f\|_{\mathcal{D}} \preceq \|f\|_{\mathcal{D}^{2,p}}^{\frac{1}{1-p}} \preceq \|f\|_{\mathcal{Q}_p,1}^{\frac{1}{1-p}}$$

holds for any inner function f on $\mathbf{D}$. Thus, we achieve

$$n^{\frac{1-p}{2}} \preceq \left\| \prod_{k=1}^{n} \sigma_{w_k} \right\|_{\mathcal{Q}_p,1},$$

hence producing

$$\left\| \prod_{k=1}^{n} \sigma_{w_k} \right\|_{\mathcal{Q}_p,1} \approx n^{\frac{1-p}{2}}, \quad n \in \mathbf{N}.$$

Theorems 7.1.3 and 7.1.4 are the special cases of the Korenblum-type estimates in [AlSi, Theorems 3.1 and 3.2] which are more general than [Kor, Theorem 1].

Note 7.5.2. Section 7.2 comes from [Xi5] and [BetBrJa, Proposition 1] plus a combination of [ChaSMar], [Mar] and [Kor]. In fact, there are two facts related to Theorem 7.2.5:

(i) If $1 < \alpha < \infty$, then

$$\sup \left\{ \int_{\mathbf{T}} \exp \left(\alpha |f(\zeta)|^2 \right) |d\zeta| : \ f \in \mathcal{D} \text{ with } f(0) = 0 \text{ and } \|f\|_{\mathcal{D}}^2 \leq \pi \right\} = \infty.$$

(ii) If $0 < \alpha < 1$, then

$$\sup \left\{ \int_{\mathbf{T}} \exp \left(\alpha |f(\zeta)|^2 \right) |d\zeta| : \ f \in \mathcal{D} \text{ with } f(0) = 0 \text{ and } \|f\|_{\mathcal{D}}^2 \leq \pi \right\} < \infty.$$

To check (i), we simply take (cf. [ChaSMar])

$$f(z) = \frac{-\log(1 - az)}{\sqrt{-\log(1 - a^2)}}, \quad a \in (0, 1).$$

And, we may employ the Littlewood–Paley form of Green's Theorem (cf. [Koo, p. 224] to get

$$\int_{\mathbf{T}} \exp\left(\alpha|f(\zeta)|^2\right)|d\zeta|$$

$$= 2\pi + 4\alpha \int_{\mathbf{D}} (-\log|z|)\left(1 + \alpha|f(z)|^2\right)\left(\exp\left(\alpha|f(z)|^2\right)\right)|f'(z)|^2 dm(z)$$

hence proving (ii) — see [PaVu].

The proofs of Theorems 7.2.4 and 7.2.5 are taken from [Mar] which extracts the technique in both Adams' Lemma in [Ad1] and Moser's Theorem in [Mos]. Here it is also worth pointing out that [ChaSMar] provides a capacity-free proof via the well-known Beurling inequality:

$$|\{\zeta \in \mathbf{T} : |f(\zeta)| > t\}| \le \exp(-t^2 + 1) \quad \text{for} \quad t > 0 \quad \text{and} \quad \|f\|_{D}^2 \le \pi.$$

Furthermore, [Es] shows some sharp inequalities on the uniform harmonic majorants that extend Chang–Marshall's Theorem. In addition, some weighted area integral estimates of the exponential type for the diagonal Besov spaces can be found in [BuFeVu]. Theorem 7.2.8 reveals the law of the iterated logarithm for the Bloch space — see also [Mak] for more information.

Note 7.5.3. Section 7.3 is an adaptation of [Pa2]. It is worth comparing Theorem 7.3.1 with [Xi3, Theorem 4.4.2] — both may be regarded as mutually reversed propositions in some sense.

Note 7.5.4. Section 7.4 is taken from the major part of [GoMa]. For the Bloch part of Theorem 7.4.1 see also [Gir3]. Beyond the radial growth of derivatives we can say something on the radial variation of functions in $\mathcal{Q}_p$. For $f \in \mathcal{H}$ and $\zeta \in \mathbf{T}$ let

$$V(f, \zeta) = \int_0^1 |f'(r\zeta)| dr$$

be the radial variation of f along the segment from 0 to $\zeta \in \mathbf{T}$. Then the exceptional set of f is defined by

$$E(f) = \{\zeta \in \mathbf{T} : V(f, \zeta) = \infty\}.$$

As an immediate consequence of Theorem 7.4.2, we get that if $f \in \mathcal{Q}_p$, $p \in (0, 1)$, then $|E(f)| = 0$. But this result cannot extend to $p \ge 1$ since

$$f(z) = \sum_{k=1} k^{-1} z^{2^k}$$

is an element of $\mathcal{Q}_1 = \mathcal{BMOA}$ and has the property

$$V(f, \zeta) = \infty \quad \text{for all} \quad \zeta \in \mathbf{T}.$$

On the other hand, according to Beurling–Zygmund's Theorem: if $f \in \mathcal{D}^{2,p}$, $p \in [0,1)$, then $cap_p(E(f)) = 0$ (cf. [Beu] for $p = 0$ and [KahSa, Chapter 4]), we can trivially get

$$f \in \mathcal{Q}_p, \ p \in [0,1) \Rightarrow cap_p(E(f)) = 0.$$

In addition to this, we can also derive that if $f \in \mathcal{Q}_p$ where $p \in [0,1)$, then $\lim_{r \to 1}(1-r)|f'(r\zeta)| = 0$ for every $\zeta \in \mathbf{T}$ except perhaps for a set with p-capacity 0 – see [GoMa, Section 4]. Here, it is worthy of special remark that [JoMu] gives an affirmative answer to Anderson's conjecture: for any conformal map $f \in \mathcal{H}$ there is $\zeta \in \mathbf{T}$ such that

$$\int_0^1 |f''(r\zeta)|\, dr < \infty.$$

Chapter 8

Holomorphic Q-Classes on Hyperbolic Riemann Surfaces

Up to this point we have dealt with many essential properties of the holomorphic Q functions on the open unit disk which is the canonical example of all hyperbolic Riemann surfaces. We broaden our perspective now to consider their generalizations on hyperbolic Riemann surfaces. Since this study is far from complete, the very first step is for us to introduce such a $\mathcal{Q}_p$ class, naturally and geometrically. This idea leads to the main topic of this chapter and will be detailed through the following five sections:

- Basics about Riemann Surfaces;
- Area and Seminorm Inequalities;
- Intermediate Setting – BMOA Class;
- Sharpness;
- Limiting Case – Bloch Classes.

8.1 Basics about Riemann Surfaces

In complex analysis, a Riemann surface is a one-dimensional complex manifold. Riemann surfaces may be thought of as deformed versions of the complex plane — locally near every point they look like patches of the complex plane, but the global topology can be quite different. For example, they can look like a sphere or a torus or a couple of sheets glued together. Below is the precise definition.

Definition 8.1.1. A Riemann surface is a connected Hausdorff space $\mathbf{X}$ together with a collection of charts $\{\mathbf{U}_\alpha, z_\alpha\}$ satisfying the following three conditions:

(i) The $\mathbf{U}_\alpha$ form an open covering of $\mathbf{X}$.

(ii) Each z_α is a homeomorphic mapping of $\mathbf{U}_\alpha$ onto an open subset of $\mathbf{C}$.

(iii) If $\mathbf{U}_\alpha \cap \mathbf{U}_\beta \neq \emptyset$, then $z_{\alpha\beta} = z_\beta \circ z_\alpha^{-1}$ is holomorphic on $z_\alpha(\mathbf{U}_\alpha \cap \mathbf{U}_\beta)$.

In order to get a better understanding of Riemann surfaces, we consider some typical examples as follows.

Example 8.1.2.   (i) $\mathbf{C}$ is the most trivial Riemann surface. The identity map $z_\alpha(z) = z$ defines a chart for $\mathbf{C}$, but also the conjugate map $w_\beta(z) = \bar{z}$ defines another chart on $\mathbf{C}$. These two charts are not compatible, so this endows $\mathbf{C}$ with two distinct Riemann surface structures.

(ii) In an analogous fashion, every open subset of $\mathbf{C}$ can be viewed as a Riemann surface in a natural way. More generally, every open subset of a Riemann surface is a Riemann surface.

(iii) Let $\mathbf{S} = \mathbf{C} \cup \{\infty\}$ and let $z_\alpha(z) = z$ where $z \in \mathbf{S} \setminus \{\infty\}$ and $w_\beta(z) = 1/z$ where $z \in \mathbf{S} \setminus \{0\}$ and $1/\infty$ is defined to be 0. Then we obtain two compatible charts, making $\mathbf{S}$ into a Riemann surface which is called the Riemann sphere because it can be interpreted as wrapping $\mathbf{C}$ around the sphere. Unlike $\mathbf{C}$, it is compact.

On the other hand, it is worth remarking a few more facts. First of all, the system $\{\mathbf{U}_\alpha, z_\alpha\}$ defines a conformal structure on $\mathbf{X}$, and if it is understood which conformal structure we are referring to, then we will directly speak of the Riemann surface $\mathbf{X}$. Secondly, the topology of $\mathbf{X}$ is completely determined via the mappings $\{z_\alpha\}$, and a point $u \in \mathbf{U}_\alpha$ is uniquely determined by the complex number $z_\alpha(u)$ — due to this, z_α is regarded as a local variable. The subscript is often dropped and $z(u)$ is identified with u — for example, $\{z \in \mathbf{C} : |z - z_0| < r\}$ can refer either to an open disk on $\mathbf{C}$ or to its inverse image on $\mathbf{X}$. Thirdly, the identification of a point on the Riemann surface with the corresponding value of a local variable produces no difficulty whenever one deals with concepts such as holomorphic/harmonic/subharmonic functions and analytic arcs that are invariant under conformal mappings. Finally, we want to say that the main point of Riemann surfaces is that holomorphic functions may be defined between them. So Riemann surfaces are nowadays considered the natural setting for studying the global behavior of these functions, especially multi-valued functions such as the logarithm. In the meantime, we know that geometrical facts about Riemann surfaces are as nice as possible, and they often provide the intuition and motivation for generalizations to other curves, manifolds or varieties

Here we recall the following definition.

Definition 8.1.3.   (i) A mapping $f : \mathbf{X} \mapsto \mathbf{Y}$ between two Riemann surfaces $\mathbf{X}$ and $\mathbf{Y}$ is called holomorphic/harmonic/subharmonic provided for every chart $\{\mathbf{U}_\alpha, z_\alpha\}$ of $\mathbf{X}$ and every chart $\{\mathbf{V}_\beta, w_\beta\}$ of $\mathbf{Y}$, the map $w_\beta \circ f \circ z_\alpha^{-1}$ is holomorphic/harmonic/subharmonic as a function from $\mathbf{C}$ to $\mathbf{C}$ wherever it is defined. In particular, if $\mathbf{Y} = \mathbf{C}$, then the corresponding f is called a holomorphic/harmonic/subharmonic function on $\mathbf{X}$.

(ii) Two Riemann surfaces **X** and **Y** are called conformally equivalent provided there exists a bijective holomorphic function from **X** to **Y** whose inverse is also holomorphic.

The well-known uniformization theorem for Riemann surfaces says that every simply connected Riemann surface is conformally equivalent to **C** or to the Riemann sphere $\mathbf{C}^e$ or to the open disk **D**. From this, Riemann surfaces can be classified into compact and noncompact ones. Traditionally, a noncompact Riemann surface is also said to be open. But, the open Riemann surfaces can be compactified by adding a single point — the ideal boundary: if u_n lies outside any given compact set for all sufficiently large $n \in \mathbf{N}$, then we say the sequence $\{u_n\}$ of points tends to ∞. To establish a further classification of Riemann surfaces, we need to discuss the existence of the Green function on a Riemann surface via the so-called Perron family of subharmonic functions. To be more specific, given a Riemann surface **X**, let $\mathcal{F}$ be a family of real-valued subharmonic functions on **X** satisfying the following two conditions:

(i) If $f_1, f_2 \in \mathcal{F}$ then $\max\{f_1, f_2\} \in \mathcal{F}$.

(ii) Let **J** be a Jordan region on **X**. Assume $f \in \mathcal{F}$ and suppose F is a real-valued harmonic function on **J** with the same boundary values as f and $F = f$ on $\mathbf{X} \setminus \mathbf{J}$. Then $F \in \mathcal{F}$.

Such an $\mathcal{F}$ is called a Perron family on **X**. The essential property of the Perron families is that $\sup\{f : f \in \mathcal{F}\}$ is either harmonic or identical with ∞.

Given a point $x_0 \in \mathbf{X}$, let z be a local variable at x with $z(x_0) = 0$ and $\mathcal{F}_{x_0}$ be the family of real-valued functions f with the following three properties:

(i) f is defined and subharmonic on $\mathbf{X} \setminus \{x_0\}$;

(ii) f is identically 0 outside a compact set;

(iii) $\limsup_{x \to x_0} \big(f(x) + \log|z(x)|\big) < \infty$.

Clearly $\mathcal{F}_{x_0}$ is a Perron family. Moreover, if

$$g_{\mathbf{X}}(x, x_0) = \sup\{f(x) : f \in \mathcal{F}_{x_0}\} < \infty, \quad x \in \mathbf{X},$$

then we say that **X** has the Green function $g_{\mathbf{X}}(x, x_0)$ with a pole at x_0. Clearly, this function is harmonic on $\mathbf{X} \setminus \{x_0\}$ and independent of the choice of local variable $z(x)$ at x_0. It is not hard to see that $\lim_{x \to x_0} g_{\mathbf{X}}(x, x_0) = \infty$ and so that $g_{\mathbf{X}}(x, x_0)$ is not constant. Note that if **X** is compact, then it has no Green function because otherwise it follows that $g_{\mathbf{X}}(x, x_0)$ would have a minimum, contradicting the previous properties of the Green function. So, a further classification of Riemann surfaces follows.

Definition 8.1.4. An open Riemann surface is called hyperbolic if it has the Green function; otherwise it is called parabolic.

For example, $\mathbf{D}$ is hyperbolic and $\mathbf{C}$ is parabolic. Of course, $\mathbf{S}$ is neither hyperbolic nor parabolic. Parabolic surfaces share many properties with compact Riemann surfaces — in particular, a positive harmonic function on a parabolic Riemann surface is a constant.

Given a hyperbolic Riemann surface $\mathbf{X}$ we can introduce two conformally invariant metrics. First, we can use the Green function to define the Robin function and hence the logarithmic capacitary density as follows: fix $z_0 \in \mathbf{X}$ and let

$$\gamma_{\mathbf{X}}(z_0) = \lim_{z \to z_0} \left(g_{\mathbf{X}}(z, z_0) + \log|z - z_0| \right),$$

where z and z_0 also represent the values of local variable at the points z and z_0. The number $\gamma(z_0)$ is the well-known Robin's constant at z_0 with respect to the local variable z. This generates a definition of the logarithmic capacitary density at z_0 via

$$c_{\mathbf{X}}(z_0) = \exp\left(-\gamma_{\mathbf{X}}(z_0)\right).$$

Naturally, $c_{\mathbf{X}}(z)|dz|$ is called the logarithmic capacity metric. Note that if $\mathbf{X} = \mathbf{D}$ then

$$c_{\mathbf{D}}(z)|dz| = (1 - |z|^2)^{-1}|dz|, \quad z \in \mathbf{D}.$$

This actually suggests a consideration of the hyperbolic or Poincaré metric on $\mathbf{X}$. To be more specific, let $\mathbf{X}$ and $\mathbf{Y}$ be two Riemann surfaces, and consider a holomorphic mapping $\tau : \mathbf{Y} \mapsto \mathbf{X}$. We say that τ is a local homeomorphism if every point on $\mathbf{Y}$ has a neighborhood $\mathbf{V}$ such that the restriction $\tau|_{\mathbf{V}}$ of τ to $\mathbf{V}$ is a homeomorphism. In this case, the pair $(\mathbf{Y}, \tau)$ is called a covering surface of $\mathbf{X}$, $\tau(y)$ is the projection of $y \in \mathbf{Y}$ and y is said to lie over $\tau(y)$. As is well known, a hyperbolic Riemann surface $\mathbf{X}$ can be modeled by a Fuchsian model $\mathbf{D}/Fuc(\mathbf{D})$ where $Fuc(\mathbf{D})$ is a Fuchsian group — a discrete subgroup of $Aut(\mathbf{D})$. In this case, the open unit disk $\mathbf{D}$ is called the universal covering surface of $\mathbf{X}$ and so there is a holomorphic mapping, i.e., universal covering mapping, τ from $\mathbf{D}$ onto $\mathbf{X}$. Now, the Poincaré metric density of a hyperbolic Riemann surface $\mathbf{X}$ is defined by

$$\lambda_{\mathbf{X}}(z) = \inf\{(1 - |w|^2)^{-1} : \tau(w) = z\},$$

which certainly does not depend on the choice of a universal covering mapping since $\tau \circ \gamma = \tau$ for all $\gamma \in Fuc(\mathbf{D})$. The hyperbolic metric $\lambda_{\mathbf{X}}(z)|dz|$ on $\mathbf{X}$ is real-analytic and has constant Gaussian curvature -4. It is the unique metric on $\mathbf{X}$ obeying

$$\tau^*\left(\lambda_{\mathbf{X}}(z)|dz|\right) = \lambda_{\mathbf{D}}(w)|dw| = (1 - |w|^2)^{-1}|dw|,$$

where the left-hand-side is the pull-back metric. Moreover, there is associated to $Fuc(\mathbf{D})$ the Poincaré normal polygon determined below. Fix $z_0 \in \mathbf{D}$ with $\gamma(z_0) \neq z_0$ unless $\gamma = id$ — the identity element of $Fuc(\mathbf{D})$. Recall that

$$d_{\mathbf{D}}(z_1, z_2) = \log\left(\frac{1 + |\sigma_{z_1}(z_2)|}{1 - |\sigma_{z_1}(z_2)|}\right)^{\frac{1}{2}}, \quad z_1, z_2 \in \mathbf{D}$$

is the hyperbolic distance on $\mathbf{D}$. And let

$$\Omega_{z_0} = \{z \in \mathbf{D} : \; d_{\mathbf{D}}(z, z_0) < d_{\mathbf{D}}(z, \gamma(z_0)) \quad \text{for all} \quad \gamma \in Fuc(\mathbf{D}) \setminus \{id\}\}$$

be a fundamental region for $Fuc(\mathbf{D})$ which is an open domain $\Omega \subset \mathbf{D}$ satisfying

$$\gamma(\Omega) \cap \Omega = \emptyset \quad \text{for all} \quad \gamma \in Fuc(\mathbf{D}) \setminus \{id\} \quad \text{and} \quad \mathbf{D} = \bigcup_{\gamma \in Fuc(\mathbf{D})} \gamma(\overline{\Omega}).$$

Nevertheless, the important properties are the facts that any fundamental region Ω is a copy of the hyperbolic Riemann surface $\mathbf{X}$, any universal covering mapping $\tau : \Omega \mapsto \mathbf{X}$ is surjective and the boundary $\partial\Omega$ has zero area. Furthermore, the Green function of Ω is defined by the well-known Myrberg's formula:

$$g_\Omega(z, w) = g_{\mathbf{X}}(\tau(z), \tau(w)) = \sum_{\gamma \in Fuc(\mathbf{D})} g_{\mathbf{D}}(z, \gamma(w)), \quad z, w \in \Omega.$$

8.2 Area and Seminorm Inequalities

A careful look at the a priori estimate in Example 1.1.1, along with the foregoing introduction of Green functions, leads to a consideration of the so-called holomorphic Q classes over general hyperbolic Riemann surfaces.

Definition 8.2.1. Let $p \in [0, \infty)$ and $\mathbf{X}$ be a hyperbolic Riemann surface. We say that a holomorphic function $f : \mathbf{X} \mapsto \mathbf{C}$ belongs to $\mathcal{Q}_p(\mathbf{X})$ provided

$$\|f\|_{\mathcal{Q}_p(\mathbf{X})} = \sup_{w \in \mathbf{X}} E_p(f, w, \mathbf{X}) < \infty,$$

where

$$E_p(f, w, \mathbf{X}) = \left(\frac{i}{2} \int_{\mathbf{X}} |f'(z)|^2 \big(g_{\mathbf{X}}(z, w)\big)^p dz \wedge d\bar{z} \right)^{\frac{1}{2}}$$

and $dz \wedge d\bar{z} = 2idxdy$ for a local variable $z = x + iy$. In particular, we denote $\mathcal{Q}_0(\mathbf{X})$ and $\mathcal{Q}_1(\mathbf{X})$ by $\mathcal{D}(\mathbf{X})$ — the Dirichlet space on $\mathbf{X}$ and $\mathcal{BMOA}(\mathbf{X})$ — the BMOA space on $\mathbf{X}$, respectively.

Here and henceforth, the derivative of a function f on $\mathbf{X}$ is determined in the following way: if $z \in \mathbf{X}$ and t is a local parameter in a neighborhood of z such that $t(z) = 0$, then $f'(z)$ represents the usual derivative of $f \circ t^{-1}$ at 0.

A very basic problem is to clarify the relationship among these Q classes. To do so, we need to have a deep understanding of the isoperimetric inequality for Riemann surfaces.

Lemma 8.2.2. *Suppose* $\mathbf{X}$ *is a Riemann surface. Assume* Ω *is a relatively compact subdomain of* $\mathbf{X}$ *and has a piecewise smooth boundary* $\partial\Omega$. *If* $f : \mathbf{X} \mapsto \mathbf{C}$ *is holomorphic, then the following isoperimetric inequality holds:*

$$\frac{i}{2} \int_\Omega |f'(z)|^2 dz \wedge d\bar{z} \leq (4\pi)^{-1} Length\big(f(\partial\Omega)\big) = (4\pi)^{-1} \int_{f(\partial\Omega)} |dz|.$$

Proof. Assuming $f(\partial\Omega)$ is positively oriented, we use the Green formula to derive

$$\int_\Omega |f'(z)|^2 dz \wedge d\bar{z} = \int_\Omega df(z) \wedge \overline{df(z)} = \int_{f(\partial\Omega)} f(z)\overline{df(z)} = \int_{f(\partial\Omega)} w d\bar{w}.$$

Suppose that $\mathrm{Length}(C)$ stands for the length of a curve C. By the polygonal approximation, we may assume that $f(\partial\Omega)$ is a polygonal closed curve. Accordingly, $f(\partial\Omega)$ may be written as a sum of finite polygonal Jordan closed curves $\{C_j\}_{j=1}^n$ so that

$$\sum_{j=1}^n \mathrm{Length}(C_j) \le \mathrm{Length}(C) \quad \text{and} \quad \int_C w d\bar{w} = \sum_{j=1}^n \int_{C_j} w d\bar{w}.$$

If Ω_j represents the Jordan region bounded by C_j, then it follows from the Green formula and the elementary isoperimetric inequality that

$$\left| \int_{C_j} w d\bar{w} \right| = \left| \int_{\Omega_j} dw d\bar{w} \right| \le (2\pi)^{-1}\mathrm{Length}(C_j), \quad j = 1, \ldots, n.$$

Consequently,

$$\begin{aligned}
\frac{i}{2}\int_\Omega |f'(z)|^2 dz \wedge d\bar{z} &\le 2^{-1}\sum_{j=1}^n \left| \int_{C_j} w d\bar{w} \right| \\
&\le (4\pi)^{-1}\sum_{j=1}^n \mathrm{Length}(C_j) \\
&\le (4\pi)^{-1}\mathrm{Length}(C),
\end{aligned}$$

as desired. $\square$

This lemma is used to derive the following result.

Theorem 8.2.3. *Let $\mathbf{X}$ be a hyperbolic Riemann surface and $g_{\mathbf{X}}(z, z_0)$ be its Green function with pole at $z_0 \in \mathbf{X}$. For $t \ge 0$ let $\mathbf{X}_t = \{z \in \mathbf{X} : g_{\mathbf{X}}(z, z_0) > t\}$. If $f : \mathbf{X} \mapsto \mathbf{C}$ is holomorphic, then the function*

$$A(t) = \frac{i}{2}\int_{\mathbf{X}_t} |f'(z)|^2 dz \wedge d\bar{z}$$

has the following three properties:

(i) *$A(t)$ is continuous and decreasing with increasing $t \ge 0$.*

(ii) *$e^{2s}A(s) \le e^{2t}A(t)$ for $s \ge t \ge 0$.*

(iii) *For $p, t \in [0, \infty)$,*

$$\left(E_p(f, z_0, \mathbf{X}_t)\right)^2 = \int_0^\infty A(s)ds^p = -\int_t^\infty s^p dA(s).$$

Here and hereafter, the right-hand-side integral (as well as its like) will be understood under Riemann–Stieltjes integration.

Proof. First, we prove that those three properties are valid under the assumption that $\mathbf{X}$ is a finite Riemann surface and f is holomorphic on the closure $\bar{\mathbf{X}}$ of $\mathbf{X}$. For the sake of simplicity, we write

$$G(z) = g_{\mathbf{X}}(z, z_0) + i g_{\mathbf{X}}^*(z, z_0) = g(z) + i g^*(z),$$

where $g^*(z) = g_{\mathbf{X}}^*(z, z_0)$ is the harmonic conjugate of $g(z) = g_{\mathbf{X}}(z, z_0)$ and is locally defined up to an additive constant. For $t \geq 0$, set

$$\Gamma_t = \{z \in \bar{\mathbf{X}} : g_{\mathbf{X}}(z, z_0) = t\} \quad \text{and} \quad A_0(t) = \int_{\Gamma_t} \left|\frac{f'(z)}{G'(z)}\right|^2 \frac{\partial g(z)}{\partial n} ds.$$

Here and afterwards, ds denotes the arc length measure on Γ_t and $\frac{\partial g(z)}{\partial n}$ the derivative in the inner normal direction with respect to $\mathbf{X}_t$.

Under $p, t \in [0, \infty)$, we make the substitutions

$$dg(z) \wedge dg^*(z) = |G'(z)|^2 dz \wedge d\bar{z}, \quad \lambda = g(z) \quad \text{and} \quad d\lambda = \left(\frac{\partial g(z)}{\partial n}\right) dn$$

to get a chain of equalities:

$$\frac{i}{2} \int_{\mathbf{X}_t} |f'(z)|^2 \left(g_{\mathbf{X}}(z, z_0)\right)^p dz \wedge d\bar{z}$$

$$= \frac{i}{2} \int_{\mathbf{X}_t} \left|\frac{f'(z)}{G'(z)}\right|^2 \left(g_{\mathbf{X}}(z, z_0)\right)^p |G'(z)|^2 dz \wedge d\bar{z}$$

$$= \frac{i}{2} \int_{\mathbf{X}_t} \left|\frac{f'(z)}{G'(z)}\right|^2 \left(g_{\mathbf{X}}(z, z_0)\right)^p dg(z) \wedge dg^*(z)$$

$$= \int_t^\infty \int_{\Gamma_\lambda} \left|\frac{f'(z)}{G'(z)}\right|^2 \left(g_{\mathbf{X}}(z, z_0)\right)^p \left(\frac{\partial g(z)}{\partial n}\right)^2 ds dn$$

$$= \int_t^\infty \int_{\Gamma_\lambda} \left|\frac{f'(z)}{G'(z)}\right|^2 \left(g_{\mathbf{X}}(z, z_0)\right)^p \left(\frac{\partial g(z)}{\partial n}\right) ds d\lambda$$

$$= \int_t^\infty \lambda^p A_0(\lambda) d\lambda.$$

In particular,

$$A(t) = \int_t^\infty A_0(\lambda) d\lambda, \quad t \geq 0.$$

Thus, (i) and (iii) follow.

As before, let $\text{Length}\big(f(\Gamma_t)\big)$ be the length of $f(\Gamma_t)$. Then by the Cauchy–Schwarz inequality, we get

$$
\begin{aligned}
\Big(\text{Length}\big(f(\Gamma_t)\big)\Big)^2 &\leq \left(\int_{\Gamma_t} |f'(z)||dz|\right)^2 \\
&= \left(\int_{\Gamma_t} \left|\frac{f'(z)}{G'(z)}\right|\left(\frac{\partial g(z)}{\partial n}\right)ds\right)^2 \\
&\leq \left(\int_{\Gamma_t} \left|\frac{f'(z)}{G'(z)}\right|^2 \left(\frac{\partial g(z)}{\partial n}\right)ds\right)\left(\int_{\Gamma_t} \left(\frac{\partial g(z)}{\partial n}\right)ds\right) \\
&= 2\pi A_0(t).
\end{aligned}
$$

Note that the isoperimetric inequality in Lemma 8.2.2 ensures

$$
A(t) \leq (4\pi)^{-1}\Big(\text{Length}\big(f(\Gamma_t)\big)\Big)^2.
$$

So it follows that

$$
2A(t) \leq A_0(t) \quad \text{and} \quad \frac{d\big(e^{2t}A(t)\big)}{dt} = e^{2t}\big(2A(t) - A_0(t)\big) \leq 0,
$$

and consequently, $e^{2t}A(t)$ decreases with increasing $t \geq 0$. This proves (ii).

Now, let us handle the case that $\mathbf{X}$ is a general hyperbolic Riemann surface. For this, let $\{\mathbf{X}^j\}_{j\in\mathbf{N}}$ be a regular exhaustion of $\mathbf{X}$ and $A_j(t)$ be defined for each $\mathbf{X}^j$. If f and z_0 are given as above, then from what we have verified for finite Riemann surfaces, we can read off

$$
s \geq t \geq 0 \Rightarrow e^{2s}A_j(s) \leq e^{2t}A_j(t)
$$

and

$$
p, t \geq 0 \Rightarrow \frac{i}{2}\int_{\mathbf{X}_t^j} |f'(z)|^2 \big(g_{\mathbf{X}^j}(z, z_0)\big)^p dz \wedge d\bar{z} = -\int_t^\infty \lambda^p dA_j(\lambda).
$$

Here, $g_{\mathbf{X}^j}(z, z_0)$ is the Green function of $\mathbf{X}^j$ and

$$
\mathbf{X}_t^j = \{z \in \mathbf{X} : g_{\mathbf{X}^j}(z, z_0) > t\}.
$$

Clearly, $\lim_{j\to\infty} A_j(t) = A(t)$ for every $t \geq 0$. Accordingly, (ii) follows. Furthermore, via integrating by parts we derive that for $p, t \geq 0$,

$$
\Big(E_p\big(f, z_0, \mathbf{X}_t^j\big)\Big)^2 = t^p A_j(t) + p\int_t^\infty \lambda^{p-1} A_j(\lambda)d\lambda.
$$

This, together with letting $j \to \infty$ and integrating by parts once again, produces:

$$
\Big(E_p\big(f, z_0, \mathbf{X}_t\big)\Big)^2 = t^p A(t) + p\int_t^\infty \lambda^{p-1} A(\lambda)d\lambda = -\int_t^\infty \lambda^p dA(\lambda), \quad t > 0,
$$

thereupon proving (iii) for $p \geq 0$ and $t > 0$. Finally, letting $t \to 0$ completes the argument. □

To compare one Q class with another, we need the following elementary lemma.

Lemma 8.2.4. *Given a nonnegative function $A(t)$ on $(0, \infty)$ with the following two properties:*

(i) *$A(t)$ is continuous and decreasing with increasing $t > 0$.*

(ii) *$e^{2t_2} A(t_2) \leq e^{2t_1} A(t_1)$ when $t_2 \geq t_1 > 0$.*

For $p, t \in [0, \infty)$ let $B_p(t) = -\int_t^\infty s^p dA(s)$. If $p \geq q \geq 0$, then

$$B_p(0) \leq \frac{2^q \Gamma(p+1)}{2^p \Gamma(q+1)} B_q(0)$$

for which

$$B_p(0) = \frac{2^q \Gamma(p+1)}{2^p \Gamma(q+1)} B_q(0) < \infty$$

if and only if

$$A(0) = \lim_{t \to 0} A(t) < \infty \quad and \quad A(t) = e^{-2t} A(0), \quad t > 0.$$

Proof. Suppose $p \geq q \geq 0$ and $B_q(0) < \infty$. Integrating by parts and using the above condition (ii), we have that for $t > 0$,

$$
\begin{aligned}
B_q(t) &= t^q A(t) + q \int_t^\infty s^{q-1} A(s) ds \\
&\leq A(t) \left(t^q + q e^{2t} \int_t^\infty s^{q-1} e^{-2s} ds \right) \\
&= 2A(t) e^{2t} \int_t^\infty s^q e^{-2s} ds.
\end{aligned}
$$

Using again (ii) above, we also find

$$\frac{dB_q(t)}{B_q(t)} \leq -\frac{2t^q A(t) dt}{B_q(t)} \leq -\frac{t^q e^{-2t} dt}{\int_t^\infty s^q e^{-2s} ds}, \quad t > 0.$$

Integrating this inequality from 0 to t, we get a further inequality

$$B_q(t) \leq \frac{2^{q+1} B_q(0)}{\Gamma(q+1)} \int_t^\infty s^q e^{-2s} ds, \quad t \geq 0.$$

Consequently,

$$
\begin{aligned}
B_p(0) &= -\int_0^\infty t^{p-q} t^q dA(t) \\
&= (p-q)\int_0^\infty t^{p-q-1} B_q(t) dt \\
&\leq \frac{2^{q+1}(p-q)B_q(0)}{\Gamma(q+1)} \int_0^\infty t^{p-q-1} \int_t^\infty s^q e^{-2s} ds\, dt \\
&= \frac{2^q \Gamma(p+1)}{2^p \Gamma(q+1)} B_q(0).
\end{aligned}
$$

This proves the first part of the lemma. Below is the argument for the second part.
Suppose

$$
A(0) = \lim_{t\to 0} A(t) \quad \text{and} \quad A(t) = A(0)e^{-2t} \quad \text{for} \quad t > 0.
$$

Then

$$
B_p(0) = 2^{-p}\Gamma(p+1)A(0) \quad \text{for} \quad p \geq 0,
$$

and hence the desired equality holds with $p \geq q \geq 0$. Conversely, suppose

$$
B_p(0) = \frac{2^q \Gamma(p+1)}{2^p \Gamma(q+1)} B_q(0) < \infty, \quad p > q \geq 0,
$$

but there exists $s_0 > t_0 > 0$ such that $A(s_0) < e^{-2(s_0-t_0)} A(t_0)$. Then the continuity of $A(\cdot)$ implies that there is a $\delta > 0$ such that

$$
A(s_0) < e^{-2(s_0-t_0)} A(t) \quad \text{as} \quad t \in (t_0 - \delta, t_0].
$$

Hence by the condition (ii),

$$
A(s) < e^{-2(s-t)} A(t) \quad \text{as} \quad t \in (t_0 - \delta, t_0] \quad \text{and} \quad s \geq s_0.
$$

As a consequence, we obtain

$$
B_q(t) < 2A(t)e^{2t} \int_t^\infty s^q e^{-2s} ds \quad \text{as} \quad t \in (t_0 - \delta, t_0],
$$

and so

$$
B_p(0) < \frac{2^q \Gamma(p+1)}{2^p \Gamma(q+1)} B_q(0) < \infty, \quad p > q \geq 0.
$$

Clearly, this contradicts the hypothesis. □

Now, we can establish the following area and seminorm inequalities associated with the holomorphic Q classes over hyperbolic Riemann surfaces.

Theorem 8.2.5. *Let $0 \leq q < p$ and $\mathbf{X}$ be a hyperbolic Riemann surface with $w \in \mathbf{X}$. Then:*

(i) $E_p(f, w, \mathbf{X}) \leq \left(\dfrac{2^q \Gamma(p+1)}{2^p \Gamma(q+1)} \right)^{\frac{1}{2}} E_q(f, w, \mathbf{X})$ *for any holomorphic* $f : \mathbf{X} \mapsto \mathbf{C}$.

(ii) $\mathcal{Q}_q(\mathbf{X}) \subseteq \mathcal{Q}_{\mathbf{p}}(\mathbf{X})$ *with*

$$\|f\|_{\mathcal{Q}_p(\mathbf{X})} \leq \left(\frac{2^q \Gamma(p+1)}{2^p \Gamma(q+1)} \right)^{\frac{1}{2}} \|f\|_{\mathcal{Q}_q(\mathbf{X})}, \quad f \in \mathcal{Q}_q(\mathbf{X}).$$

Proof. It follows immediately from Theorem 8.2.3 and Lemma 8.2.4. $\square$

8.3   Intermediate Setting – BMOA Class

The case $q = 0$ and $p = 1$ of Theorem 8.2.5 is of independent interest since it is a critical case and indeed has a root in the BMOA-theory on hyperbolic Riemann surfaces.

Theorem 8.3.1. *Let $\mathbf{X}$ be a hyperbolic Riemann surface and $f : \mathbf{X} \mapsto \mathbf{C}$ be holomorphic.*

(i) *Given $w \in \mathbf{X}$, let $h_{\mathbf{X}}(z, w)$ be the least harmonic majorant of $|f(z) - f(w)|^2$ on $\mathbf{X}$, then*

$$\left(E_1(f, w, \mathbf{X}) \right)^2 = \frac{\pi}{2} h_{\mathbf{X}}(w, w).$$

(ii) $\|f\|_{\mathcal{BMOA}(\mathbf{X})} \leq \sqrt{\dfrac{Area\big(f(\mathbf{X})\big)}{2}}.$

Proof. (i) Taking a regular exhaustion of $\mathbf{X}$ into account, we see that it is enough to check the formula under the hypothesis that $\mathbf{X}$ is a finite Riemann surface and f is holomorphic on the closure $\overline{\mathbf{X}}$ of $\mathbf{X}$.

Suppose $\mathbf{W}$ is the interior of a compact bordered Riemann surface $\overline{\mathbf{W}}$ and $\partial \mathbf{W}$ stands for the boundary of $\mathbf{W}$. If u and v are $\mathcal{C}^2$ functions on $\overline{\mathbf{W}}$, then the Green formula just says

$$\frac{i}{2} \int_{\mathbf{W}} (v \Delta u - u \Delta v) dz \wedge d\bar{z} = \int_{\partial \mathbf{W}} \left(u \frac{\partial v}{\partial n} - v \frac{\partial u}{\partial n} \right) ds,$$

where Δ, $\partial / \partial n$ and ds stand for the Laplacian, differentiation in the inner normal direction and arc length measure on $\partial \mathbf{W}$, respectively.

Applying this formula to $u(z) = |f(z) - f(w)|^2$ and $v(z) = g_{\mathbf{X}}(z, w)$ in the domain obtained by removing from $\mathbf{X}$ a small open disk centered at w, shrinking the disk, and noticing $\Delta u(z) = 4|f'(z)|^2$ and $\Delta v(z) = 0$, we derive the formula in (i) since the integrals along the boundary of the small disk approach zero and

$$\int_{\partial \mathbf{X}} \frac{\partial g_{\mathbf{X}}(z, w)}{\partial n} ds = 2\pi.$$

(ii) The argument for the inequality is split into three steps.

Step 1. Suppose $\tau_1 : \mathbf{D} \mapsto \mathbf{X}$ is a universal covering mapping. Then the least harmonic majorant of a subharmonic function is preserved by τ_1, that is to say, if h is the least harmonic majorant of a subharmonic function u on $\mathbf{X}$, then $h \circ \tau_1$ is the least harmonic majorant of $u \circ \tau_1$ on $\mathbf{D}$. This general fact implies

$$\|f\|_{\mathcal{BMOA}(\mathbf{X})} = \|f \circ \tau_1\|_{\mathcal{BMOA}(\mathbf{D})}.$$

In fact, for $\zeta, \eta \in \mathbf{D}$ let $h_{\mathbf{D}}(\zeta, \eta)$ be the least harmonic majorant of the subharmonic function $|f \circ \tau_1(\zeta) - f \circ \tau_1(\eta)|^2$ on $\mathbf{D}$, then the above-mentioned fact gives

$$h_{\mathbf{D}}(\zeta, \eta) = h_{\mathbf{X}}\big(\tau_1(\zeta), \tau_1(\eta)\big), \quad \zeta, \eta \in \mathbf{D}.$$

This, along with (i), implies

$$
\begin{aligned}
2\pi^{-1}\|f\|^2_{\mathcal{BMOA}(\mathbf{X})} &= \sup_{w \in \mathbf{X}} h_{\mathbf{X}}(w, w) \\
&= \sup_{\eta \in \mathbf{D}} h_{\mathbf{X}}\big(\tau_1(\eta), \tau_1(\eta)\big) \\
&= \sup_{\eta \in \mathbf{D}} h_{\mathbf{D}}(\eta, \eta) \\
&= 2\pi^{-1}\|f \circ \tau_1\|^2_{\mathcal{BMOA}(\mathbf{D})},
\end{aligned}
$$

as required.

Step 2. In order to verify the inequality in (ii), we choose a universal covering mapping τ_2 from $\mathbf{D}$ onto the range set $f(\mathbf{X})$ of f, and show

$$\|f\|_{\mathcal{BMOA}(\mathbf{X})} \leq \|\tau_2\|_{\mathcal{BMOA}(\mathbf{D})}.$$

Clearly, it suffices to prove

$$\|f \circ \tau_1\|_{\mathcal{BMOA}(\mathbf{D})} \leq \|\tau_2\|_{\mathcal{BMOA}(\mathbf{D})}.$$

Note that $\tau_1^{-1} \circ (f \circ \tau_1)$ has a single-valued branch, say, ϕ on $\mathbf{D}$. So, it follows that $\phi \in \mathcal{H}^\infty$ with $\|\phi\|_{\mathcal{H}^\infty} \leq 1$ and so that $f \circ \tau_1 = \tau_2 \circ \phi$ is valid. With this, we see that if $b = \phi(a)$, then $f \circ \tau_1(a) = \tau_2(b)$. Now, let $h_{\mathbf{D},1}(z, a)$ and $h_{\mathbf{D},2}(z, a)$ be the least harmonic majorants of

$$u(z) = |f \circ \tau_1(z) - f \circ \tau_1(a)|^2 \quad \text{and} \quad v(z) = |\tau_2(z) - \tau_2(b)|^2$$

respectively. Then $u(z)$ is subordinate to $v(z)$ and hence by the Littlewood Subordination Principle [Hil, p. 421],

$$h_{\mathbf{D},1}(z,a) \leq h_{\mathbf{D},2}(\phi(z),b), \quad z \in \mathbf{D}.$$

Using this inequality as well as (i), we derive

$$\frac{2}{\pi}\|f \circ \tau_1\|^2_{\mathcal{BMOA}(\mathbf{D})} = \sup_{a \in \mathbf{D}} h_{\mathbf{D},1}(a,a) \leq \sup_{b \in \mathbf{D}} h_{\mathbf{D},2}(b,b) \leq \frac{2}{\pi}\|\tau_2\|^2_{\mathcal{BMOA}(\mathbf{D})},$$

hence reaching the desired inequality.

Step 3. To complete the proof, let I be the identity mapping from $f(\mathbf{X})$ onto itself. Substituting $f(\mathbf{X})$ and I for $\mathbf{X}$ and f in Step 1, we get

$$\|I\|_{\mathcal{BMOA}(f(\mathbf{X}))} = \|I \circ \tau_2\|_{\mathcal{BMOA}(\mathbf{D})} = \|\tau_2\|_{\mathcal{BMOA}(\mathbf{D})}.$$

This, together with Step 2 and Theorem 8.2.5, implies

$$
\begin{aligned}
\|f\|^2_{\mathcal{BMOA}(\mathbf{X})} &\leq \|\tau_2\|^2_{\mathcal{BMOA}(\mathbf{D})} \\
&= \|I\|^2_{\mathcal{BMOA}(f(\mathbf{X}))} \\
&\leq 2^{-1}\|I\|^2_{\mathcal{D}(f(\mathbf{X}))} \\
&= 2^{-1}\text{Area}\big(I(f(\mathbf{X}))\big) \\
&= 2^{-1}\text{Area}(f(\mathbf{X})),
\end{aligned}
$$

as required. $\qquad\qquad\square$

To handle equality of Theorem 8.3.1 and to check when

$$\|f\|_{\mathcal{BMOA}(\mathbf{X})} = \sqrt{2^{-1}}\|f\|_{\mathcal{D}(\mathbf{X})}$$

happens, we need the following lemma.

Lemma 8.3.2. *Let $\mathbf{X} \subset \mathbf{C}$ be a hyperbolic Riemann surface with finite area.*

(i) *Given $w \in \mathbf{X}$, let $1_{\mathbf{X}}(z,w)$ be the least harmonic majorant of $|z-w|^2$ on $\mathbf{X}$, then*

$$1_{\mathbf{X}}(w,w) \leq \frac{\text{Area}(\mathbf{X})}{\pi}$$

for which equality holds if and only if $\mathbf{X}$ is a domain of the form $\mathbf{X} = \{z \in \mathbf{C} : |z-w| < r\} \setminus \mathbf{E}$, where $r > 0$ and $\mathbf{E}$ is a closed set with $\text{cap}_0(\mathbf{E}) = 0$.

(ii) *If $I(z) = z$ for $z \in \mathbf{X}$, then*

$$\|I\|_{\mathcal{BMOA}(\mathbf{X})} \leq \sqrt{\frac{\text{Area}(\mathbf{X})}{2}}$$

for which equality holds if and only if $\mathbf{X}$ is a domain of the form $\mathbf{X} = \{z \in \mathbf{C} : |z-w| < r\} \setminus \mathbf{E}$, where $r > 0$ and $\mathbf{E}$ is a closed set with $\text{cap}_0(\mathbf{E}) = 0$.

Proof. (i) Step 1. Suppose that $\mathbf{X}$ is a domain in $\mathbf{C}$ with smooth boundary $\partial\mathbf{X}$. Using the Green formula, we derive

$$4\mathrm{Area}(\mathbf{X}) = \int_{\mathbf{X}} \Delta(|z - w|^2) dm(z) = \int_{\partial\mathbf{X}} \frac{\partial|z - w|^2}{\partial n} ds.$$

If $u(z) = |z - w|^2 \exp\left(2g_{\mathbf{X}}(z, w)\right)$, then another application of the Green formula produces

$$
\begin{aligned}
\int_{\mathbf{X}} \Delta u(z) dm(z) &= \int_{\partial\mathbf{X}} \frac{\partial u(z)}{\partial n} ds \\
&= \int_{\partial\mathbf{X}} \frac{\partial|z - w|^2}{\partial n} ds + 2 \int_{\partial\mathbf{X}} |z - w|^2 \frac{\partial g_{\mathbf{X}}(z, w)}{\partial n} ds \\
&= 4\mathrm{Area}(\mathbf{X}) - 4\pi 1_{\mathbf{X}}(w, w),
\end{aligned}
$$

and so the inequality in (i) follows.

Step 2. If $\mathbf{X}$ is a general domain in $\mathbf{C}$, then there is a smooth exhaustion $\{\mathbf{X}_j\}$ of $\mathbf{X}$ such that $w \in \cap_{j=1}^{\infty}\mathbf{X}_j$. Writing $1_{\mathbf{X}_j}(z, w)$ and $u_j(z)$ respectively for the functions for $\mathbf{X}_j$ that correspond to $1_{\mathbf{X}}(z, w)$ and $u(z)$ as in Step 1, we get from Step 1 that

$$1_{\mathbf{X}_j}(w, w) + \frac{1}{4\pi} \int_{\mathbf{X}_j} \Delta u_j(z) dm(z) = \frac{\mathrm{Area}(\mathbf{X}_j)}{\pi}, \quad j \in \mathbf{N}.$$

Letting $j \to \infty$ and using the Lebesgue Monotone Convergence Theorem and Fatou's Lemma, we find

$$1_{\mathbf{X}}(w, w) + \frac{1}{4\pi} \int_{\mathbf{X}} \Delta u(z) dm(z) = \frac{\mathrm{Area}(\mathbf{X})}{\pi},$$

thus establishing the required inequality.

Step 3. We check the equivalent condition for equality. Assuming

$$1_{\mathbf{X}}(z, w) = \frac{\mathrm{Area}(\mathbf{X})}{\pi},$$

we work out from Step 2 that $\Delta u(z) = 0$ on $\mathbf{X}$ and so that

$$g_{\mathbf{X}}(z, w) = \log \frac{r}{|z - w|}$$

for some constant $r > 0$. This forces that $\mathbf{X}$ is a domain as determined in (i). Conversely, if $\mathbf{X}$ is such a domain, then $1_{\mathbf{X}}(z, w) = r^2$ on $\mathbf{X}$ and hence equality occurs since $cap_0(\mathbf{E}) = 0$ implies $\mathrm{Area}(\mathbf{E}) = m(\mathbf{E}) = 0$.

(ii) The inequality is a special case of Theorem 8.3.1 (ii). It remains to check the equality condition. On the one hand, if

$$\|I\|_{\mathcal{BMOA}(\mathbf{X})} = \sqrt{\frac{\mathrm{Area}(\mathbf{X})}{2}},$$

then there is a sequence of points $\{w_j\}$ in $\mathbf{X}$ such that

$$\|I\|^2_{\mathcal{BMOA}(\mathbf{X})} = \frac{\pi}{2} \lim_{j \to \infty} 1_{\mathbf{X}}(w_j, w_j).$$

Without loss of generality, we may assume that w_j is convergent to some point $w \in \overline{\mathbf{X}}$ and $g_{\mathbf{X}}(z, w_j)$ converges uniformly on compact subsets of $\mathbf{X} \setminus \{w\}$. If

$$v_j(z) = |z - w_j|^2 \exp\left(2g_{\mathbf{X}}(z, w_j)\right),$$

then an application of Fatou's Lemma yields

$$\frac{2}{\pi}\|I\|^2_{\mathcal{BMOA}(\mathbf{X})} + \frac{1}{4\pi} \int_{\mathbf{X}} \lim_{j \to \infty} \Delta v_j(z) dm(z) \le \frac{\text{Area}(\mathbf{X})}{\pi}.$$

Consequently, $\lim_{j \to \infty} \Delta v_j(z) = 0$ for all $z \in \mathbf{X}$ thanks to the above-assumed equality. Therefore we obtain

$$\lim_{j \to \infty} g_{\mathbf{X}}(z, w_j) = \log \frac{r}{|z - w|}$$

for some constant $r > 0$. This in turns implies that $\mathbf{X}$ is contained in the open disk $\{z \in \mathbf{C} : |z - w| < r\}$. If $\mathbf{E} = \{z \in \mathbf{C} : |z - w| < r\} \setminus \mathbf{X}$ and $cap_0(\mathbf{E}) > 0$, then we can choose a small number $\epsilon > 0$ such that

$$cap_0(\mathbf{E}_\epsilon) = cap_0\left(\mathbf{E} \cap \{z \in \mathbf{C} : |z - w| \ge \epsilon\}\right) > 0.$$

Letting $\mathbf{X}_\epsilon = \mathbf{X} \cup \{z \in \mathbf{C} : |z - w| < \epsilon\}$, we read

$$\mathbf{X} \subseteq \mathbf{X}_\epsilon \subseteq \{z \in \mathbf{C} : |z - w| < r\}$$

and consequently,

$$\log \frac{r}{|z - w|} = \lim_{j \to \infty} g_{\mathbf{X}}(z, w_j) \le \lim_{j \to \infty} g_{\mathbf{X}_\epsilon}(z, w_j) = g_{\mathbf{X}_\epsilon}(z, w) \le \log \frac{r}{|z - w|}.$$

This tells us that $cap_0(\mathbf{E}_\epsilon) = 0$, a contradiction. Accordingly, $cap_0(\mathbf{E}) = 0$ and $\mathbf{X}$ must be a domain as described in the equality condition of (ii). Conversely, if $\mathbf{X}$ is as above, then it is easy to calculate

$$\lim_{z \to w} 1_{\mathbf{X}}(z, z) = r^2 \quad \text{as well as} \quad \|I\|_{\mathcal{BMOA}(\mathbf{X})} = \sqrt{\frac{\text{Area}(\mathbf{X})}{2}}.$$

We are done. $\qquad \square$

Theorem 8.3.3. *Let $\mathbf{X}$ be a hyperbolic Riemann surface and $f : \mathbf{X} \mapsto \mathbf{C}$ be holomorphic.*

(i) *If*

$$\|f\|_{\mathcal{BMOA}(\mathbf{X})} = \sqrt{\frac{Area(f(\mathbf{X}))}{2}},$$

then $f(\mathbf{X}) = \{w \in \mathbf{C} : |w - c| < r\} \setminus \mathbf{E}$ *with* $c \in \mathbf{C}$, $r > 0$ *and* $cap_0(\mathbf{E}) = 0$, *but not conversely.*

(ii) $\|f\|_{\mathcal{BMOA}(\mathbf{X})} = \dfrac{\|f\|_{\mathcal{D}(\mathbf{X})}}{\sqrt{2}}$ *if and only if there is a simply connected hyperbolic Riemann surface* $\mathbf{Y}$ *such that* $\mathbf{X}$ *equals* $\mathbf{Y}$ *set-minus at most* $\mathbf{E}$ *with* $cap_0(\mathbf{E}) = 0$ *and* f *is extended to a conformal mapping from* $\mathbf{Y}$ *onto an open disk in* $\mathbf{C}$.

Proof. (i) This follows from the argument for Theorem 8.3.1 (ii) and the equality condition of Lemma 8.3.2 (ii). Concerning the converse, we select the conformal mapping $f_1 : \mathbf{D} \mapsto \mathbf{D} \cap \{z \in \mathbf{C} : \Re z > -2^{-1}\}$ with $f_1(0) = 0$ and put $f = f_1^2$. Clearly, we have $f(\mathbf{D}) = \mathbf{D}$ which yields $Area(f(\mathbf{D})) = \pi$. However, for any $\epsilon \in (0, 1)$ there is a $\delta \in (0, 1)$ such that

$$|w| \leq \epsilon \Rightarrow \int_{\mathbf{T}} |f \circ \sigma_{-w}(\zeta)|^2 |d\zeta| \leq 1 - \delta \quad \text{and} \quad |w| > \epsilon \Rightarrow |f(w)| \geq \delta.$$

Accordingly, it follows from Theorem 8.3.1 (i) and a routine computation that if $h_{\mathbf{X}}(z, w)$ still stands for the least harmonic majorant of $|f(z) - f(w)|^2$ on $\mathbf{D}$, then

$$
\begin{aligned}
2\pi^{-1}\|f\|^2_{\mathcal{BMOA}(\mathbf{X})} &= \sup_{w \in \mathbf{X}} h_{\mathbf{X}}(w, w) \\
&= \sup_{w \in \mathbf{X}} \left((2\pi)^{-1} \int_{\mathbf{T}} |f \circ \sigma_{-w}(\zeta)|^2 |d\zeta| - |f(w)|^2 \right) \\
&\leq 1 - \delta,
\end{aligned}
$$

and hence the desired equality is not valid.

(ii) If

$$\|f\|_{\mathcal{BMOA}(\mathbf{X})} = \sqrt{2^{-1}}\|f\|_{\mathcal{D}(\mathbf{X})},$$

then by Theorem 8.3.1 we have

$$\|f\|_{\mathcal{BMOA}(\mathbf{X})} = \sqrt{2^{-1}Area(f(\mathbf{X}))} = \sqrt{2^{-1}}\|f\|_{\mathcal{D}(\mathbf{X})}.$$

This, together with (i), implies that $f(\mathbf{X}) = \{w \in \mathbf{C} : |w - c| < r\} \setminus \mathbf{E}$ for some $c \in \mathbf{C}$, some $r > 0$ and some $\mathbf{E}$ with $cap_0(\mathbf{E}) = 0$, but also that f is univalent on $\mathbf{X}$. Note that $cap_0(\mathbf{E}) = 0$ implies that $\mathbf{E}$ is removable for $\mathcal{BMOA}$ functions — see also the forthcoming Lemma 8.4.1. So f and $\mathbf{X}$ must be as above. The converse follows from the fact that $h_{\mathbf{X}}(w, w)$ tends to r^2 as $w \to \partial \mathbf{X}$ and $f(w) \to c$. $\square$

8.4 Sharpness

In this section, we determine when two equalities in Theorem 8.2.5 occur. To this end, we recall that the Green function is a conformal invariant — this means that if $f : \mathbf{X} \mapsto \mathbf{Y}$ is a conformal mapping from the hyperbolic Riemann surface $\mathbf{X}$ to the other hyperbolic Riemann surface $\mathbf{Y}$, then

$$g_{\mathbf{Y}}\big(f(z), f(w)\big) = g_{\mathbf{X}}(z, w) \quad \text{for all} \quad (z, w) \in \mathbf{X} \times \mathbf{X}.$$

This result can be generalized in the following form.

Lemma 8.4.1. *Let* $\mathbf{X}, \mathbf{Y} \subset \mathbf{C}$ *be two hyperbolic Riemann surfaces and* $f : \mathbf{X} \mapsto \mathbf{Y}$ *holomorpic. If* f *is injective and* $\mathbf{Y} \setminus f(\mathbf{X})$ *is a closed set of logarithmic capacity zero, then*

$$g_{\mathbf{Y}}\big(f(z), f(w)\big) = g_{\mathbf{X}}(z, w) \quad \text{for all} \quad (z, w) \in \mathbf{X} \times \mathbf{X}.$$

Conversely, if there are two points $z, w \in \mathbf{X}$ *such that* $g_{\mathbf{Y}}\big(f(z), f(w)\big) = g_{\mathbf{X}}(z, w)$, *then* f *is injective and* $\mathbf{Y} \setminus f(\mathbf{X})$ *is a closed set of logarithmic capacity zero.*

Proof. It suffices to check the second part because the first part follows from the fact that $\mathbf{Y}$ and $\mathbf{Y} \setminus f(\mathbf{X})$ has the same Green function in this situation.

In so doing, we assume $g_{\mathbf{Y}}\big(f(z), f(w)\big) = g_{\mathbf{X}}(z, w)$ for distinct points $z, w \in \mathbf{X}$. Then the Lindelöf Principle gives

$$g_{\mathbf{Y}}\big(f(\zeta), f(w)\big) = \sum_{f(\eta)=f(w)} n(\eta, f) g_{\mathbf{X}}(\zeta, w) + u_{f(w)}(\zeta), \quad \zeta \in \mathbf{X},$$

where $n(\eta, f)$ is the order of f at $\eta \in \mathbf{X}$ and u_y is a nonnegative harmonic function attached to $y \in \mathbf{Y}$ and defined on $\mathbf{X}$. For $\zeta = z$ the last formula yields

$$g_{\mathbf{Y}}\big(f(z), f(w)\big) \geq n(w, f) g_{\mathbf{X}}(z, w) + u_{f(w)}(z) \geq g_{\mathbf{X}}(z, w).$$

This, along with the assumption, implies the equality holds throughout, so

$$n(w, f) = 1, \quad u_{f(w)}(z) = 0 \quad \text{and} \quad f(\zeta) \neq f(w) \quad \text{for all} \quad \zeta \in \mathbf{X} \setminus \{w\}.$$

Consequently,

$$g_{\mathbf{Y}}\big(f(\zeta), f(w)\big) = g_{\mathbf{X}}(\zeta, w) \quad \text{for all} \quad \zeta \in \mathbf{X}.$$

Furthermore, by the symmetry of the Green function, it follows that $f(\eta) \neq f(\zeta)$ for all $\eta \in \mathbf{X} \setminus \{\zeta\}$, $n(\eta, f) = 1$ and

$$g_{\mathbf{Y}}\big(f(\eta), f(\zeta)\big) = g_{\mathbf{X}}(\eta, \zeta) \quad \text{for all} \quad \eta \in \mathbf{X}.$$

But this also holds for all $\zeta \in \mathbf{X}$. As a consequence, f is injective. Moreover, from the last equation we conclude that $u_{f(\omega)} = 0$ for all $\omega \in \mathbf{X}$. According to the dichotomy associated with the Lindelöf Principle (cf. [Hei]), we must have $u_y(x) = 0$ for all $x \in \mathbf{X}$ and $y \in \mathbf{Y}$. Since f is injective, we find that

$$\sup_{y \in \mathbf{Y}} \sum_{f(x)=y} n(x, f) = 1,$$

and so that

$$\mathbf{Y} \setminus f(\mathbf{X}) = \left\{ y \in \mathbf{Y} : \sum_{f(x)=y} n(x, f) < 1 \right\}$$

is a closed set of logarithmic capacity zero.   □

In what follows, we say that a subset $\mathbf{E}$ of a simply connected hyperbolic Riemann surface $\mathbf{X}$ is of logarithmic capacity zero provided the conformal map from $\mathbf{X}$ onto $\mathbf{D}$ sends $\mathbf{E}$ onto a set of logarithmic capacity zero on $\mathbf{D}$.

Theorem 8.4.2. *Let $\mathbf{X}$ be a hyperbolic Riemann surface, $w \in \mathbf{X}$, and f be a non-constant holomorphic function on $\mathbf{X}$. Then the following two statements are equivalent:*

(i) $E_p(f, w, \mathbf{X}) = \left(\dfrac{2^q \Gamma(p+1)}{2^p \Gamma(q+1)} \right)^{\frac{1}{2}} E_q(f, w, \mathbf{X}) < \infty, \quad p > q \geq 0.$

(ii) $\mathbf{X}$ *is obtained from a simply connected hyperbolic Riemann surface $\mathbf{Y}$ by removing at most a set of logarithmic capacity zero and f is extended onto a conformal mapping from $\mathbf{Y}$ to an open disk in $\mathbf{C}$ centered at $f(w) \in \mathbf{C}$.*

Proof. Suppose (i) is valid. Upon continuing to use the notation in Theorem 8.2.3 and Lemma 8.2.4, we find that $A(t) = e^{-2t} A(0)$ for any $t > 0$, and that $G(z)$ and $F(z) = \exp\left(-G(z) \right)$ are multiple-valued but $\Re G(z)$ and $|F(z)|$ are single-valued. Then $F(z)$ is single-valued near w and $F'(w) = \exp\left(-\gamma_{\mathbf{X}}(w) \right) \neq 0$ where $\gamma_{\mathbf{X}}(w)$ is the Robin constant. Taking $\xi = \zeta + i\eta = F(z)$ as a local variable near w, we have $\xi(w) = 0$ and $g_{\mathbf{X}}(z, w) = -\log|\xi|$ near w. Upon writing $f(\xi) = \sum_{j=0}^{\infty} b_j \xi^j$ where $b_0 = f(z_0)$, we get that for a large $t > 0$,

$$e^{-2t} A(0) = A(t) = \int_{|\xi| < e^{-t}} |f'(\xi)|^2 dm(\xi) = \pi |b_1|^2 e^{-2t} + 2\pi |b_2|^2 e^{-4t} + \cdots,$$

so that $A(0) = \pi |b_1|^2$ and $b_j = 0$ for $j - 1 \in \mathbf{N}$. This actually means that $f(\xi) = b_0 + b_1 \xi = b_0 + b_1 F(z)$ near z_0 and then $f = b_0 + b_1 F$ — where F is single-valued on $\mathbf{X}$ thanks to the uniqueness theorem. Consequently, $f(\mathbf{X})$ is contained in the open disk $\mathbf{D}(b_0, |b_1|)$ centered at $b_0 = f(z_0)$ with radius $|b_1| = \sqrt{A(0)/\pi}$ due to $|F| < 1$ on $\mathbf{X}$.

Using Theorem 8.3.1 and making a simple calculation with $e^{-2t}A(0) = A(t)$ and Theorem 8.2.3 (iii), we derive

$$\begin{aligned}
2^{-1}\mathrm{Area}\big(f(\mathbf{X})\big) &\geq \big(E_1(f, z_0, \mathbf{X})\big)^2 \\
&= 2^{-1}A(0) \\
&= \frac{i}{4}\int_{\mathbf{X}}|f'(z)|^2 dz \wedge d\bar{z} \\
&\geq 2^{-1}\mathrm{Area}\big(f(\mathbf{X})\big),
\end{aligned}$$

hence finding that inequalities become equalities, and consequently f is univalent with f^{-1} as its inverse.

Note again that the Green function is conformal invariant. So the above argument yields

$$\begin{aligned}
g_{f(\mathbf{X})}(w, b_0) &= g_{f(\mathbf{X})}\big(w, f(z_0)\big) \\
&= g_{\mathbf{X}}\big(f^{-1}(w), z_0\big) \\
&= -\log|F\big(f^{-1}(w)\big)| \\
&= \log\frac{|b_1|}{|w - b_0|} \\
&= g_{\mathbf{D}(b_0,|b_1|)}(w, 0), \quad w \in f(\mathbf{X}).
\end{aligned}$$

These equalities plus Lemma 8.4.1 imply that $\mathbf{D}(b_0, |b_1|) \setminus f(\mathbf{X})$ is of logarithmic capacity 0. This proves (ii).

Conversely, if (ii) holds. Then we may assume that f conformally maps $\mathbf{X}$ into an open disk $\mathbf{D}\big(f(z_0), r_0\big)$ with radius r_0 about $f(z_0)$ and $\mathbf{D}\big(f(z_0), r_0\big) \setminus f(\mathbf{X})$ has logarithmic capacity 0. Thus, for $p \geq 0$ we have

$$\begin{aligned}
\big(E_p(f, z_0, \mathbf{X})\big)^2 &= \int_{f(\mathbf{X})}\Big(g_{f(\mathbf{X})}(\xi, f(z_0))\Big)^p dm(\xi) \\
&= \int_{\mathbf{D}\big(f(z_0),r_0\big)}\Big(g_{\mathbf{D}\big(f(z_0),r_0\big)}(\xi, f(z_0))\Big)^p dm(\xi) \\
&= \int_{\mathbf{D}(b_0,|b_1|)}\Big(\log\frac{r_0}{|\xi - f(z_0)|}\Big)^p dm(\xi) \\
&= \frac{\pi\Gamma(p+1)}{2^p},
\end{aligned}$$

hence verifying (i). So, we are done. $\qquad\square$

To deal with equality in Theorem 8.2.3 (ii), we need to have an insight into the Green function near the ideal boundary of a Riemann surface.

Lemma 8.4.3. *Let* $\mathbf{X}$ *be a hyperbolic Riemann surface and* $\partial\mathbf{X}$ *its ideal boundary; let* $z_j \to \partial\mathbf{X}$ *and* $g_{\mathbf{X}}(z, z_j)$ *converge to a harmonic function* $g_{\mathbf{X}}(z)$ *locally uniformly in* $\mathbf{X}$, *and let* $f : \mathbf{X} \mapsto \mathbf{C}$ *be a nonconstant holomorphic function with* $\|f\|_{\mathcal{Q}_0(\mathbf{X})} < \infty$. *For* $t \geq 0$ *and* $j \in \mathbf{N}$, *define*

$$\mathbf{X}_t^j = \{z \in \mathbf{X} :\ g_{\mathbf{X}}(z, z_j) > t\} \quad and \quad A_j(t) = \frac{i}{2} \int_{\mathbf{X}_t^j} |f'(z)|^2\, dz \wedge d\bar{z}.$$

Then the following statements are valid:

Case 1: $g_{\mathbf{X}} \equiv 0$ *implies*

(i) $\lim_{j \to \infty} A_j(t) = 0$ *for* $t > 0$;

(ii) $\lim_{j \to \infty} E_p(f, z_j, \mathbf{X}) = 0$ *for* $p > 0$.

Case 2: $g_{\mathbf{X}} \not\equiv 0$ *implies that if*

$$p, t \geq 0, \quad \mathbf{X}_t = \{z \in \mathbf{X} :\ g_{\mathbf{X}}(z) > t\} \quad and \quad A(t) = \frac{i}{2} \int_{\mathbf{X}_t} |f'(z)|^2 dz \wedge d\bar{z},$$

then

(iii) $\lim_{j \to \infty} A_j(t) = A(t)$ *for* $t \geq 0$;

(iv) $A(t)$ *is a continuous and decreasing function with* $e^{2t_2} A(t_2) \leq e^{2t_1} A(t_1)$ *for* $t_2 \geq t_1 \geq 0$;

(v)

$$\lim_{j \to \infty} \int_0^\infty t^p dA_j(t) = \int_0^\infty t^p dA(t)$$

and

$$\lim_{j \to \infty} E_p(f, z_j, \mathbf{X}) = \frac{i}{2} \int_{\mathbf{X}} |f'(z)|^2 \big(g_{\mathbf{X}}(z)\big)^p dz \wedge d\bar{z}.$$

Proof. Case 1: $g_{\mathbf{X}} \equiv 0$.

Fix $t > 0$. Since $\|f\|_{\mathcal{Q}_0(\mathbf{X})} < \infty$, for any $\epsilon > 0$ there is a compact set $\mathbf{K} \subset \mathbf{X}$ such that

$$\frac{i}{2} \int_{\mathbf{X} \backslash \mathbf{K}} |f'(z)|^2 dz \wedge d\bar{z} < \epsilon.$$

Using the hypothesis that $g_{\mathbf{X}}(z, z_n)$ tends to 0 uniformly on $\mathbf{K}$, we get an $N \in \mathbf{N}$ such that

$$n - N \in \mathbf{N}, \quad z \in \mathbf{K} \Rightarrow g_{\mathbf{X}}(z, z_n) \leq t.$$

Thus, $n - N \in \mathbf{N}$ implies $\mathbf{X}_t^n \subseteq \mathbf{X} \setminus \mathbf{K}$ and then

$$A_n(t) = \frac{i}{2} \int_{\mathbf{X}_t^n} |f'(z)|^2\, dz \wedge d\bar{z} \leq \frac{i}{2} \int_{\mathbf{X} \backslash \mathbf{K}} |f'(z)|^2\, dz d\bar{z} < \epsilon,$$

proving (i).

To see (ii), we follow the second part of the argument for Theorem 8.2.3 to get

$$
\begin{aligned}
\lim_{j \to \infty} \left(E_p(f, z_j, \mathbf{X}_t^n) \right)^2 &= \lim_{j \to \infty} \left(t^p A_j(t) + p \int_t^\infty \lambda^{p-1} A_j(\lambda) d\lambda \right) \\
&\leq 2 e^{2t} \lim_{j \to \infty} A_j(t) \int_t^\infty s^p e^{-2s} ds \\
&\leq 2 \Gamma(p+1) e^{2t} \lim_{j \to \infty} A_j(t) \\
&= 0.
\end{aligned}
$$

In the meantime, for $p, t > 0$,

$$
\left(E_p(f, z_j, \mathbf{X} \setminus \mathbf{X}_t^j) \right)^2 \leq \frac{t^p i}{2} \int_{\mathbf{X} \setminus \mathbf{X}_t^j} |f'(z)|^2 \, dz \wedge d\bar{z} < t^p \|f\|_{\mathcal{Q}_0(\mathbf{X})}^2.
$$

The foregoing two estimates imply (ii).

Case 2: $g \not\equiv 0$.

In this case, we have $g_{\mathbf{X}}(z) > 0$ for $z \in \mathbf{X}$, but we claim that $g_{\mathbf{X}}(z)$ takes arbitrarily small value. Assume, on the contrary, that there is a positive number t_0 such that $g_{\mathbf{X}}(z) \geq t_0 > 0$ for $z \in \mathbf{X}$. Then, for $0 < t < t_0$, we have

$$
\mathbf{X}_t \subseteq \mathbf{X} = \liminf_{j \to \infty} \mathbf{X}_t^j = \bigcup_{k=1}^\infty \left(\bigcap_{j=k}^\infty \mathbf{X}_t^j \right).
$$

Consequently,

$$
\|f\|_{\mathcal{Q}_0(\mathbf{X})}^2 \leq \liminf_{j \to \infty} A_j(t) \leq \limsup_{j \to \infty} A_j(t) \leq \|f\|_{\mathcal{Q}_0(\mathbf{X})}^2,
$$

a contradiction to the inequality $e^{2t_2} A_j(t_2) \leq e^{2t_1} A_j(t_1)$ for $t_2 \geq t_1 \geq 0$. This verifies the claim. Accordingly, $g_{\mathbf{X}}(z)$ cannot be a constant and $A(t)$ is continuous. Note that if $\overline{\mathbf{X}_t}$ stands for the closure of $\mathbf{X}_t$ for $t > 0$, then

$$
\mathbf{X}_t \subseteq \liminf_{j \to \infty} \mathbf{X}_t^j = \bigcup_{k=1}^\infty \left(\bigcap_{j=k}^\infty \mathbf{X}_t^j \right) \subseteq \limsup_{j \to \infty} \mathbf{X}_t^j = \bigcap_{k=1}^\infty \left(\bigcup_{j=k}^\infty \mathbf{X}_t^j \right) \subseteq \overline{\mathbf{X}_t}
$$

implies that $A(t) \leq \liminf_{j \to \infty} A_j(t)$ and

$$
\limsup_{j \to \infty} A_j(t) \leq \frac{i}{2} \int_{\overline{\mathbf{X}_t}} |f'(z)|^2 \, dz \wedge d\bar{z} = \frac{i}{2} \int_{\mathbf{X}_t} |f'(z)|^2 \, dz \wedge d\bar{z}.
$$

Thus, $\lim_{j \to \infty} A_j(t) = A(t)$ for which

$$
e^{2t_2} A(t_2) \leq e^{2t_1} A(t_1) \quad \text{for} \quad t_2 \geq t_1 \geq 0.
$$

This verifies (iii).

Of course, (iv) follows from (iii).

It is clear that (v) is valid for $p = 0$. When $p > 0$, integrating by parts twice, we obtain

$$\int_0^\infty t^p dA_j(t) = p \int_0^\infty t^{p-1} A_j(t) dt \to p \int_0^\infty t^{p-1} A(t) dt = \int_0^\infty t^p dA(t),$$

hence obtaining the first limit of (v). To check the second one, let $p > 0$ and $\epsilon > 0$ be given. Then Theorem 8.2.5 (i) gives

$$\big(E_p(f, z_j, \mathbf{X})\big)^2 \le 2^{-p} \Gamma(p+1) \big(E_0(f, z_j, \mathbf{X})\big)^2 < \infty.$$

And hence Fatou's Lemma asserts that

$$\frac{i}{2} \int_{\mathbf{X}} |f'(z)|^2 \big(g_{\mathbf{X}}(z)\big)^p \, dz \wedge d\bar{z} < \infty$$

so that, by the absolute continuity of integrals, there is a $\delta > 0$ such that

$$\frac{i}{2} \int_{\mathbf{E}} |f'(z)|^2 \, dz \wedge d\bar{z} < \delta \Rightarrow \frac{i}{2} \int_{\mathbf{E}} |f'(z)|^2 \big(g_{\mathbf{X}}(z)\big)^p \, dz \wedge d\bar{z} < \epsilon/3$$

holds for any measurable set $\mathbf{E}$. Taking $t > 0$ such that

$$2A(t) < \delta \quad \text{and} \quad \|f\|^2_{\mathcal{Q}_0(\mathbf{X})} \left(\int_t^\infty s^p e^{-2s} ds + 2t^p e^{-2t} \right) < \epsilon/3,$$

and letting

$$\mathbf{Y}_j^t = \bigcup_{k=j}^\infty \mathbf{X}_t^k \quad \text{and} \quad \mathbf{Y}_t = \limsup_{j \to \infty} \mathbf{X}_t^j = \cap_{j=1}^\infty \mathbf{Y}_j^t,$$

we have

$$A(t) = \frac{i}{2} \int_{\mathbf{Y}_t} |f'(z)|^2 \, dz \wedge d\bar{z} = \frac{i}{2} \lim_{j \to \infty} \int_{\mathbf{Y}_j^t} |f'(z)|^2 \, dz \wedge d\bar{z}$$

hence getting an $N \in \mathbf{N}$ such that

$$\frac{i}{2} \int_{\mathbf{Y}_N} |f'(z)|^2 \, dz \wedge d\bar{z} < 2A(t) < \delta$$

implies

$$\frac{i}{2} \int_{\mathbf{Y}_N} |f'(z)|^2 \big(g_{\mathbf{X}}(z)\big)^p \, dz \wedge d\bar{z} < \epsilon/3.$$

From the preceding estimates we can conclude that for $j > N$,

$$
\begin{aligned}
\left(E_p(f, z_j, \mathbf{Y}_N)\right)^2 &= \left(E_p(f, z_j, \mathbf{X}_t^j)\right)^2 + \left(E_p(f, z_j, \mathbf{Y}_N \setminus \mathbf{X}_t^j)\right)^2 \\
&< A_j(t)e^{2t} \int_t^\infty s^p e^{-2s}\, ds + 2t^p A(t) \\
&\leq \|f\|_{\mathcal{Q}_0(\mathbf{X})}^2 \left(\int_t^\infty s^p e^{-2s}\, ds + 2t^p e^{-2t} \right) \\
&< \epsilon/3.
\end{aligned}
$$

On the other hand, since $g_\mathbf{X}(z, z_j) \leq t$ for $j \geq N$ and $z \in \mathbf{X} \setminus \mathbf{Y}_N$, by the Bounded Convergence Theorem, there is an $N_1 > N$ such that as $j > N_1$,

$$
\left| \left(E_p(f, z_j, \mathbf{X} \setminus \mathbf{Y}_N)\right)^2 - \frac{i}{2} \int_{\mathbf{X} \setminus \mathbf{Y}_N} |f'(z)|^2 (g_\mathbf{X}(z))^p\, dz \wedge d\bar{z} \right| < \epsilon/3.
$$

Putting together the last three $\epsilon/3$-estimates and using the triangle inequality, we conclude that if $j > N_1$, then

$$
\left| \left(E_p(f, z_j, \mathbf{X})\right)^2 - \frac{i}{2} \int_{\mathbf{X}} |f'(z)|^2 (g_\mathbf{X}(z))^p\, dz \wedge d\bar{z} \right| < \epsilon,
$$

and the second limit of (iv) follows. $\qquad\square$

The following assertion characterizes when equality in Theorem 8.2.5 (ii) happens.

Theorem 8.4.4. Let $\mathbf{X}$ be a hyperbolic Riemann surface, $z_0 \in \mathbf{X}$, and $f : \mathbf{X} \mapsto \mathbf{C}$ a nonconstant holomorphic function with $\|f\|_{\mathcal{Q}_0(\mathbf{X})} < \infty$. Then the following two statements are equivalent:

(i) $\|f\|_{\mathcal{Q}_p(\mathbf{X})} = \left(\dfrac{2^q \Gamma(p+1)}{2^p \Gamma(q+1)} \right)^{\frac{1}{2}} \|f\|_{\mathcal{Q}_q(\mathbf{X})}, \quad p > q \geq 0.$

(ii) $\mathbf{X}$ is obtained from a simply connected hyperbolic Riemann surface $\mathbf{Y}$ by removing at most a set of logarithmic capacity zero and f is extended onto a conformal mapping from $\mathbf{Y}$ to an open disk in $\mathbf{C}$ centered at $f(w) \in \mathbf{C}$.

Proof. (i)$\Rightarrow$(ii) Suppose (i) is valid. Then there exists a sequence $\{z_j\}$ of points in $\mathbf{X}$, which converges to a point z_0 in either $\mathbf{X}$ or $\partial\mathbf{X}$, such that

$$
\lim_{j \to \infty} \left(E_p(f, z_j, \mathbf{X})\right)^2 = \|f\|_{\mathcal{Q}_p(\mathbf{X})}^2.
$$

If $z_0 \in \mathbf{X}$, then

$$
\left(E_p(f, z_0, \mathbf{X})\right)^2 = \|f\|_{\mathcal{Q}_p(\mathbf{X})}^2 = \frac{2^q \Gamma(p+1)}{2^p \Gamma(q+1)} \|f\|_{\mathcal{Q}_q(\mathbf{X})}^2,
$$

and hence

$$\left(E_p(f, z_0, \mathbf{X})\right)^2 = \frac{2^q \Gamma(p+1)}{2^p \Gamma(q+1)} \left(E_q(f, z_0, \mathbf{X})\right)^2$$

which produces (ii) thanks to Theorem 8.4.2.

If $z_0 \in \partial \mathbf{X}$, then there exists z_j in $\mathbf{X}$ such that $z_j \to z_0$ as $j \to \infty$. Since $\lim_{j \to \infty} \sup_{z \in \mathbf{E}} g_{\mathbf{X}}(z, z_j) < \infty$ for any compact set $\mathbf{E} \subset \mathbf{X}$, an application of the Montel Theorem produces a subsequence of $\{g_{\mathbf{X}}(z, z_j)\}$ converging to a harmonic function $g_{\mathbf{X}}(z)$ locally uniformly in $\mathbf{X}$. This $g_{\mathbf{X}}$ is not identical with 0. In fact, if not, then by Lemma 8.4.3 (ii), $\lim_{j \to \infty} E_p(f, z_j, \mathbf{X}) = 0$ and so $\|f\|_{\mathcal{Q}_p(\mathbf{X})} = 0$ which forces that f is a constant — this is a contradiction to the hypothesis.

Now, by Lemma 8.4.3, we have

$$\|f\|_{\mathcal{Q}_p(\mathbf{X})}^2 = -\lim_{j \to \infty} \int_0^\infty t^p dA_j(t) = -\int_0^\infty t^p dA(t).$$

Meanwhile, we also have

$$\|f\|_{\mathcal{Q}_q(\mathbf{X})}^2 \geq -\lim_{j \to \infty} \int_0^\infty t^q dA_j(t) = -\int_0^\infty t^q dA(t).$$

These, plus Lemma 8.4.1, yield

$$-\int_0^\infty t^p dA(t) \geq -\frac{2^q \Gamma(p+1)}{2^p \Gamma(q+1)} \int_0^\infty t^q dA(t) \geq -\int_0^\infty t^p dA(t),$$

and so equality holds and implies $A(t) = A(0)e^{-2t}$ for $t > 0$. Consequently,

$$\frac{A(0)}{2} \geq \left(E_1(f, z_0, \mathbf{X})\right)^2 \geq \lim_{j \to \infty} \int_0^\infty t dA_j(t) = \int_0^\infty t dA(t) = \frac{A(0)}{2}.$$

This implies (ii) thanks to Theorem 8.3.3.

(ii)⇒(i) If (ii) holds, then by the uniformization theorem we may assume that $\mathbf{X}$ is obtained from the unit disk by removing at most a set of logarithmic capacity zero and f is the identity. Now the calculation in Example 1.1.1 implies $\|f\|_{\mathcal{Q}_p(\mathbf{X})}^2 = 2^{-p} \pi \Gamma(p+1)$, giving (i). We are done.   □

Looking over Theorem 8.4.4, we have a natural question below: is the hypothesis $\|f\|_{\mathcal{Q}_0(\mathbf{X})} = \|f\|_{\mathcal{D}(\mathbf{X})} < \infty$ necessary? The forthcoming example shows that we cannot drop this condition at least for $p \geq 1$.

Example 8.4.5. Let $\{z_j\}$ be an increasing sequence on $[0, 1)$ with $z_1 = 0$ and $|\sigma_{z_{j-1}}(z_j)| = 1 - j^{-2}$ for $j - 1 \in \mathbf{N}$, and let $B(z) = \prod_{j=1}^\infty \sigma_{z_j}(z)$. Then

$$\|B\|_{\mathcal{D}} = \infty \quad \text{and} \quad \|B\|_{\mathcal{Q}_p} = \left(\pi 2^{-p} \Gamma(p+1)\right)^{\frac{1}{2}}, \quad p \geq 1$$

but Theorem 8.4.4 (ii) is not valid for such $\mathbf{D}$ and B.

Proof. Clearly, the above-defined function B is not conformal on $\mathbf{D}$. So Theorem 8.4.4 (ii) fails in this case. Now, for $j \in \mathbf{N}$ and $z \in \mathbf{D}$, let

$$B_j(z) = B\big(\sigma_{z_j}(z)\big) = z \prod_{k \neq j} \sigma_{z_k}\big(\sigma_{z_j}(z)\big).$$

Then, $B_j(0) = 0$ and $B_j(\mathbf{D}) \subset \mathbf{D}$ for $j \in \mathbf{N}$. A simple calculation gives

$$1 > B'_{2j+1}(0) \geq \big(1 - (2j+1)^{-2}\big)^{2j} \prod_{k > 2j+1} \big(1 - k^{-2}\big) \to 1 \quad \text{as} \quad j \to \infty,$$

and so $\lim_{j \to \infty} B'_{2j+1}(0) = 1$. This, along with a normal family argument and an application of the Schwarz Lemma, implies that $B_{2j+1}(z)$ converges z locally uniformly in $\mathbf{D}$. Now, for $p > 0$ and $0 < r < 1$, we have

$$\lim_{j \to \infty} \int_{|\sigma_{z_{2j+1}}(z)| < r} |B'(z)|^2 \big(g_\mathbf{D}(z, z_{2j+1})\big)^p dm(z) = \int_{|z| < r} (-\log|z|)^p \, dm(z).$$

This implies that if $p \geq 1$, then

$$\begin{aligned}
\big(\pi 2^{-p} \Gamma(p+1)\big)^{\frac{1}{2}} &\leq \|B\|_{\mathcal{Q}_p} \\
&\leq \big(2^{-p} \Gamma(p+1)\big)^{\frac{1}{2}} \|B\|_{\mathcal{BMOA}(\mathbf{D})} \\
&\leq \big(2^{-p} \Gamma(p+1)\big)^{\frac{1}{2}} \big(2^{-1} \text{Area}(f(\mathbf{D}))\big)^{\frac{1}{2}} \\
&\leq \big(\pi 2^{-p} \Gamma(p+1)\big)^{\frac{1}{2}}
\end{aligned}$$

and hence the equalities hold. Note also that B is an infinite Blaschke product. So it cannot belong to $\mathcal{D}$, i.e., $\|B\|_\mathcal{D} = \infty$. We are done. $\qquad \square$

8.5 Limiting Case – Bloch Classes

In this section we deal with the limiting case $p \to \infty$ of Theorem 8.2.3.

Lemma 8.5.1. *Given a nonnegative function $A(t)$ on $(0, \infty)$ with the following two properties:*

(i) *$A(t)$ is continuous and decreasing with increasing $t > 0$.*

(ii) *$e^{2t_2} A(t_2) \leq e^{2t_1} A(t_1)$ when $t_2 \geq t_1 > 0$.*

Furthermore, for $p, t \in [0, \infty)$ let $B_p(t) = -\int_t^\infty s^p dA(s)$. If there exists $p_0 \in [0, \infty)$ such that $B_{p_0}(0) < \infty$, then

$$\lim_{p \to \infty} \frac{2^p B_p(0)}{\Gamma(p+1)} = \lim_{t \to \infty} e^{2t} A(t).$$

Proof. By Lemma 8.2.4 we know that $B_{p_0}(0) < \infty$ ensures $B_p(0) < \infty$ as $p \geq p_0$. Moreover, integrating by parts twice and using the condition (ii) we have

$$
\begin{aligned}
B_p(0) &= p \int_0^\infty t^{p-1} A(t) dt \\
&= p \int_0^\infty e^{2t} A(t) t^{p-1} e^{-2t} dt \\
&= p e^{2t} A(t) \int_0^t s^{p-1} e^{-2s} ds \Big|_0^\infty - p \int_0^\infty \int_0^t s^{p-1} e^{-2s} ds\, d(e^{2t} A(t)) \\
&= \frac{\Gamma(p+1)}{2^p} \lim_{t\to\infty} e^{2t} A(t) - p \int_0^\infty \int_0^t s^{p-1} e^{-2s} ds\, d(e^{2t} A(t)).
\end{aligned}
$$

Therefore, it suffices to check the limit

$$
I_p = \frac{2^p p}{\Gamma(p+1)} \int_0^\infty \int_0^t s^{p-1} e^{-2s} ds\, d(e^{2t} A(t)) \to 0 \quad \text{as} \ \ p \to \infty.
$$

To this end, we observe that the condition (ii) yields that for any $\epsilon > 0$ there is a $t_0 > 0$ such that $-\epsilon < \int_{t_0}^\infty d(e^{2t} A(t)) \leq 0$, and consequently,

$$
I_{p,1} = \frac{2^p p}{\Gamma(p+1)} \int_{t_0}^\infty \int_0^t s^{p-1} e^{-2s} ds\, d(e^{2t} A(t)) \geq \int_{t_0}^\infty d(e^{2t} A(t)) > -\epsilon.
$$

At the same time, integrating by parts we get

$$
\begin{aligned}
I_{p,2} &= \frac{2^p p}{\Gamma(p+1)} \int_0^{t_0} \int_0^t s^{p-1} e^{-2s} ds\, d(e^{2t} A(t)) \\
&\geq \frac{2^p}{\Gamma(p+1)} \int_0^{t_0} t^p\, d(e^{2t} A(t)) \\
&\geq \frac{2^p}{\Gamma(p+1)} \int_0^{t_0} t^p e^{2t}\, dA(t) \\
&\geq \frac{2^p e^{2t_0} t_0^{p-p_0}}{\Gamma(p+1)} \int_0^{t_0} t^{p_0}\, dA(t) \\
&\geq -\frac{2^p e^{2t_0} t_0^{p-p_0} B_{p_0}(0)}{\Gamma(p+1)} \\
&\to \ 0 \quad \text{as} \ p \to \infty.
\end{aligned}
$$

Accordingly, $-2\epsilon < I_p \leq 0$ for sufficiently large p. This proves the lemma. $\square$

Theorem 8.5.2. *Let* $\mathbf{X}$ *be a hyperbolic Riemann surface,* $z_0 \in \mathbf{X}$, *and* $f : \mathbf{X} \mapsto \mathbf{C}$ *holomorphic and nonconstant.*

(i) *If*

$$\mathbf{X}_t = \{z \in \mathbf{X} : g_{\mathbf{X}}(z, z_0) > t\} \quad and \quad A(t) = \frac{i}{2} \int_{\mathbf{X}_t} |f'(z)|^2 dz \wedge d\bar{z} \quad for \quad t > 0,$$

then

$$\lim_{t\to\infty} e^{2t} A(t) = \pi \Big(\frac{|f'(z_0)|}{c_{\mathbf{X}}(z_0)} \Big)^2.$$

(ii) *If $E_{p_0}(f, z_0, \mathbf{X}) < \infty$ for some $p_0 \in [0, \infty)$, then*

$$\lim_{p\to\infty} 2^p \big(\pi\Gamma(p+1) \big)^{-1} \big(E_p(f, z_0, \mathbf{X}) \big)^2 = \Big(\frac{|f'(z_0)|}{c_{\mathbf{X}}(z_0)} \Big)^2.$$

(iii)

$$\Big(\frac{|f'(z_0)|}{c_{\mathbf{X}}(z_0)} \Big)^2 \le 2^p \big(\pi\Gamma(p+1) \big)^{-1} \big(E_p(f, z_0, \mathbf{X}) \big)^2, \quad p \in [0, \infty),$$

where equality holds if and only if $\mathbf{X}$ is obtained from a simply connected hyperbolic Riemann surface $\mathbf{Y}$ by removing at most a set of logarithmic capacity zero and f is extended onto a conformal mapping from $\mathbf{Y}$ to an open disk in $\mathbf{C}$ centered at $f(w) \in \mathbf{C}$.

(iv)

$$\|f\|_{\mathcal{CB}(\mathbf{X})} = \sup_{w \in \mathbf{X}} \frac{|f'(w)|}{c_{\mathbf{X}}(w)} \le \sqrt{2^p \big(\pi\Gamma(p+1) \big)^{-1}} \|f\|_{\mathcal{Q}_p(\mathbf{X})}, \quad p \in [0, \infty),$$

where equality holds under $\|f\|_{\mathcal{D}(\mathbf{X})} < \infty$ if and only if $\mathbf{X}$ is obtained from a simply connected hyperbolic Riemann surface $\mathbf{Y}$ by removing at most a set of logarithmic capacity zero and f is extended onto a conformal mapping from $\mathbf{Y}$ to an open disk $\mathbf{D}\big(f(w)\big)$ centered at $f(w) \in \mathbf{C}$.

Proof. (i) Note that

$$\xi = \exp\big(-(g_{\mathbf{X}}(z, z_0) + ig_{\mathbf{X}}^*(z, z_0)) \big)$$

may be taken as a local variable at z_0. So it follows that $\xi(z_0) = 0$, $\gamma_{\mathbf{X}}(0) = 0$ and $c_{\mathbf{X}}(0) = 1$ under the variable ξ. For a sufficiently large $t > 0$, $\mathbf{X}_t$ is the parameter disk $\{\xi \in \mathbf{D} : |\xi| < e^{-t}\}$, and therefore

$$e^{2t} A(t) = e^{2t} \int_{|\xi| < e^{-t}} |f'(\xi)|^2 dm(\xi) \to \pi|f'(0)|^2 \quad as \quad t \to \infty,$$

as desired.

(ii) follows from (i) and Lemma 8.5.1.

(iii) follows from (ii), Theorem 8.2.5 (i), Theorem 8.4.2 and the uniformalization theorem.

(iv) follows from (iii), Theorem 8.2.5 (ii), Theorem 8.4.4 and the uniformalization theorem. $\square$

This theorem suggests that we introduce a Bloch-type space based on the capacitary density. More explicitly, we say that a holomorphic function f on a given hyperbolic Riemann surface $\mathbf{X}$ is of the capacitary Bloch class, written $f \in \mathcal{CB}(\mathbf{X})$ provided $\|f\|_{\mathcal{CB}(\mathbf{X})} < \infty$. It is clear that if $\mathbf{X}$ is simply connected and $\tau : \mathbf{D} \mapsto \mathbf{X}$ is surjectively conformal, then

$$\|f\|_{\mathcal{CB}(\mathbf{X})} = \|f \circ \tau\|_{\mathcal{B}}$$

and hence $\mathcal{CB}(\mathbf{X})$ is a natural generalization of the classical Bloch space from the open unit disk $\mathbf{D}$ to the hyperbolic Riemann surface $\mathbf{X}$.

Meanwhile, we can define another type of Bloch space through the hyperbolic metric. Given a hyperbolic Riemann surface $\mathbf{X}$ with the Poincaré metric $\lambda_{\mathbf{X}}$, we say that a holomorphic function $f : \mathbf{X} \mapsto \mathbf{C}$ is of Bloch class, written $f \in \mathcal{B}(\mathbf{X})$, provided

$$\|f\|_{\mathcal{B}(\mathbf{X})} = \sup_{z \in \mathbf{X}} \frac{|f'(z)|}{\lambda_{\mathbf{X}}(z)} < \infty.$$

It is interesting to compare the above two Bloch classes.

Theorem 8.5.3. *Let $\mathbf{X}$ be a hyperbolic Riemann surface, $Fuc(\mathbf{D})$ a Fuchsian group such that $\mathbf{D}/Fuc(\mathbf{D})$ is conformally equivalent to $\mathbf{X}$, and Ω the fundamental region of $Fuc(\mathbf{D})$. Then:*

(i) $\mathcal{CB}(\mathbf{X}) \subseteq \mathcal{B}(\mathbf{X})$. *But there is a hyperbolic Riemann surface $\mathbf{Y}$ such that*

$$\mathcal{CB}(\mathbf{Y}) \neq \mathcal{B}(\mathbf{Y}).$$

(ii) $\delta(\mathbf{X}) = \inf_{w \in \Omega} \prod_{\gamma \in Fuc(\mathbf{D})} |\sigma_w(\gamma(w))| > 0$ *implies* $\mathcal{CB}(\mathbf{X}) = \mathcal{B}(\mathbf{X})$, *but not conversely.*

Proof. (i) The first part just follows from the inequality $c_{\mathbf{X}}(z) \leq \lambda_{\mathbf{X}}(z)$ for any hyperbolic Riemann surface $\mathbf{X}$. So, it is enough to check this inequality. To this end, given $z \in \mathbf{X}$ and a local variable t in a neighborhood of z with $t(z) = 0$, let $\mathcal{A}_z^*(\mathbf{X})$ be the family of all multiple-valued holomorphic functions $F : \mathbf{X} \mapsto \mathbf{C}$ with $|F|$ being single-valued, $F(z) = 0$ and $F'(z) = 1$ for one of the branches. From [SariO, pp. 177–178] it turns out

$$c_{\mathbf{X}}(z) = \max \left\{ \sup_{w \in \mathbf{X}} |F(w)| : F \in \mathcal{A}_z^*(\mathbf{X}) \right\} > 0.$$

Assume $F_{\mathbf{X}} \in \mathcal{A}_z^*(\mathbf{X})$ is the unique extremal function with respect to the local coordinate t, and $\tau : \mathbf{D} \mapsto \mathbf{X}$ is a holomorphic universal covering mapping such that $\tau(0) = z$. Then $c_{\mathbf{X}}(F_{\mathbf{X}} \circ \tau)$ is a single-valued holomorphic self-map of $\mathbf{D}$ that has 0 as its fixed point. Note that this function vanishes at each point of the set $\tau^{-1}(z)$. So, the Schwarz Lemma is employed to deduce

$$c_{\mathbf{X}}(z)|F_{\mathbf{X}}'(z)||\tau'(0)| \leq 1,$$

where $\tau'(0)$ stands for the derivative of $t \circ \tau$ at 0. Consequently,

$$c_{\mathbf{X}}(z) \le |\tau'(0)|^{-1} = \lambda_{\mathbf{X}}(z),$$

as required.

As for the second part, let

$$f(z) = (1+z)/(1-z) \quad \text{and} \quad \mathbf{L} = \{m + in : \ m, n \in \mathbf{Z}\},$$

where $\mathbf{Z}$ stands for the set of all integers. Then $f \notin \mathcal{B}$ and $f^{-1}(\mathbf{L})$ is a countable set and hence a set of logarithmic capacity zero. If $\mathbf{Y} = \mathbf{D} \setminus f^{-1}(\mathbf{L})$, then $g_{\mathbf{D}}(z, w) = g_{\mathbf{Y}}(z, w)$ for $z, w \in \mathbf{Y}$ and consequently, $c_{\mathbf{D}}(z) = c_{\mathbf{Y}}(z)$ for each $z \in \mathbf{Y}$. Putting $h = f|_{\mathbf{Y}}$, i.e., the restriction of f to $\mathbf{Y}$, we see that $h \in \mathcal{B}(\mathbf{Y})$ since h omits all values in $\mathbf{L}$. But, because of $f \notin \mathcal{B}$, it follows that $f \notin \mathcal{CB}(\mathbf{D})$. Also, since $f'(z) = h'(z)$ and $c_{\mathbf{D}}(z) = c_{\mathbf{Y}}(z)$ for each $z \in \mathbf{Y}$, we conclude that $h \in \mathcal{CB}(\mathbf{Y})$ amounts to $f \in \mathcal{CB}(\mathbf{D})$, thereupon getting $h \in \mathcal{B}(\mathbf{Y}) \setminus \mathcal{CB}(\mathbf{Y})$, as desired.

(ii) We begin with claiming that if $\tau : \mathbf{D} \mapsto \mathbf{X}$ is a universal covering mapping, then

$$\frac{c_{\mathbf{X}}\big(\tau(w)\big)}{\lambda_{\mathbf{X}}\big(\tau(w)\big)} = \prod_{\gamma \in Fuc(\mathbf{D})} |\sigma_w\big(\gamma(w)\big)|, \quad w \in \mathbf{\Omega}.$$

Since

$$g_{\mathbf{\Omega}}(z, w) = \sum_{\gamma \in Fuc(\mathbf{D})} \Big(-\log|\sigma_z\big(\gamma(w)\big)| \Big),$$

it is easy to derive

$$\lim_{z \to w} \big(g_{\mathbf{\Omega}}(z, w) + \log|z - w| \big) = \log(1 - |w|^2) + \sum_{\gamma \ne id} \Big(-\log|\sigma_w\big(\gamma(w)\big)| \Big).$$

This, plus the conformal invariance of $c_{\mathbf{X}}(\cdot)$ and $\lambda_{\mathbf{X}}(\cdot)$, yields

$$
\begin{aligned}
c_{\mathbf{X}}\big(\tau(w)\big)|\tau'(w)| &= |\tau'(w)| \exp\Big(-\lim_{z \to w} \big(g_{\mathbf{\Omega}}(z, w) + \log|z - w| \big) \Big) \\
&= |\tau'(w)| c_{\mathbf{D}}(w) \prod_{\gamma \in Fuc(\mathbf{D})} |\sigma_w\big(\gamma(w)\big)| \\
&= \lambda_{\mathbf{X}}\big(\tau(w)\big) \prod_{\gamma \in Fuc(\mathbf{D})} |\sigma_w\big(\gamma(w)\big)|.
\end{aligned}
$$

Note once again that $\tau'(w)$ denotes the derivative of $t \circ \tau$ at w where t is a local variable at $\tau(w) \in \mathbf{X}$ with $t\big(\tau(w)\big) = 0$. Thus, the last sequence of equalities verifies the desired claim.

Next, it follows from the above claim that

$$0 < \delta(\mathbf{X}) \le \frac{c_{\mathbf{X}}\big(\tau(w)\big)}{\lambda_{\mathbf{X}}\big(\tau(w)\big)} \le 1, \quad w \in \mathbf{\Omega}.$$

Since $\tau : \Omega \mapsto \mathbf{X}$ is injective and $\tau : \overline{\Omega} \mapsto \mathbf{X}$ is surjective and $\partial\Omega$ has area zero, we conclude that $\mathcal{CB}(\mathbf{X}) = \mathcal{B}(\mathbf{X})$.

To check the converse, we consider the punctured unit disk $\mathbf{Y} = \mathbf{D} \setminus \{0\}$ which is certainly a hyperbolic Riemann surface. In this case, we have $g_\mathbf{Y}(z, w) = g_\mathbf{D}(z, w)$ for all $z, w \in \mathbf{Y}$, ensuring $c_\mathbf{Y}(w) = c_\mathbf{D}(w)$ for each $w \in \mathbf{Y}$, which in turn means that $\mathcal{CB}(\mathbf{Y}) = \mathcal{CB}(\mathbf{D})$. Since $\mathcal{B}(\mathbf{Y}) = \mathcal{B}$, it follows that $\mathcal{CB}(\mathbf{Y}) = \mathcal{B}(\mathbf{Y})$. Now, if

$$\phi(z) = \frac{z+1}{z-1} \quad \text{and} \quad \tau(z) = \exp\phi(z),$$

then ϕ and τ are a conformal mapping and a universal covering mapping from $\mathbf{D}$ onto $\mathbf{LH} = \{\zeta \in \mathbf{C} : \Re\zeta < 0\}$ and $\mathbf{Y}$ respectively. Hence, if $\psi(\zeta) = e^\zeta$, then $\psi(\zeta + 2k\pi i) = \psi(\zeta)$ for each $\zeta \in \mathbf{LH}$ and each $k \in \mathbf{Z}$, and hence, if Γ is a Fuchsian group such that $\mathbf{D}/\Gamma$ is conformally equivalent to $\mathbf{Y}$, then for $\gamma \in \Gamma$ we have $\phi(\gamma(z)) = \phi(z) + 2k\pi i$ for some integer k. For the definiteness, we use the notation $\phi(\gamma_k(z)) = \phi(z) + 2k\pi i$ for each integer k. Nevertheless, the hyperbolic distance $d_\mathbf{LH}(\zeta, \zeta + 2\pi i)$ on $\mathbf{LH}$ approaches 0 as $\Re\zeta \to -\infty$. Thus, we get

$$\lim_{0<x\to1} d_\mathbf{D}\big(x, \gamma_1(x)\big) = 0 \quad \text{whence} \quad \lim_{0<x\to1} |\sigma_x\big(\gamma_1(x)\big)| = 0.$$

Since

$$|\sigma_x\big(\gamma_1(x)\big)| \geq \prod_{k=1}^{\infty} |\sigma_x\big(\gamma_k(x)\big)| \geq \prod_{\gamma\in\Gamma} |\sigma_x\big(\gamma(x)\big)| \quad \text{for} \quad x \in (0,1),$$

we must have $\delta(\mathbf{Y}) = 0$. This proves the desired assertion. □

Finally, we compare $\mathcal{Q}_p(\mathbf{X})$, $p > 1$ with $\mathcal{B}(\mathbf{X})$.

Theorem 8.5.4. *Let $\mathbf{X}$ be a hyperbolic Riemann surface, $\tau : \mathbf{D} \mapsto \mathbf{X}$ a universal covering mapping, and*

$$d_\mathbf{X}(z, w) = \inf\{d_\mathbf{D}(\zeta, \eta) : \quad z = \tau(\zeta), \ w = \tau(\eta)\}$$

the hyperbolic distance between $z, w \in \mathbf{X}$. Then:

(i) *$\mathcal{Q}_p(\mathbf{X}) \subseteq \mathcal{B}(\mathbf{X})$ for $p \in [0, \infty)$, but there is a hyperbolic Riemann surface $\mathbf{Y}$ such that $\mathcal{Q}_p(\mathbf{X}) \neq \mathcal{B}(\mathbf{X})$ for $p \in (1, \infty)$.*

(ii) *A holomorphic function $f : \mathbf{X} \mapsto \mathbf{C}$ belongs to $\mathcal{B}(\mathbf{X})$ if and only if f is Lipschitz continuous in the hyperbolic distance; that is,*

$$\|f\|_{\mathcal{B}(\mathbf{X}),1} = \sup_{z,w\in\mathbf{X}} \frac{|f(z) - f(w)|}{d_\mathbf{X}(z, w)} < \infty.$$

(iii) *If*

$$C(\mathbf{X}) = \sup_{z,w\in\mathbf{X}} d_\mathbf{X}(z, w) \left(\log\sqrt{\frac{\exp\big(g_\mathbf{X}(z, w)\big) + 1}{\exp\big(g_\mathbf{X}(z, w)\big) - 1}}\right)^{-1} < \infty,$$

then $\mathcal{B}(\mathbf{X}) = \mathcal{Q}_p(\mathbf{X})$ for any $p \in (1, \infty)$. But there is a hyperbolic Riemann surface $\mathbf{Y}$ such that $C(\mathbf{Y}) = \infty$ and $\mathcal{B}(\mathbf{Y}) = \mathcal{Q}_p(\mathbf{Y})$.

Proof. (i) The inclusion follows from Theorems 8.5.2 and 8.5.3. On the other hand, let again

$$f(z) = (1 + z)/(1 - z) \quad \text{and} \quad \mathbf{L} = \{m + in : \ m, n \in \mathbf{Z}\}.$$

Then $f \notin \mathcal{B} = \mathcal{Q}_p$ for $p > 1$, and $f^{-1}(\mathbf{L})$ is a countable set and hence a set of logarithmic capacity zero. If $\mathbf{Y} = \mathbf{D} \setminus f^{-1}(\mathbf{L})$, then $g_{\mathbf{D}}(z, w) = g_{\mathbf{Y}}(z, w)$ for $z, w \in \mathbf{Y}$. Putting $h = f|_{\mathbf{Y}}$, i.e., the restriction of f to $\mathbf{Y}$, we see that $h \in \mathcal{B}(\mathbf{Y})$ since h omits all values in $\mathbf{L}$. However, $f \notin \mathcal{Q}_p$ as $p > 1$ and

$$E_p(h, w, \mathbf{Y}) = E_p(f, w, \mathbf{D}) = E_p(f, w), \quad w \in \mathbf{Y}$$

lead to $h \notin \mathcal{Q}_p(\mathbf{Y})$. This proves $\mathcal{B}(\mathbf{Y}) \neq \mathcal{Q}_p(\mathbf{Y})$.

(ii) Suppose that τ is a universal covering mapping from $\mathbf{D}$ onto $\mathbf{X}$. If $f \in \mathcal{B}(\mathbf{X})$, then $f \circ \tau \in \mathcal{B}$ and hence

$$\left|f\big(\tau(\zeta)\big) - f\big(\tau(\eta)\big)\right| = d_{\mathbf{D}}(\zeta, \eta) \frac{\left|f\big(\tau(\zeta)\big) - f\big(\tau(\eta)\big)\right|}{d_{\mathbf{D}}(\zeta, \eta)} \preceq \|f \circ \tau\|_{\mathcal{B}} d_{\mathbf{D}}(\zeta, \eta),$$

thus giving $\|f\|_{\mathcal{B}(\mathbf{X}),1} < \infty$. Conversely, since

$$\frac{\left|f\big(\tau(\zeta)\big) - f\big(\tau(\eta)\big)\right|}{d_{\mathbf{D}}(\zeta, \eta)} \leq \frac{|f(z) - f(w)|}{d_{\mathbf{X}}(z, w)}, \quad z = \tau(\zeta), \ w = \tau(\eta),$$

we conclude that $\|f\|_{\mathcal{B}(\mathbf{X}),1} < \infty$ implies $f \circ \tau \in \mathcal{B}$, namely, $f \in \mathcal{B}(\mathbf{X})$.

(iii) Suppose $\{\mathbf{X}_j\}$ is a regular exhaustion of $\mathbf{X}$ and

$$h_j(z, w) = g_{\mathbf{X}_j}(z, w) + i g_{\mathbf{X}_j}^*(z, w)$$

where $g_{\mathbf{X}_j}^*(z, w)$ is a harmonic conjugate of $g_{\mathbf{X}_j}(z, w)$, being locally defined up to an additive constant. Applying the formulas

$$\Delta u(z) = 4|f'(z)|^2 \quad \text{and} \quad \Delta g_{\mathbf{X}_j}(z, w) = p(p-1)\big(g_{\mathbf{X}_j}(z, w)\big)^{p-2}|\nabla g_{\mathbf{X}_j}(z, w)|^2$$

to the Green formula

$$\frac{i}{2} \int_{\mathbf{X}_j} \big(v(z)\Delta u(z) - u(z)\Delta v(z)\big) dz \wedge d\bar{z} = 2 \int_{\partial \mathbf{X}_j} \left(u \frac{\partial v}{\partial n} - v \frac{\partial u}{\partial n}\right) ds$$

with

$$u(z) = |f(z) - f(w)|^2 \quad \text{and} \quad v(z) = \big(g_{\mathbf{X}_j}(z, w)\big)^p,$$

and taking the limit as the radius of the disk at w shrinks to 0, we derive

$$\big(E_p(f, w, \mathbf{X}_j)\big)^2 = \frac{ip(p-1)}{2^2} \int_{\mathbf{X}_j} |f(z) - f(w)|^2 \frac{|\nabla g_{\mathbf{X}_j}(z, w)|^2}{\big(g_{\mathbf{X}_j}(z, w)\big)^{2-p}} dz \wedge d\bar{z}.$$

Let $d^{(j)}(z,w)$ be the restriction of $d_{\mathbf{X}}(z,w)$ to $\mathbf{X}_j$ and

$$L_j(z,w) = \log\sqrt{\frac{\exp\left(g_{\mathbf{X}_j}(z,w)\right)+1}{\exp\left(g_{\mathbf{X}_j}(z,w)\right)-1}}.$$

Then the inequality $g_{\mathbf{X}_j}(z,w) \leq g_{\mathbf{X}}(z,w)$ for $z,w \in \mathbf{X}_j$ implies

$$
\begin{aligned}
C(\mathbf{X}_j) &= \sup\left\{\frac{d^{(j)}(z,w)}{L_j(z,w)} : z,w \in \mathbf{X}_j\right\} \\
&\leq \sup_{z,w\in\mathbf{X}_j} d_{\mathbf{X}}(z,w)\left(\log\sqrt{\frac{\exp\left(g_{\mathbf{X}}(z,w)\right)+1}{\exp\left(g_{\mathbf{X}}(z,w)\right)-1}}\right)^{-1} \\
&\leq C(\mathbf{X}).
\end{aligned}
$$

Now if

$$f \in \mathcal{B}(X) \quad\text{and}\quad \|f\|_{\mathcal{B}(\mathbf{X}_j),1} = \sup_{z,w\in\mathbf{X}_j}\frac{|f(z)-f(w)|}{d^{(j)}(z,w)},$$

then by (i),

$$
\begin{aligned}
&\left(E_p(f,w,\mathbf{X}_j)\right)^2 \\
&\leq \left(\|f\|_{\mathcal{B}(\mathbf{X}_j),1}\right)^2\frac{ip(p-1)}{2^2}\int_{\mathbf{X}_j}\left(d^{(j)}(z,w)\right)^2\frac{|\nabla g_{\mathbf{X}_j}(z,w)|^2}{\left(g_{\mathbf{X}_j}(z,w)\right)^{2-p}}dz\wedge d\bar{z} \\
&\leq \left(C(\mathbf{X}_j)\right)^2\left(\|f\|_{\mathcal{B}(\mathbf{X}_j),1}\right)^2\frac{ip(p-1)}{2^2}\int_{\mathbf{X}_j}\left(L_j(z,w)\right)^2\frac{|\nabla g_{\mathbf{X}_j}(z,w)|^2}{\left(g_{\mathbf{X}_j}(z,w)\right)^{2-p}}dz\wedge d\bar{z} \\
&\preceq \left(C(\mathbf{X}_j)\right)^2\left(\|f\|_{\mathcal{B}(\mathbf{X}_j),1}\right)^2\int_0^\infty\frac{\left(\int_{\{z\in\mathbf{X}_j:g_{\mathbf{X}_j}(z,w)=t\}}\frac{\partial g_{\mathbf{X}_j}(z,w)}{\partial n}ds\right)}{\left(\log\frac{e^t+1}{e^t-1}\right)^{-2}}dt^{p-1} \\
&\preceq \left(C(\mathbf{X}_j)\right)^2\left(\|f\|_{\mathcal{B}(\mathbf{X}_j),1}\right)^2\int_0^\infty\left(\log\frac{e^t+1}{e^t-1}\right)^2 dt^{p-1} \\
&\preceq \left(C(\mathbf{X}_j)\right)^2\left(\sup_{z,w\in\mathbf{X}_j}\frac{|f(z)-f(w)|}{d_{\mathbf{X}}(z,w)}\right)^2 \\
&\preceq \left(C(\mathbf{X}_j)\right)^2\|f\|_{\mathcal{B}(\mathbf{X}),1}^2.
\end{aligned}
$$

Note that $g_{\mathbf{X}_j}(z,w) \to g_{\mathbf{X}}(z,w)$ as $j \to \infty$. So we have

$$\left(E_p(f,w,\mathbf{X}_j)\right)^2 \preceq \left(C(\mathbf{X}_j)\right)^2\|f\|_{\mathcal{B}(\mathbf{X}),1}^2.$$

This implies $f \in \mathcal{Q}_p(\mathbf{X})$, $p > 1$. Consequently, $\mathcal{B}(\mathbf{X}) = \mathcal{Q}_p(\mathbf{X})$, $p > 1$ thanks to $\mathcal{Q}_p(\mathbf{X}) \subseteq \mathcal{CB}(\mathbf{X})$.

For the converse, let $\mathbf{Y} = \mathbf{D} \setminus \{0\}$, then $g_{\mathbf{Y}}(z, w) = g_{\mathbf{D}}(z, w)$ for $z, w \in \mathbf{Y}$ and hence $\mathcal{Q}_p(\mathbf{Y}) = \mathcal{Q}_p$. Note that $\lim_{z \to 0} g_{\mathbf{Y}}(z, 2^{-1}) = \log 2$. So

$$\lim_{z \to 0} \log \sqrt{\frac{\exp \left(g_{\mathbf{Y}}(z, 2^{-1})\right) + 1}{\exp \left(g_{\mathbf{Y}}(z, 2^{-1})\right) - 1}} = \log \sqrt{2}.$$

However, by using the mapping $z = e^{\zeta}$ from $\mathbf{LH} = \{\zeta \in \mathbf{C} : \Re \zeta < 0\}$ onto $\mathbf{Y}$, it is not hard to see that $d_{\mathbf{Y}}(z, 2^{-1})$ grows asymptotically to $\log(-\log |z|)$ as $z \to 0$, so $C(\mathbf{Y}) = \infty$. But, we will see that $\mathcal{B}(\mathbf{Y}) = \mathcal{Q}_p(\mathbf{Y})$ for $p > 1$. In fact, suppose $f \in \mathcal{B}(\mathbf{Y})$. If $F(\zeta) = f(e^{\zeta})$ belongs to $\mathcal{B}(\mathbf{LH})$ then

$$\sup_{z \in \mathbf{Y}} |z|(-\log |z|)|f'(z)| = \sup_{\zeta \in \mathbf{LH}} |\Re \zeta||F'(\zeta)| < \infty.$$

But $f \in \mathcal{B}(\mathbf{Y})$ forces that it has an isolated singularity at $z = 0$ which means that either f' is bounded near $z = 0$ or $\limsup_{z \to 0} |z|^{-2}|f'(z)| > 0$. From the last equality it turns out that for $f \in \mathcal{B}(\mathbf{Y})$ we must have that f' is bounded near $z = 0$ and so that we can define $f(0)$ and $f'(0)$ as finite values. Consequently, $f \in \mathcal{B} = \mathcal{Q}_p$ when $p > 1$. This gives $\mathcal{B}(\mathbf{Y}) \subseteq \mathcal{Q}_p(\mathbf{Y}) \subseteq \mathcal{CB}(\mathbf{Y}) \subseteq \mathcal{B}(\mathbf{Y})$, as desired. $\qquad \square$

8.6 Notes

Note 8.6.1. Section 8.1 is a small (but necessary for our aim) portion of [Ahl, Chapters 9 and 10]. Some basic properties on the hyperbolic and capacitary metrics can be found in [Min1], [Min2] and [Min3].

Note 8.6.2. Section 8.2 is a combination of [AuCh, Sections 2 and 3]. Here, it is worth pointing out that the inclusion chain:

$$\mathcal{Q}_0(\mathbf{X}) \subseteq \mathcal{Q}_q(\mathbf{X}) \subseteq \mathcal{Q}_p(\mathbf{X}), \quad \text{for} \quad p > q > 0,$$

was first verified by [AuHRZ] in a way extending [Met] and [Kob1] which proved $\mathcal{Q}_0(\mathbf{X}) \subseteq \mathcal{BMOA}(\mathbf{X})$. In addition, the left-hand-side inclusion can be improved under the condition that the hyperbolic Riemann surface $\mathbf{X}$ is regular; that is,

$$\lim_{w \to \partial \mathbf{X}} g_{\mathbf{X}}(z, w) = 0, \quad z \in \mathbf{X}.$$

More precisely, if $f \in \mathcal{Q}_0(\mathbf{X})$ and $\mathbf{X}$ is regular, then

$$f \in \mathcal{Q}_{p,0}(\mathbf{X}), \quad \text{i.e.,} \quad \lim_{w \to \partial \mathbf{X}} E_p(f, w, \mathbf{X}) = 0;$$

see also [AuHRZ, Theorem 3.2]. Theorem 8.2.5 has been extended to $\mathcal{Q}_p$ spaces on the unit ball of $\mathbf{C}^n$ — see also [Che].

Note 8.6.3. Section 8.3 just describes the main ideas and results in [Kob1] and [Kob2]. A survey on the study of BMOA on Riemann surfaces is given by [Met]. Here it is perhaps appropriate to mention that Theorem 8.3.1 (ii) cannot extend to $p \in (0,1)$. This is because $\exp\big((z+1)/(z-1)\big)$ does not belong to $\mathcal{Q}_p$ where $p \in [0,1)$ while its image area is clearly less than π. On the other hand, we can obtain an interesting consequence of Theorem 8.3.1 as follows: if $\mathbf{G}$ is a hyperbolic subdomain of $\mathbf{C}$, then

$$\int_{\mathbf{G}} g_{\mathbf{G}}(z,w)dm(z) \leq 2^{-1}\mathrm{Area}(\mathbf{G});$$

see also [Kob1, Corollary 7]. This leads to the concept of a BMOA-domain $\mathbf{G}$ which means that any holomorphic function sending $\mathbf{D}$ into $\mathbf{G}$ lies in $\mathcal{BMOA}$. According to [Met, Theorem 7.3], we have that $\mathbf{G}$ is a BMOA-domain if and only if the identity map I belongs to $\mathcal{BMOA}(\mathbf{G})$, namely,

$$\sup_{w \in \mathbf{G}} \int_{\mathbf{G}} g_{\mathbf{G}}(z,w)dm(z) < \infty.$$

Furthermore, it has been proved in [HayPo] and [Ste2] that $\mathbf{G}$ is a BMOA-domain when and only when there are positive constants κ_1 and κ_2 such that

$$cap_0\big((\mathbf{C} \setminus \mathbf{G}) \cap \mathbf{D}(w,\kappa_1)\big) \geq \kappa_2, \quad w \in \mathbf{G},$$

where $\mathbf{D}(w,\kappa_1)$ stands for the open disk centered at w with radius κ_1. Of course, there are many examples to show that this condition does not characterize the range of a function in $\mathcal{BMOA}$ — see [Met, p. 90]. Naturally, we can define a $\mathcal{Q}_p$-domain, but find that the cases $p \geq 1$ produce the BMOA-domains and Bloch-domains while the cases $p \in [0,1)$ do not make sense — in other words, there are no $\mathcal{Q}_p$-domains whenever $p \in [0,1)$ — see [EsXi] — but we can talk about conformal domains in this case — see Section 1.3. In addition, a John–Nirenberg type estimate for $\mathcal{BMOA}$ on the hyperbolic Riemann surface was established in [Zha1] — more explicitly, a holomorphic function f on the hyperbolic Riemann surface $\mathbf{X}$ belongs to $\mathcal{BMOA}(\mathbf{X})$ if and only if there are positive constants κ_1 and κ_2 such that for every $w \in \mathbf{X}$ and $t > 0$,

$$\frac{i}{2}\int_{\{z \in \mathbf{X} : |f(z)-f(w)|>t\}} |f'(z)|^2 g_{\mathbf{X}}(z,w)dz \wedge d\bar{z} \leq \kappa_1 \exp(-\kappa_2 t).$$

Furthermore, for another investigation of holomorphic maps between Riemann surfaces which preserve BMO, see also [Got1], [Got2], and [Got3].

Note 8.6.4. Section 8.4 is a modified combination of [AuCh, Sections 4 and 5]. Lemma 8.4.1 is a sort of special formulation of [Min3, Theorem 1] and its proof.

Note 8.6.5. Section 8.5 is taken from [AuCh, Section 6] and [AuLXZ]. The argument for $c_{\mathbf{X}}(z)|dz| \leq \lambda_{\mathbf{X}}(z)|dz|$ in Theorem 8.5.3 is extracted from [Min3, Theorem 2] which also tells us that both metrics coincide at a single point if and only if $\mathbf{X}$ is simply connected. Furthermore, it would be of interest to study when $\mathcal{CB}(\mathbf{X})$ is contained in $\mathcal{Q}_p(\mathbf{X})$ whenever $p > 1$. In [AuCh, Section 7] the following area inequality has been proved: if $\mathbf{X}$ is a finite Riemann surface with the Green function $g_{\mathbf{X}}(\cdot, \cdot)$ and $f : \mathbf{X} \mapsto \mathbf{C}$ is holomorphic, then $\|f\|_{\mathcal{CB}(\mathbf{X})} \leq \sqrt{\pi^{-1}\text{Area}(f(\mathbf{X}))}$, which is established via verifying

$$e^{2(t-t_0)} \leq \frac{\text{Area}\big(f(\{z \in \mathbf{X} : g_{\mathbf{X}}(z, w) > t_0\})\big)}{\text{Area}\big(f(\{z \in \mathbf{X} : g_{\mathbf{X}}(z, w) > t\})\big)}, \quad t > t_0 \geq 0.$$

Bibliography

[Ad1]   D. R. Adams, *A sharp inequality of J. Moser for higher order deriva-tives*. Ann. of Math. **128** (1988), 385–398.

[Ad2]   D. R. Adams, *A note on the Choquet integrals with respect to Haus-dorff capacity*. Function spaces and applications, Proc. Lund 1986, Lecture Notes in Math. **1302**. Springer-Verlag, Berlin, 1988, pp. 115–124.

[Ad3]   D. R. Adams, *The classification problem for the capacities associated with the Besov and Tribel-Lizorkin spaces*. Approximation and Func-tion Spaces, Banach Center Publications **22** (1989), 9–24.

[Ad4]   D. R. Adams, *Choquet integrals in potential theory*. Publ. Mat. **42** (1998), 3–66.

[AdFr]   D. R. Adams and M. Frazier, *Composition operators on potential spaces*. Proc. Amer. Math. Soc. **114** (1992), 155-165.

[AdHe]   D. R. Adams and L. I. Hedberg, *Function Spaces and Potential The-ory*. Springer-Verlag, 1996.

[AgPa]   R. P. Agarwal and P. Y. H. Pang, *Opial Inequalities with Applications in Differential and Difference Equations*. Kluwer Academic Publish-ers, Dordrecht/Boston/London, 1995.

[AhJe]   P. Ahern and M. Jevtić, *Inner multipliers of the Besov space,* $0 < p < 1$. Rocky Mountain J. Math. **20**(1990), 753–764.

[Ahl]   L. V. Ahlfors, *Complex Analysis*. 2nd Edition, McGraw-Hill Book Company, New York, 1966.

[AlCaSi] A. Aleman, M. Carlsson and A. M. Simbotin, *Preduals of Q_p spaces and Carleson imbeddings of weighted Dirichlet spaces*. Preprint, 2005.

[AlSi]   A. Aleman and A. M. Simbotin, *Estimates in Möbius invariant spaces of analytic functions*. Complex Var. Theory Appl. **49**(2004), 487–510.

[AlvJMi] J. Alvarez and M. Milman, *Spaces of Carleson measures: duality and interpolation*. Ark. Mat. **25**(1987), 155–174.

[AlvVMV]   V. Alvarez, M. A. Marquez and D. Vukotić, *Composition and superposition operators between the Bloch space and Bergman spaces.* Preprint, 2004.

[An]   J. M. Anderson, *Bloch functions: the basic theory.* Operators and Function Theory (Lancaster, 1984), 1–17, NATO Adv. Sci. Inst. Ser. C Math. Phys. Sci., **153**, Reidel, Dordrecht, 1985.

[AnClPo]   J. M. Anderson, J. Clunie and Ch. Pommerenke, *On Bloch functions and normal functions.* J. Reine. Angew. Math. **270**(1974), 12–37.

[AnRo]   J. A. Antonino and S. Romaguera, *A short proof of the Cima-Wogen* $L(f) =$ circle *theorem.* Proc. Amer. Math. Soc. **92**(1984), 391–392.

[ArFi]   J. Arazy and S. Fisher, *Some aspects of the minimal, Möbius-invariant space of analytic functions on the unit disc.* Interpolation spaces and allied topics in analysis (Lund, 1983), Lecture Notes in Math. **1070**. Springer, Berlin, 1994, pp. 24–44.

[ArFiJaPe] J. Arazy, S. Fisher, S. Janson and J. Peetre, *Membership of Hankel operators on the ball in unitary ideas.* J. London Math. Soc. (2) **43**(1991), 485–508.

[ArFiPe1]   J. Arazy, S. Fisher and J. Peetre, *Möbius invariant function spaces.* J. Reine Angew. Math. **363**(1985), 110–145.

[ArFiPe2]   J. Arazy, S. Fisher and J. Peetre, *Hankel operators on weighted Bergman spaces.* Amer. J. Math. **110**(1988), 989–1054.

[ArFiPe3]   J. Arazy, S. Fisher and J. Peetre, *Besov norms of rational functions.* Lecture Notes in Math. **1032**. Springer-Verlag, Berlin, 1988, pp. 125–129.

[AuCh]   R. Aulaskari and H. Chen, *Area inequality and* Q_p *norm.* J. Funct. Anal. **221** (2005), 1–24.

[AuHRZ]   R. Aulaskari, Y. He, J. Ristioja and R. Zhao, Q_p *spaces on Riemann surfaces.* Canad. J. Math. **50**(1988), 449–464.

[AuLXZ]   R. Aulaskari, P. Lappan, J. Xiao and R. Zhao, *BMOA(R,m) and capacity density Bloch spaces on hyperbolic Riemann surfaces.* Results Math. **29**(1996), 203–226.

[AuMW]   R. Aulaskari, S. Makhmutov and H. Wulan, *On* q_p *sequences.* Finite or infinite dimensional complex analysis and applications, 117–125, Adv. Complex Anal. Appl. **2**. Kluwer Acad. Publ., Dordrecht, 2004.

[AuNoZh]   R. Aulaskari, M. Nowark and R. Zhao, *The n-th derivative characterizations of the Möbius bounded Dirichlet space.* Bull. Austral. Math. Soc. **58**(1998), 43–56.

[AuReTo] R. Aulaskari, L. F. Resendis O and L. M. Tovar S, Q_p *spaces and harmonic majorants.* Complex Var. Theory Appl. **49**(2004), 241–156.

[AxSh] S. Axler and A. L. Shields, *Extreme points in VMO and BMO.* Indiana Univ. Math. J. **31**(1982), 1–6.

[Bae1] A. Baernstein II, *Integral means, univalent functions and circular symmetrization.* Acta Math. **133**(1974), 139–169.

[Bae2] A. Baernstein II, *Analytic functions of bounded mean oscillation.* Aspects of Contemporary Complex Analysis. Academic Press, 1980, pp. 3–36.

[Ban] C. Bandle, *Isoperimetric Inequalities and Applications.* Monographs and Studies in Mathematics 7. Pitman, 1980.

[BanFlu] C. Bandle and M. Flucher, *Harmonic radius and concentration of energy, hyperbolic radius and Liouville's equations* $\Delta u = e^u$ *and* $\Delta u = u^{\frac{n+2}{n-2}}$. SIAM Rev. **38**(1996), 191–238.

[Bec] W. Beckner, *Geometric asymptotics and the logarithmic Sobolev inequality.* Forum Math. **11**(1999), 105–137.

[BergSc] S. Bergman and M. Schiffer, *Kernel Functions and Elliptic Differential Equations in Mathematical Physics.* Pure and Applied Mathematics 4. Academic Press, 1970.

[BernGRT] S. Bernstein, K. Gürlebeck, L. F. Reséndis O. and L. M. Tovar S, *Dirichlet and Hardy spaces of harmonic and monogenic functions.* Z. Anal. Anwendungen **24** (2005), 763–789.

[BetBrJa] D. Betebenner, L. Brown and F. Jafari, *On the range of functions belonging to weighted Dirichlet spaces.* Preprint, 1997.

[Beu] A. Beurling, *Ensembles exceptionnels.* Acta Math. **72**(1940), 1–13.

[BisJo] C. J. Bishop and P. W. Jones, *Harmonic measure,* L^2 *estimates and the Schwarzian derivative.* J. Anal. Math. **62**(1994), 77–113.

[Bo] B. Böe, *A norm on the holomorphic Besov space.* Proc. Amer. Math. Soc. **131**(2003), 235–241.

[BoNi] B. Böe and A. Nicolau, *Interpolation by functions in the Bloch space.* J. Anal. Math. **94**(2004), 171–194.

[BroSh] L. Brown and A. L. Shields, *Cyclic vectors in the Dirichlet space.* Trans. Amer. Math. Soc. **285**(1984), 269–304.

[BuFeVu] S. Buckley, J. Fernández and D. Vukotić, *Superposition operators on Dirichlet type space.* Papers on analysis, Rep. Univ. Jyväskylä Dep. Math. Stat. **83**. Univ. Jyväskylä, Jyväskylä, 2001, pp. 41–61.

[CeKaSo]   P. Cerejeiras, U. Kähler and F. Sommen, *Clifford analysis over projective hyperbolic space II*. Math. Meth. Appl. Sci. **25**(2002), 1465–1478.

[ChaDXi]   D. C. Chang and J. Xiao, *A note on isoperimetric and Sobolev inequalities*. Arch. Inequal. Appl. **2**(2004), 427–434.

[ChaSMar] S. -Y. A. Chang and D. E. Marshall, *On a sharp inequality concerning the Dirichlet integral*. Amer. J. Math. **107**(1985), 1015–1033.

[Che]   Z. Chen, Q_p *and* M_p *norm in the unit ball of* $\mathbb{C}^n$. Preprint, 2005.

[CiWo]   J. Cima and W. Wogen, *Extreme points of the unit ball of the Bloch space* $\mathcal{B}_0$. Michigan Math. J. **25**(1978), 213–222.

[CnDe]   J. Cnops and R. Delanghe, *Möbius invariant spaces in the unit ball*. Appl. Anal. **73**(1999), 45–64.

[CoheCol] J. M. Cohen and F. Colonna, *Extreme points of the Bloch space of a homogeneous tree*. Israel J. Math. **94**(1996), 247–271.

[Cohn]   W. Cohn, *A factorization theorem for the derivative of a function in* H^p. Proc. Amer. Math. Soc. **127**(1999), 509–517.

[CohnVe]   W. Cohn and I. E. Verbitsky, *Factorization of tent spaces and Hankel operators*. J. Funct. Anal. **175**(2000), 308–329.

[CoiMeSt] R. R. Coifman, Y. Meyer and E. Stein, *Some new function spaces and their applications to harmonic analysis*. J. Funct. Anal. **62**(1985), 304–335.

[CoiRoc]   R. R. Coifman and R. Rochberg, *Representation theorems for holomorphic and harmonic functions in* L^p. Asterisque **77**(1980), 11–66.

[Col]   F. Colonna, *Extreme points of a convex set of Bloch functions*. Seminars in Complex Analysis and Geometry, (Arcavacata, 1988), 23–59.

[Con]   J. B. Conway, *Functions of One Complex Variable II*. Graduate Texts in Mathematics **159**. Springer-Verlag, New York, 1995.

[CuRu]   J. G. Cuerva and J. L. Rubio, *Weighted Norm Inequalities and Related Topics*. Amsterdam, North-Holland, 1985.

[DaXi]   G. Dafni and J. Xiao, *Some new tent spaces and duality theorems for fractional Carleson measures and* $Q_\alpha(\mathbf{R}^n)$. J. Funct. Anal. **208**(2004), 377–422.

[DeV]   C. L. DeVito, *Functional Analysis*. Academic Press, 1978.

[DoGiVu]   J. Donaire, D. Girela and D. Vukotić, *On univalent functions in some Möbius invariant spaces*. J. Reine Angew. Math. **553**(2002), 43–72.

[Du]   P. Duren, *Theory of* H^p *Spaces*. Academic Press, 1970.

[DuRoSh] P. Duren, B. W. Romberg and A. L. Shields, *Linear functionals on H^p spaces with $0 < p < 1$*. J. Reine Angew. Math. **238**(1969), 32–60.

[DuSc] P. Duren and A. Schuster, *Bergman Spaces*. Mathematical Surveys and Monographs, **100**. Amer. Math. Soc., Providence, RI, 2004.

[Dy] K. M. Dyakonov, *Absolute values of BMOA functions*. Rev. Mat. Iberoamericana **15**(1999), 451–473.

[ElGRT] A. El-Sayed Ahmed, K. Gürlebeck, L. R. Reséndis and L. M. Tovar S, *Characterizations for Bloch space by $\mathbf{B}^{p,q}$ spaces in Clifford analysis*. Complex Var. Elliptic Equ. **51** (2006), 119–136.

[Eng] M. Englis, *Q_p-spaces: generalizations to bounded symmetric domains*. Preprint, 2005.

[Es] M. Essén, *Sharp estimates of uniform harmonic majorants in the plane*. Ark. Mat. **25**(1987), 15–28.

[EsSh] M. Essén and D. Shea, *On some questions of uniqueness in the theory of symmetrization*. Ann. Acad. Sci. Fenn. Ser. A I Math. 4(1978/79), 311–340.

[EsWuXi] M. Essén, H. Wulan and J. Xiao, *Several function-theoretic characterizations of Möbius invariant Q_K spaces*. J. Funct. Anal. **230**(2006), 78–115.

[EsXi] M. Essén and J. Xiao, *Q_p spaces – a survey*. Complex function spaces (Mekrijärvi, 1999), Univ. Joensuu Dept. Math. Rep. Ser. **4**. Univ. Joensuu, 2001, pp. 41–60.

[Ga] J. Garnett, *Bounded Analytic Functions*. Academic Press, 1981.

[GhZ] P. G. Ghatage and D. Zheng, *Analytic functions of bounded mean oscillation and the Bloch space*. Integral Equations Operator Theory **17** (1993), 501–515.

[Gir1] D. Girela, *Integral means and BMOA-norms of logarithms of univalent functions*. J. London Math. Soc. (2) **33**(1986), 117–132.

[Gir2] D. Girela, *BMO, A_2-weights and univalent functions*. Analysis **7**(1987), 129–143.

[Gir3] D. Girela, *On analytic functions with finite Dirichlet integral*. Complex Var. Theory Appl. **12**(1989), 9–15.

[GoMa] C. Gonzalez and M. A. Marquez, *On the growth of the derivative of Q_p functions*. Ann. Univ. Mariae Curie-Sklodowska Sect. A **55**(2001), 23–38.

[Got1] Y. Gotoh, *On BMO functions on Riemann surfaces*. J. Math. Kyoto Univ. **25**(1985), 331–339.

[Got2]   Y. Gotoh, *On holomorphic maps between Riemann surfaces which preserve BMO*. J. Math. Kyoto Univ. **35**(1995), 299–324.

[Got3]   Y. Gotoh, *On uniform and relative uniform domains*. Preprint, 1999.

[GuCDS]   K. Gürlebeck, J. Cnops, R. Delanghe and M. V. Shapiro, Q_p-*spaces in Clifford analysis*. Adv. Appl. Clifford Algebras **11**(2001), 201–218.

[GuMa]   K. Gürlebeck and H. R. Malonek, *On strict inclusions of weighted Dirichlet spaces of monogenic functions*. Bull. Austral. Math. Soc. **64**(2001), 33–50.

[Hay]   W. K. Hayman, *Multivalent Functions, second edition*. Cambridge University Press, Cambridge, 1994.

[HayPo]   W. K. Hayman and Ch. Pommerenke, *On analytic functions of bounded mean oscillation*. Bull. London Math. Soc. **10**(1978), 219–224.

[HedKZ]   H. Hedenmalm, B. Korenblum and K. Zhu, *The Theory of Bergman Spaces*. GTM **199**. Springer, 2000.

[Hei]   M. Heins, *On the Lindelöf principle*. Ann. of Math. **61**(1955), 440–473.

[Hil]   E. Hille, *Analytic Function Theory II*. Chelsea, 1962.

[Hu]   Z. J. Hu, Q_p *spaces in the unit ball of* $\mathbf{C}^n$ *with* $\frac{n-1}{n} < p \leq 1$. J. Huzhou Teachers College **25:3**(2003), 1–10.

[Hun]   G. A. Hunt, *Some theorems concerning Brownian motion*. Trans. Amer. Math. Soc. **81**(1956), 294–319.

[JaPeSe]   S. Janson, J. Peetre and S. Semmes, *On the action of Hankel and Toeplitz operators on some function spaces*. Duke Math. J. **51**(1984), 937–958.

[Jo]   P. W. Jones, L^∞-*estimates for the* $\bar{\partial}$ *problem in a half plane*. Acta Math. **150**(1983), 137–152.

[JoMu]   P. W. Jones and P. Müller, *Radial variation of Bloch functions*. Math. Res. Lett. **4**(1997), 395–400.

[KahSa]   J. P. Kahane and R. Salem, *Ensembles Parfaits et Séries Trigonométriques*. 2nd Edition, Hermann, Paris, 1994.

[Kak]   S. Kakutani, *Two dimensional Brownian motion and harmonic functions*. Proc. Imp. Acad. Tokyo **20**(1944), 706–714.

[KimKw]   H. O. Kim and E. G. Kwon, *Weighted subspaces of Hardy spaces*. Canad. J. Math. **40**(1988), 1074–1083.

[Kno]   K. Knopp, *Theory and Applications of Infinite Series*. Hafner Publishing Co., New York, 1971.

[Kob1]   S. Kobayashi, *Range sets and BMO norms of analytic functions.* Canad. J. Math. **36**(1984), 747–755.

[Kob2]   S. Kobayashi, *Image areas and BMO norms of analytic functions.* Kodai Math. J. **8**(1985), 163–170.

[Koo]   P. Koosis, *Introduction to H_p Spaces.* 2nd Edition, Cambridge University Press, Cambridge, 1998.

[Kor]   B. Korenblum, *BMO estimates and radial growth of Bloch functions.* Bull. Amer. Math. Soc. **12**(1985), 99–102.

[KotLR]   M. Kotilainen, V. Latvala and J. Rättyä, *Carleson measures and conformal self-mappings in the real unit ball.* Math. Nachr. (to appear).

[Kr]   S. G. Krantz, *Fractional integration on Hardy spaces.* Studia Math. **73**(1982), 87–94.

[Kw]   E. G. Kwon, *A characterization of Bloch space and Besov space.* J. Math. Anal. Appl. (2006) (to appear).

[Leu]   H. Leutwiler, *Lectures on BMO.* University of Joensuu Publications in Sci. **14**(1989), 121–148.

[LiO]   B. Li and C. Ouyang, *Randomization of Q_p spaces on the unit ball of $\mathbf{C}^n$.* Sci. China Ser. A **48** (suppl.)(2005), 306–317.

[Lis]   S. Li, *Composition operators on Q_p spaces.* Georgian Math. J. **12**(2005), 505–514.

[Lix]   X. Li, *On Hyperbolic Q Classes.* Ann. Acad. Sci. Fenn. Math. Diss. **145**, 2005.

[Lue]   D. H. Luecking, *A new proof of an inequality of Littlewood and Paley.* Proc. Amer. Math. Soc. **103**(1988), 887–893.

[Mak]   N. G. Makarov, *Probability methods in the theory of conformal mappings.* Leningrad Math. J. **1:1**(1990), 1–56.

[Mar]   D. E. Marshall, *A new proof of a sharp inequality concerning the Dirichlet integral.* Ark. Mat. **27**(1989), 131–137.

[Mel]   M. S. Melnikov, *Metric properties of analytic α-capacity and approximation of analytic functions with a Hölder condition by rational functions.* Math. USSR Sb. **8**(1969), 115–124.

[Met]   T. A. Metzger, *On BMOA for Riemann surfaces.* Canad. J. Math. **33**(1981), 1255–1260.

[Min1]   C. D. Minda, *The hyperbolic metric and coverings of Riemann surfaces.* Pacific J. Math. **84**(1979), 171–182.

[Min2]   C. D. Minda, *The Hahn metric on Riemann surfaces*. Kodai Math.
            J. **6**(1983), 57–69.

[Min3]   C. D. Minda, *The capacity metric on Riemann surfaces*. Ann. Acad.
            Sci. Fenn. Ser. A I Math. **12**(1987), 25–32.

[Mos]   J. Moser, *A sharp form of an inequality by N. Trudinger*. Indiana
            Math. J. **20**(1971), 1077–1091.

[Ne]   R. Nevanlinna, *Analytic Functions*. Springer-Verlag, 1970.

[Or]   J. Orobitg, *On spectral synthesis in some Hardy-Sobolev spaces*. Proc.
            Roy. Irish Acad. Sect. A **92**(1982), 205–223.

[OrVe]   J. Orobitg and J. Verdera, *Choquet integrals, Hauddorff content and
            the Hardy-Littlewood maximal operator*. Bull. London Math. Soc.
            30(1998), 145–150.

[PauS]   J. Pau and D. Suarez, *On the stable rank of $Q_p \cap H^\infty$*. Manuscript,
            2006.

[Pa1]   M. Pavlovic, *Introduction to Function Spaces on the Disk*. Posebna
            Izdanja [Special Editions] **20**. Matematički Institut SANU, Belgrade,
            2004.

[Pa2]   M. Pavlovic, *Hadamard product in Q_p spaces*. J. Math. Anal. Appl.
            305(2005), 589–598.

[PaVu]   M. Pavlovic and D. Vukotić, *A weak form of the Chang-Marshall in-
            equality via Green's formula*. Rocky Mountain J. Math. (2006) (to
            appear).

[PaXi]   M. Pavlovic and J. Xiao, *Splitting planar isoperimetric inequality
            through preduality of $Q_p, 0 < p < 1$*. J. Funct. Anal. **233**(2006), 40–59.

[Pee]   J. Peetre, *Invariant function spaces and Hankel operators – a rapid
            survey*. Expo. Math. **5**(1987), 3-16.

[PerR]   F. Pérez-Gonzlez and J. Rättyä, *Forelli-Rudin estimates, Carleson
            measures and $F(p, q, s)$-functions*. J. Math. Anal. Appl. **315** (2006),
            394–414.

[Pet]   K. E. Petersen, *Brownian Motion, Hardy Spaces and Bounded Mean
            Oscillation*. London Math. Soc. LNS **28**, 1977.

[RaReTo]   E. Ramírez de Arellano, L. F. Resendis O and L. M. Tovar S, Zhao
            $F(p, q, s)$ function spaces and harmonic majorants. Advances in anal-
            ysis, World Sci. Publ., Hacjensack, NJ. 2005, pp. 121–137.

[Rat]   J. Rättyä, *On Some Complex Function Spaces and Classes*. Ann.
            Acad. Sci. Fenn. Math. Diss. **124**, 2001.

[ReKa]   G. Ren and U. Kähler, *Hardy-Littlewood inequalities and Q_p-spaces*. Z. Anal. Anwendungen **24**(2005), 375–388.

[Roc1]   R. Rochberg, *Interpolation by functions in Bergman spaces*. Michigan Math. J. **29**(1982), 229–236.

[Roc2]   R. Rochberg, *Decomposition theorems for Bergman spaces and their applications*. S.C. Power (ed.), Operators and Function Theory, NATO ASI Series, Series C: Math. and Physical Sci., Vol. **153**(1985), 225–277.

[RocWu]   R. Rochberg and Z. Wu, *A new characterization of Dirichlet type spaces and applications*. Illinois J. Math. **37**(1993), 101-122.

[RosSh]   W. T. Ross and H. S. Shapiro, *Generalized Analytic Continuation*. University Lecture Series **25**. Amer. Math. Soc., Providence, RI, 2002.

[Rud]   W. Rudin, *Functional Analysis*. McGraw-Hill, N.J., 1973.

[Sara]   D. Sarason, *Functions of vanishing mean oscillation*. Trans. Amer. Math. Soc. **207**(1975), 391–405.

[SariO]   L. Sario and K. Oikawa, *Capacity Functions*. Springer Verlag, Berlin – Heidelberg – New York, 1969.

[Sch]   A. Schuster, *Interpolation by Bloch functions*. Illinois J. Math. **43**(1999), 677–691.

[Sha]   R. Shamoyan, *On diagonal map in spaces of analytic functions in the unit polydisk*. Manuscript, 2006.

[Sei]   K. Seip, *Interpolation and Sampling in Spaces of Analytic Functions*. University Lecture Series **33**. Amer. Math. Soc., Providence, RI, 2004.

[Sis1]   A. Siskakis, *Semigroups of composition operators on spaces of analytic functions, a review*. Studies on composition operators (Laramie, WY, 1996), 229–252, Contemp. Math. **213**, Amer. Math. Soc., Providence, RI, 1998.

[Sis2]   A. Siskakis, *Volterra operators on spaces of analytic functions - a survey*. Proceedings of First Advanced Course in Operator Theory and Complex Analysis, University of Seville, 2004, pp. 51–68.

[Ste1]   D. A. Stegenga, *Multipliers of the Dirichlet space*. Illinois J. Math. **24**(1980), 113–139.

[Ste2]   D. A. Stegenga, *A geometric condition which implies BMOA*. Michigan Math. J. **27**(1980), 247–252.

[Sun]   C. Sundberg, *Values of BMOA functions on interpolating sequences*. Michigan Math. J. **31**(1984), 21–30.

[Tr]   H. Triebel, *Theory of Function Spaces II*. Birkhäuser Verlag, 1992.

[Ve]   I. E. Verbitskii, *Multipliers in spaces with "fractional" norms, and inner functions*. Siberian Math. J. **26**(1985), 198–216.

[Wi1]   K. J. Wirths, *On an extremum problem for Bloch functions*. J. Anal. **1**(1993), 1–5.

[Wi2]   K. J. Wirths, *Extreme Bloch functions and summation methods for Bloch functions*. Constr. Approx. **15**(1999), 427–440.

[WiXi1]   K. J. Wirths and J. Xiao, *An image-area inequality for some planar holomorphic maps*. Results Math. **38**(2000), 172–179.

[WiXi2]   K. J. Wirths and J. Xiao, *Recognizing $Q_{p,0}$ functions per Dirichlet space structure*. Bull. Belg. Math. Soc. **8**(2001), 47–59.

[WiXi3]   K. J. Wirths and J. Xiao, *Extreme points in spaces between Dirichlet and vanishing mean oscillation*. Bull. Austral. Math. Soc. **67**(2003), 365–375.

[Wu1]   Z. Wu, *The predual and second predual of W_α*. J. Funct. Anal. **109**(1993), 314–334.

[Wu2]   Z. Wu, *A class of bilinear forms on Dirichlet type spaces*. J. London Math. Soc. (2) **54** (1996), 498–514.

[Wu3]   Z. Wu, *Carleson measures and multipliers for Dirichlet spaces*. J. Funct. Anal. **169**(1999), 148–163.

[Wu4]   Z. Wu, *s-Carleson measures and predual of $F(p,q,s)$*. Manuscript, 2006.

[WuXie1]   Z. Wu and C. Xie, *Decomposition theorems for Q_p spaces*. Ark. Mat. **40**(2002), 383–401.

[WuXie2]   Z. Wu and C. Xie, *Q spaces and Morrey spaces*. J. Funct. Anal. **201**(2003), 282–297.

[Wu1]   H. Wulan, *Möbius invariant Q_p spaces: results, techniques and questions*. Adv. Math. (China) **34**(2005), 385–404.

[WulZh1]   H. Wulan and K. Zhu, *Q_K spaces via higher order derivatives*. Rocky Mountain J. Math. (2006) (to appear).

[WulZh2]   H. Wulan and K. Zhu, *Derivative-free characterizations for Q_K spaces*. J. Austral. Math. Soc. (2006) (to appear).

[Xi1]   J. Xiao, *Carleson measure, atomic decomposition and free interpolation from Bloch space*. Ann. Acad. Sci. Fenn. Ser. A I Math. **19**(1994), 35–46.

[Xi2] J. Xiao, *The $\bar{\partial}$-problem for multipliers of the Sobolev space.* Manusc. Math. **97**(1998), 217-232.

[Xi3] J. Xiao, *Holomorphic Q Classes.* Lecture Notes in Math. **1767**. Springer-Verlag, Berlin, 2001.

[Xi4] J. Xiao, *Some results on Q_p spaces, $0 < p < 1$, continued.* Forum Math. **47**(2005), 637–668.

[Xi5] J. Xiao, *Some results on Q_p spaces, $0 < p < 1$, extended.* Manuscript, 2006.

[Xio] C. J. Xiong, *Superposition operators between Q_p spaces and Bloch-type spaces.* Complex Var. Theory Appl. **50**(2005), 935–938.

[Xu] W. Xu, *Superposition operators on Bloch-type spaces.* Preprint, 2006.

[Ya] S. Yamashita, *Dirichlet-finite functions and harmonic majorants.* Illinois J. Math. **25**(1981), 626–631.

[Zha1] R. Zhao, *An exponential decay characterization of BMOA on Riemann surfaces.* Arch. Math. **79**(2002), 61–66.

[Zha2] R. Zhao, *Distances from Bloch functions to some Möbius invariant spaces.* Preprint, 2006.

[Zhu1] K. Zhu, *Operator Theory in Function Spaces.* Pure and Applied Math., Marcel Dekker, New York, 1990.

[Zhu2] K. Zhu, *Spaces of Holomorphic Functions in the Unit Ball.* Graduate Texts in Mathematics **226**. Springer-Verlag, New York, 2005.

[Zhu3] K. Zhu, *A class of Möbius invariant function spaces.* Illinois J. Math. (to appear).

Index